NANOMATERIALS-BASED COMPOSITES FOR ENERGY APPLICATIONS

Emerging Technology and Trends

NANOMATERIALS-BASED COMPOSITES FOR ENERGY APPLICATIONS

Emerging Technology and Trends

Edited by
Keka Talukdar, PhD

Apple Academic Press Inc.
4164 Lakeshore Road
Burlington ON L7L 1A4
Canada

Apple Academic Press Inc.
1265 Goldenrod Circle NE
Palm Bay, Florida 32905
USA

© 2020 by Apple Academic Press, Inc.

No claim to original U.S. Government works

Printed and bound by CPI Group (UK) Ltd, Croydon, CR0 4YY

International Standard Book Number-13: 978-1-77188-806-6 (Hardcover)
International Standard Book Number-13: 978-0-42926-505-1 (eBook)

All rights reserved. No part of this work may be reprinted or reproduced or utilized in any form or by any electronic, mechanical or other means, now known or hereafter invented, including photocopying and recording, or in any information storage or retrieval system, without permission in writing from the publisher or its distributor, except in the case of brief excerpts or quotations for use in reviews or critical articles.

This book contains information obtained from authentic and highly regarded sources. Reprinted material is quoted with permission and sources are indicated. Copyright for individual articles remains with the authors as indicated. A wide variety of references are listed. Reasonable efforts have been made to publish reliable data and information, but the authors, editors, and the publisher cannot assume responsibility for the validity of all materials or the consequences of their use. The authors, editors, and the publisher have attempted to trace the copyright holders of all material reproduced in this publication and apologize to copyright holders if permission to publish in this form has not been obtained. If any copyright material has not been acknowledged, please write and let us know so we may rectify in any future reprint.

Trademark Notice: Registered trademark of products or corporate names are used only for explanation and identification without intent to infringe.

Library and Archives Canada Cataloguing in Publication

Title: Nanomaterials-based composites for energy applications : emerging technology and trends / edited by Keka Talukdar.

Names: Talukdar, Keka, 1975- editor.

Description: Includes bibliographical references and index.

Identifiers: Canadiana (print) 2019020222X | Canadiana (ebook) 20190202262 | ISBN 9781771888066 (hardcover) | ISBN 9780429265051 (ebook)

Subjects: LCSH: Nanostructured materials. | LCSH: Nanocomposites (Materials) | LCSH: Renewable energy sources—Technological innovations.

Classification: LCC TA418.9.N35 N36 2020 | DDC 620.1/15—dc23

CIP data on file with US Library of Congress

Apple Academic Press also publishes its books in a variety of electronic formats. Some content that appears in print may not be available in electronic format. For information about Apple Academic Press products, visit our website at **www.appleacademicpress.com** and the CRC Press website at **www.crcpress.com**

About the Editor

Keka Talukdar, PhD, is an independent researcher and also currently teaches in the Department of Physics at Nadiha High School, Durgapur, West Bengal, India. She previously worked as a Research Associate at the Institute of Genomics and Bio Medical Research, Hooghly, West Bengal, and was affiliated with the National Institute of Technology, Durgapur, India, where she earned her PhD in computational nanoscience. Her research interests include modeling and simulation of CNTs and graphene properties, nanobiophysics, and biosensor and ion channel simulation. She has published over 25 journal articles as well as several book chapters with international publishers, including one with John Wiley. She has been a reviewer for serveral professional journals, including Materials Science and Engineering B, Mechanics of Materials, Nanoscale Research Letters, and several others. She has been a reviewer for several professional journals, including *Materials Science and Engineering B, Mechanics of Materials, Nanoscale Research Letters,* and several others. She also reviewed several research papers for the DAE Solid State Physics Symposium for consecutive four years.

Contents

Contributors ix

Abbreviations xi

Foreword xvii

Preface xix

PART I: Preparation and Properties of Nanocomposites for Energy Applications 1

1. **Preparation and Properties of Nanocomposites for Energy Applications** 3
   Idowu David Ibrahim, Chewe Kambole, Azunna Agwo Eze, Adeyemi Oluwaseun Adeboje, Emmanuel Rotimi Sadiku, Williams Kehinde Kupolati, Tamba Jamiru, Babatunji Wunmi Olagbenro, and Oludaisi Adekomaya

2. **Recent Progress in Nanocrystalline Oxide Thin Films** 43
   Pallabi Gogoi, D. Pamu, and L. Robindro Singh

3. **Application of Magnetic Fluid in the Energy Sector** 65
   Kinnari Parekh and R. V. Upadhyay

4. **Nanostructured Materials and Composites for Renewable Energy** 91
   Idowu David Ibrahim, Chewe Kambole, Azunna Agwo Eze, Adeyemi Oluwaseun Adeboje, Emmanuel Rotimi Sadiku, Williams Kehinde Kupolati, Tamba Jamiru, Babatunji Wunmi Olagbenro, and Oludaisi Adekomaya

5. **Simulation and Modeling of Nanotechnology Aircraft Energy System Using MATLAB** 121
   Indradeep Kumar

6. **Promising Nanomaterials and Their Applications in Energy Technology** 145
   Tathagat Waghmare

PART II: Nanocomposites for Sustained Energy 159

7. Nano-Bio Hybrid Platform to Meet the Energy Challenge 161
   Mihir Ghosh

8. ZnS/ZnO Nanocomposite in Photovoltaics: A Computational Study on Energy Conversion 185
   Keka Talukdar

9. Bio-Nanometal-Cluster Composites for Renewable Energy Storage Applications 215
   R. Govindhan and B. Karthikeyan

10. AlNi and AuZn Nanohybrids for Capacitors: A Computational Study 233
   Keka Talukdar

11. Recycling of Polymer Nanocomposites of Carbon Fiber Reinforced Compressed Natural Gas Reservoirs 255
   Gilberto João Pavani, Sérgio Adalbeto Pavani, and Carlos Arthur Ferreira

12. Nanomaterial-Based Energy Storage and Supply System in Aircraft Systems 283
   Indradeep Kumar

Color insert of illustrations .... A – H

Index *303*

Contributors

Adeyemi Oluwaseun Adeboje
Department of Civil Engineering, Tshwane University of Technology, Pretoria, South Africa

Oludaisi Adekomaya
Department of Mechanical Engineering, Mechatronics and Industrial design, Tshwane University of Technology, Pretoria, South Africa

Azunna Agwo Eze
Department of Mechanical Engineering, Mechatronics and Industrial design, Tshwane University of Technology, Pretoria, South Africa

Carlos Arthur Ferreira
Universidade Federal do Rio Grande do Sul, Av. Bento Gonçalves – 9500, Porto Alegre, Brazil

Mihir Ghosh
Department of Organic Chemistry, Weizmann Institute of Science, 234 Herzl Street, Rehovot – 7610001, Israel, E-mail: mihir.ghosh@weizmann.ac.il

Pallabi Gogoi
Department of Nanotechnology, North Eastern Hill University, Shillong, Meghalaya, India

R. Govindhan
Department of chemistry, Annamalai University, Annamalainagar – 608002, Tamil Nadu, India

Idowu David Ibrahim
Department of Mechanical Engineering, Mechatronics and Industrial design, Tshwane University of Technology, Pretoria, South Africa, E-mail: ibrahimid@tut.ac.za, ibrahimidowu47@gmail.com

Tamba Jamiru
Department of Mechanical Engineering, Mechatronics and Industrial design, Tshwane University of Technology, Pretoria, South Africa

Chewe Kambole
Department of Civil Engineering, Tshwane University of Technology, Pretoria, South Africa

B. Karthikeyan
Department of chemistry, Annamalai University, Annamalainagar – 608002, Tamil Nadu, India

Indradeep Kumar
Chairman of Bibhuti Education and Research, Bhagalpur, Bihar, India; Research Scholar, Department of Mechanical Engineering, Vels Institute of Science, Technology and Advanced Studies (VISTAS), Chennai – 600117, India

Williams Kehinde Kupolati
Department of Civil Engineering, Tshwane University of Technology, Pretoria, South Africa

Babatunji Wunmi Olagbenro
Department of Industrial and Production Engineering, University of Ibadan, Oyo State, Nigeria

D. Pamu
Department of Physics, Indian Institute of Technology Guwahati, Guwahati, Assam, India

Kinnari Parekh
Dr. K. C. Patel R&D Center, Charotar University of Science and Technology, Changa – 388421, Dist. Anand, Gujarat, India, E-mail: kinnariparekh.rnd@charusat.ac.in

Gilberto João Pavani
Instituto Federal do Rio Grande do Sul, Av. São Vicente 785, Farroupilha, Brazil,
E-mail: gilberto.pavani@farroupilha.ifrs.edu.br

Sérgio Adalbeto Pavani
Universidade Federal de Santa Maria, Av. Roraima – 1000, Santa Maria, Brazil

Emmanuel Rotimi Sadiku
Department of Chemical, Metallurgical and Materials Engineering, Polymer Section, Tshwane University of Technology, Pretoria, South Africa

L. Robindro Singh
Department of Nanotechnology, North Eastern Hill University, Shillong, Meghalaya, India

Keka Talukdar
Department of Physics, Nadiha High School, Durgapur – 713211, West Bengal, India,
E-mail: keka.talukdar@yahoo.co.in

R. V. Upadhyay
Dr. K. C. Patel R&D Center, P. D. Patel Institute of Applied Sciences,
Charotar University of Science and Technology, Changa – 388421,
Dist. Anand, Gujarat, India

Tathagat Waghmare
Faculty of Food Bio-Engineering Technology Department of Microbiology and Biotechnology, Szent Istvan University, Budapest – 1118, Hungary, Europe

Abbreviations

ABS	acrylonitrile-butadiene-styrene
ACD	automatic cooling device
AETD	automatic energy transport device
AFM	atomic force microscopy
AI	artificial intelligence
APTES	aminopropyltriethoxysilane
ATR	attenuated total reflectance
Au/Pt/Ag TNCs	Au/Pt/Ag colloidal nanocomposites
BNPs	bimetallic nanoparticles
BR	bacteriorhodopsin
BSD	back scattered detection
CA	chronoamperometry
CDG	colloidal dispersion gels
CdTe	cadmium-telluride
CEN	European Standardization Committee
CIGS	copper indium gallium selenide
CMNC	ceramics matrix nanocomposites
CNFs	carbon nanofibers
CNG	compressed natural gas
CNT	carbon nanotube
CO_2	carbon dioxide
CPUs	central processing units
CSP	concentrating solar power
CTO	conductive transparent oxide
CV	cyclic voltammetry
CVD	chemical vapor deposition
DA	dark-adapted
DASC	direct absorption solar collectors
DC	direct current
DFT	density functional theory
DLS	dynamic light scattering
DOS	density of state
DRAM	dynamic random access memory

DSSC	dye-sensitized solar cell
DST	Department of Science and Technology
ECs	electrochemical capacitors
EDD	electron difference density
EDP	electron difference potential
EDX	energy dispersive X-ray
ESR	equivalent series resistance
EV	electric vehicle
EVA	ethylene vinyl acetate
FCNTs	functionalized carbon nanotubes
FEM	finite element simulation
FES	flywheel energy storage
FESEM	field emission scanning electron microscopy
FET	field-effect transistor
FHD	ferrohydrodynamic
FRET	Förster resonance energy transfer
FT-IR	Fourier transform infrared
FW	filament winding
GPCR	G-protein coupled receptor
HDPE	high-density polyethylene
HEVs	hybrid electric vehicles
HILS	hardware-in-the-loop simulation
HLP	hydrophilic and lipophilic polysilicon
HOMO-LUMO	highest molecular-lowest unoccupied molecular orbital
HR-SEM	high resolution scanning electron microscopy
HR-TEM	high-resolution transmission electron microscopy
HST	hot spot temperature
IFT	interfacial tension
INCOSE	International Council on Systems Engineering
ISO	International Organization for Standardization
ITO	indium tin oxide
KNN	$K_{0.5}Na_{0.5}NbO_3$
LA	light-adapted
LACO	linear combination of atomic orbitals
LCP	commercial thermotropic liquid crystalline polymers
LDA	local density approximation
LDPE	low-density polyethylene
LEDs	light-emitting diodes

Abbreviations

LIBs	lithium-ion batteries
LiPF	lithium hexafluorophosphate
LLDPE	low-density polyethylene
LSPR	localized SPR
MBSE	model-based system engineering
MEMS/NEMS	micro/nanoelectromechanical systems
MHD	magnetohydrodynamic
MIM	metal-insulator-metal
MMC	metal matrix composites
MMNC	metallic nanocomposites
MNBA	micro/nano bottom ash
MNFA	micro/nano fly ash
MNPs	monometallic nanoparticles
MOR	methanol oxidation reaction
MP	Mulliken population
MWCNT	multi-walled carbon nanotube
MWNTs	multiple walled nanotubes
NC	negative capacitance
NDT	non-destructive test technology
NEGF	non-equilibrium Green's functions
NH_3	ammonia
NIR	near-infrared
NMs	nanomaterials
NPs	nanoparticles
NWP	neutrally wet polysilicon
OH	hydroxyl
OPV	organic PVs
PA6	polyamide-6
PA66	polyamide-6,6
PAMAM	polyamidoamines
PBT	poly(butylene terephthalate)
PC	polycarbonate
PCA	photo-generated catalysis activity
PCE	power conversion efficiency
PCM	phase change materials
PEEK	polyetheretherketone
PET	poly(ethylene terephthalate)
PI	polyimide, thermoplastic

PIB	polyisobutylene
PLA	polylactic acid
PLD	pulse laser deposition
PLS	polymer layered silicates
PMMA	polymethylmethacrylate
PMNC	polymer matrix nanocomposites
POM	polyoxymethylene (homo)
PP	polypropylene
PPS	polyphenylene sulfide
PS	polystyrene
PSCs	perovskite solar cells
PSSC	protein sensitized solar cell
PSU	polysulfone
PTFE	polytetrafluoroethylene
PV	photovoltaic
PVC	polyvinyl chloride
PVD	physical vapor deposition
PVDC	polyvinylidene chloride
PVDF	polyvinylidene difluoride
QAS	quaternary ammonium salts
QD	quantum dot
QDSSCs	quantum-dot-sensitized SCs
RF	radio frequency
RT	room temperature
SAED	selected area electron diffraction
SAN	styrene-acrylonitrile copolymer
SB	styrene/polybutadiene copolymer
SCs	solar cells
SE	systems engineering
SEIRS	surface-enhanced infrared spectrum
SEM	scanning electron microscopy
SERS	surface-enhanced Raman scattering
SFPS	short fiber-reinforced polyphenylene sulfide
SMART	self-monitoring analysis and reporting technology
SPDR	split post-dielectric resonator
SPR	surface plasmon resonance
SPS	surface photovoltage spectroscopy
SWNTs	single-walled nanotubes

Abbreviations

TBC	thermal barrier coatings
T_c	curie temperature
TEM	transmission electron microscopy
TLD	the-lens detector
TNPs	trimetallic nanoparticles
TPU	urethane base TPE
TSMF	temperature-sensitive magnetic fluid
TW	terawatts
UCNPs	upconversion nanoparticles
UHMWPE	ultrahigh molecular weight polyethylene
UHR	ultra-high resolution
UPE	pure unsaturated polyester
UV	ultraviolet
XPS	x-ray photoelectron spectroscopy
XRD	x-ray diffraction
ZP	zeta potential

Foreword

Energy is a most sought thing in modern society due to the constantly increasing human population that is the cause of fossil fuel resource depletion. This necessitates the need to explore new renewable and alternative energy sources. Nanoscience and nanotechnology will help to pave the path to improve the current energy systems.

This book discusses the applications of diverse renewable and alternative energy systems. In the first chapter provides a brief overview of functional and smart materials for energy applications as well as additional basic information about nanocomposite materials and methods of preparation. It concludes with presenting the future trends in the field of nanocomposites. Chapter 2 discusses the fundamentals of thin-film technology and dielectric thin film application to optical devices and thin-film capacitors. Moreover, it talks about the ferroelectric, piezoelectric oxide thin film for sensors and tunable microwave device applications. The third chapter discusses the application of magnetic fluid in the energy sector. On this topic, in-depth information on the magnetic fluids related to energy harvester and storage with the novel flow design is discussed. It also looks at a heat transfers using convection pipe study and automatic cooling devices. In the fourth chapter, nanostructured and nanocomposite materials used in the renewable and alternative energy sector are discussed.

In Chapter 5, the idea about computational modeling of nanoparticles and aircraft energy system conceptual design is discussed. The sixth chapter discusses semiconductor photocatalytic nanomaterials with the novel structural and electronic properties for solar energy applications, which can be further extended to various field such as water and air purification. The seventh chapter discusses nature-inspired bionanomaterials for energy conversion and energy storage systems. Chapter 8 discusses theoretical simulation by a density functional approach to elucidate the structural and electronic properties of ZnS/ZnO related to solar energy conversion efficiency. The ninth chapter considers the bio-nanometal-cluster composite unique properties (morphological, electrical, optical and mechanical) that are useful for enhancing the energy conversion and storage performances,under the "bio-nanometal-cluster composites for renewable energy storage applications." Chapter 10 is concerned with the energy storage applications of AlNi and AuZn nanocrystals, which are investigated by the density functional approach. Additionally, this chapter discusses the change of capacitance of crystals in various conditions and interface properties of the nanohybrids. The penultimate chapter talks about that the recycling of polymer nanocomposite of carbon fiber reinforced compressed natural gas reservoirs. The final chapter deals with the exploration of nanomaterials and nanoengineering in the aviation industry, and the aim is to reduce fuel consumption and carbon emissions associated with air freight and air travel. Further, it talks about improvements in renewable and conventional energy source in the aircraft system.

The basic aspects along with the recent developments in nano-based materials related to diverse energy applications are well covered in this book. With the versatility of the themes in each chapter, the reader will enjoy the multidisciplinary approach given in this book. Also, the broad platform described here induces novel ideas and thought, which can then develop into a novel nano-based material systems for new energy applications.

—**Sathish-Kumar Kamaraj**
Research Professor, Technological Institute of El Llano
Aguascalientes/National Technology Institute of Mexico,
(Instituto Tecnológico El Llano Aguascalientes/Tecnológico
Nacional de México), El Llano, Aguascalientes, México

Preface

The potential of nanotechnology is completely realized by present-day scientists and researchers to find a way to meet our ever-increasing energy demand. Tremendous motivation toward the application of nanotechnology in energy demand has led scientists to develop new technology and materials for sustainable development. This is because the increasing carbon footprint due to the burning of fossil fuels is alarming for our climate, and hence, alternative sources of energy are required. So, in the energy sector, the advent and use of new functional materials are essential.

This book brings together some of the notable prospects of nanomaterials-based composites and hybrids in energy harvesting, storage, and conversion. The aim is to capture the real picture of recent technological development in the energy sector and fundamentals of the same as well. It discusses the methods of preparation of nanocomposites for energy applications, their analyzing tools and methods, and preparation of novel energy devices for different fields. The future prospect of using magnetic fluids in the energy hunt may be debated as the amount of power generation is still under question. But the prospect of combining the properties of the magnetic fluid to enhance the output power generation is very promising. This budding area of research is included in this publication.

Applications of nanocomposites for energy requirement in various parts of an airplane are another interesting contribution. Due to the vastness of the topic, not all the energy materials and methods can be covered here, but some renewable sources of energy, such as biofuel and solar power, are elaborated on. Eco-friendly photo energy conversion through bio-inspired hybrid photocatalysts (bacteriorhodopsins) is an interesting method of solar energy conversion, and it can be successfully utilized as a renewable energy source with high earth abundance. The recent progress of the same is presented in the book, emphasizing the design of solar cells using bacteriorhodopsins as a photosensitizer. The book contains an explicit discussion on the structure of thin films, their fabrication methods, and their applications in the energy sector. Thin film technology is developing fast, and new attempts are undertaken to fabricate thin films with fascinating optical properties and charge storage capability. Preserving and storing of natural resources, such as natural gas, is thoroughly reviewed in this publication and focuses on nanocomposite materials for building a gas reservoir. The book encompasses the synthesis and characterization of the bionanometal-cluster composites for energy conversion and energy storage of renewable energy. The computational studies, including energy storage and energy conversion through photovoltaics, are given special attention as modeling and simulation can tell us those unknown facts that cannot be visible experimentally. Computational outcome for energy applications in the aerospace industry is given a space in this book.

The book is a combined effort of the authors and reviewers, and I hope this will be a helpful publication for nanotechnology researchers and scholars. Beginners and also the senior researchers and engineers will be able to enhance their knowledge and develop new ideas from the contributions presented here.

While integrating the contributions of many eminent authors from different parts of the globe, all the respected authors have offered tremendous support and cooperation from the beginning to the end of the compilation of chapters. As an editor, I am thankful to them as they found some time to contribute to this book.

—**Keka Talukdar, PhD**
NIT Durgapur, India

PART I
Preparation and Properties of Nanocomposites for Energy Applications

CHAPTER 1

Preparation and Properties of Nanocomposites for Energy Applications

IDOWU DAVID IBRAHIM,[1] CHEWE KAMBOLE,[2] AZUNNA AGWO EZE,[1] ADEYEMI OLUWASEUN ADEBOJE,[2] EMMANUEL ROTIMI SADIKU,[3] WILLIAMS KEHINDE KUPOLATI,[2] TAMBA JAMIRU,[1] BABATUNJI WUNMI OLAGBENRO,[4] and OLUDAISI ADEKOMAYA[1]

[1]*Department of Mechanical Engineering, Mechatronics and Industrial design, Tshwane University of Technology, Pretoria, South Africa, E-mail: ibrahimid@tut.ac.za, ibrahimidowu47@gmail.com*

[2]*Department of Civil Engineering, Tshwane University of Technology, Pretoria, South Africa*

[3]*Department of Chemical, Metallurgical and Materials Engineering, Polymer Section, Tshwane University of Technology, Pretoria, South Africa*

[4]*Department of Industrial and Production Engineering, University of Ibadan, Oyo State, Nigeria*

ABSTRACT

The demand for functional and smart materials for energy applications has gained global interest. The inability of a single material to meet this requirement has necessitated researchers into developing nanocomposite materials by combining two or more materials, with the incorporation of nanoparticles (NPs). This chapter focuses on the methods of preparation and the basic properties of nanocomposite materials that find useful applications in the energy sector. An extensive review of the literature was undertaken with a succinct report of the summary, detailing the various properties that make the materials find significant applications in the energy industry. Finally, an outline of the future trends in the field of nanocomposites is presented.

1.1 INTRODUCTION

Energy demand for the last three decades has increased significantly with the fast development in technology, economy, and population growth. Fossil fuel has dominated the market more than any other sources of energy, but the changes in the climatic condition, carbon emission, and the gradual scarcity of fossil resources has facilitated the growing interest for other sources of energy in order to preserve the environment. Banos et al., [1] stated that the first route for reducing the demand on fossil fuel resources is by reducing the energy wastage by employing energy-saving programs. One way to do this is by utilizing energy-saving bulbs, switching-off equipment that is not in use both in the industries and in domestic homes [2, 3]. A similar program has been implemented in South Africa, and it was helpful in reducing the constraints on the national power grid.

Another major channel to meet the huge energy demand is to diversify into renewable and sustainable energy, such as solar, wind, tidal, biofuel energy, etc., which can either be on a commercial scale or for an individual (also referred to as a stand-alone system) [4]. The consideration of these options calls for a new material that has the desired properties to function perfectly irrespective of the area of application. Nanocomposites materials have proven to meet these requirements. Although, nanoparticles (NPs) sometimes, if not well dispersed, tend to reduce the properties of the composites [5]. For a nanocomposite, at least one of the dimensions of the particle must be ≤100 nm. These materials have been widely used for energy applications, due to the improved properties such as mechanical and thermal properties [6–9]. Other important reasons why such materials are seen to be suitable are light-weight, eco-friendly, renewable, biodegradable, low cost, high strength to weight, etc. This paper presents the findings on mostly polymeric nanocomposites, method of preparation, properties, and areas of application.

1.2 PROPERTIES OF NANOCOMPOSITES

Nanomaterials (NMs) are microscopic in nature that can be dispersed nanometrically. They have very large surface areas when compared to pure polymers, and they improved properties, such as increased strength, modulus, heat, and solvent resistance. They also have decreased gas permeability and flammability. Polymer nanocomposites, e.g., polymer layered silicates (PLS), have unique model systems. By exploring intercalated (insertion) and delaminated (separation in layer) hybrids, polymer statics, as well as dynamics, are confined to a wide range of distances between its radius of gyration and the statistics of its entire length [10].

The increase in the world population resulted in an increased rate of energy consumption. However, the human is faced with the challenge of developing an alternative but sufficient and renewable energy, which is environmentally sustainable for the day-to-day living. Due to its abundance, solar energy meets up the required characteristics of the renewable energy desired. In the last 10 years, nanostructured materials had been evaluated for the construction of energy harvesting panels or assemblies. There is now high interest and focus on solar energy transfer in the form of solar cells (SCs) and hydrogen energy, obtained by splitting water, photocatalytically [11]. Modern electrochemistry of materials has succeeded in the discovery of lithium-ion batteries (LIBs), which results from the intercalations of lithium-ion negative electrode or graphite and the lithium-ion positive electrode, lithium metal oxide, separated by a lithium-ion electrolyte, e.g., lithium hexafluorophosphate (LiPF) in ethylene carbonate-diethyl carbonate [12].

Migration from the utilization of bulk to NMs can make a substantial change in the properties of electrodes or electrolytes and impact positively on the ability of the materials to convert from a form of energy to another. Utilization of NPs for electrodes or electro-catalysts can increase the surface area and result in electrodes with increased reaction rate [12].

Nanostructured materials that have peculiar nano-architecture or morphologies, e.g., nanotube, may be required to attain the required properties that will be useful for the conversion of energy for use in batteries as the effects of the space to charge at the interface of small particles can lead to the great improvement of the properties [12].

Three ways in which nanostructured materials are used for converting solar energy are: (i) by mimicking photosynthesis with clusters or assembly of the donor to acceptor molecules, (ii) by using semiconductors, e.g., hydrogen, to assist photo-catalysis in the production of fuels, and (iii) utilization of nanostructure semiconductor for SCs [13]. Nanostructures utilized for energy conversion include assembly of inorganic to organic hybrid, semiconductor and metal NPs, and energy conversion schemes using carbon nanostructures [13]. In order to achieve economic and effective energy production, Kamat [13] explored solar hydrogen production and photochemical SCs.

Converting and storing energy are stiff challenges confronting the development and sustaining the future. It has, however, been limited due to the cost implications, the durability of the product as well as performance. Application of nanostructured materials is expected to ameliorate this particular challenge, if adequately researched and applied. Block copolymer is a material, which is economical and strong for the formulation and blending of NMs with strong consideration requirements for the constituents, arrangements, and dimensional matrix. Humanity has been battling with the problem of energy production and conversion for up to 10 decades [14].

The battery is a foremost energy storage device. Lithium-ion batteries are alternative storage devices for electrochemical energy with lightweight, efficient, rechargeable facility useful in cell phones, laptops, and other portable devices [14]. High-k polymer nanocomposites are easy to process, durable, flexible, and cheap. They are suitable for the storage of energy and applicable in dielectrics. The introduction of core-shell into the architecture of the nanostructures is a versatile means of designing and synthesizing high-k polymer nanocomposites [10]. Currently, there are advances to apply core-shell nanostructured polymers for the generation of energy.

The homogeneous dispersion of NPs is smoothly achieved even when the nanocomposites are extremely or highly loaded. This gives an advantage over conventional mixing process *via* melting or dispersion in solution processes [10]. Furthermore, there is the suppression of dielectric losses when NPs that are electrically conductive are used to fill nanocomposites, and this often results in a very high dielectric constant [10]. In the order of availability or abundance, TiO_2 naturally occurs in nature in tri-party polymorphs, as rutile, anatase, and brookite. Although several other high pressured polyforms have been reported, the known phases of its synthesis are $TiO_2(B)$, $TiO_2(H)$, and $TiO_2(R)$ [11].

When illuminated by light, the holes and photo-generated electrons are capable of initiating redox reactions as a result of the absorption of chemical species on the interface or the surface of the photo-catalyst. The effects of the mechanism of heterogeneous photo-catalysis on TiO_2, due to electron paramagnetic resonance (ESR) and laser flash photolysis measurements were conceptualized by Hoffmann and co-workers to include four main processes, which are:

i. The generation of charge to carrier;
ii. Charge to carrier trapping;
iii. Charge to carrier recombination; and
iv. Interfacial charge transfer.

These processes have been succinctly reported by several authors [15–21]. The photo-catalytic processes, with the use of TiO_2 or different semiconductors, have shown limitations in the production of higher photo-conversion efficiencies, which must be overcome [11].

Highly porous carbon cryogels with tunable pore structure and chemical composition were synthesized through controlled hydrolysis and polycondensation reactions by using different chemicals as precursors and either NaOH or hexamine ($C_6N_{12}N_4$) as catalysts. Gelation was followed with freeze-drying in order to preserve the highly porous structure during solvent removal, and controlled pyrolysis of the organic hydrogels and subsequent optional activation was performed. In addition, two different approaches were taken to modify the surface chemistry of the porous carbon in order to introduce nitrogen or nitrogen-boron species, leading to different porous structures and surface chemistry, as well as electrochemical properties. These carbon cryogels have been characterized and studied for energy storage applications. Specifically, they have been investigated as electrodes for electric double-layer supercapacitors, high energy and high power density LIBs with vanadium pentoxide deposited inside the pores, porous media for natural gas (methane) storage at reduced pressure and scaffolds for hydride nanocomposites, hence, resulting in improved hydrogen storage. The relationship between processing conditions, chemical composition, pore structure, and energy storage properties are discussed [22].

The structural forms of carbon: include diamond, fullerene, graphene, graphite, amorphous carbon, and carbon nanotubes (CNTs). Amorphous carbon can be classified as graphitic and diamond-like carbons. The two differ in the orbital hybridization of their molecules. Graphitic and diamond-like carbons have sp^2 and sp^3 hybridization, respectively. They both have excellent thermal conductivity, optical, and mechanical properties [22]. These properties contribute to the reasons why they find useful applications for energy generation and storage. The desired application will determine the basic properties the nanocomposites must possess, and in order for the material to be considered, properties such as mechanical, thermal, electrical conductivity and insulation characteristics are of paramount importance.

1.2.1 MECHANICAL PROPERTY

The mechanical properties of any materials are the most desired property for such a material to be considered as a potential choice in any application. This desire has since been the driving force for scientists and engineers in finding distinct and various ways to improving the mechanical properties of different materials. The improvement can be in different forms, so also is the mechanical property been considered. Various researches have been reported where properties, such as tensile strength and modulus, flexural strength and modulus, impact strength, hardness and wear resistance, have been significantly improved due to different composite modifications that were made. These modifications can be in the form of material treatment, incorporation of fibers [23], NPs [24, 25], hybridization [26], and compatibilizers [27]. For a single material on its own, the desired properties may not be possessed for a specific application. Composites materials have shown tremendous improvement in terms of mechanical properties. A composite can either be a metal-, ceramic- or polymer-based [7, 28, 29]. Among the several types of composites, based on matrices used, polymer-based composites seem to have the lowest mechanical properties. However, due to reasons, such as: easy of fabrication, lightweight, durability, low cost, and processability of polymer matrices, they find a wide range of relevant applications in recent decades [30]. The properties of composites can further be improved by the inclusion of a small amount of NPs. NPs have a high aspect ratio, which makes them unique as fillers in nanocomposites. The modification of NPs has been reported to further contribute to the mechanical properties. Pasbakhsh et al., [31] reported that 30 phr silane modification of Hal nanotube, enhanced the degree of dispersion of the nanoparticle in ethylene-propylene-diene monomer, resulting in 135, 360 and 56% increases in the: elongation-at-break, tensile strength, and modulus, respectively in comparison with the unmodified nanoparticle composites. In a separate research, Krishnaiah et al., [32] prepared Aminopropyltriethoxysilane (APTES) modified and unmodified Hal nanotubes reinforced polylactic acid (PLA) nanocomposites. The observed mechanical improvements are shown in Table 1.1, while the elongation-at-break is represented in Figure 1.1. For the nanocomposites prepared, the composites with modified nanoparticle showed improved impact strength, tensile strength, and modulus. The optimal tensile strength was recorded at 4 wt% NPs loading, while the tensile modulus and impact strength were optimal at 8 and 6 wt% loadings, respectively.

TABLE 1.1 Mechanical Properties of Unmodified and Modified Hal Nanotube Reinforced PLA Nanocomposites.

Samples	Nanoparticle Content (wt %)	Tensile Strength (MPa)	Tensile Modulus (MPa)	Impact Strength (J/m)
Pure PLA	0	49.5	1052	21.4
2uHal	2	51.8	1088	22.6
2mHal	2	55.9	1104	25.6
4uHal	4	54.8	1152	25.9
4mHal	4	62.6	1198	29.8
6uHal	6	54.7	1185	29.4
6mHal	6	55.4	1208	30.3
8uHal	8	52.7	1225	23.9
8mHal	8	53.3	1270	24.8

Adapted from Ref. [32].

Abbreviations: uHal: unmodified Hal nanotubes, mHal: modified Hal nanotubes.

Several reports have shown that NPs or filler above 5 wt% tend to experience a reduction in the mechanical properties, which is due to the improper dispersion in the composites [30]. Different researches have been conducted by using different NPs at various loadings, and the results showed significant improvement [30, 33–38]. The observed influence is due to several parameters, ranging from particle size, method of preparation, interfacial adhesion, shear stress, the degree of dispersion, etc. Sheh et al., [30] explained, by using SEM, the effect of the degree of dispersion of nanoclay on the overall properties of the prepared nanocomposites. In the analysis, LQASMMT and BQASMMT represent linear and benzyl organo-MMT modified quaternary ammonium salts (QAS), respectively. Pure montmorillonite (P-MMT) reinforced nanocomposites were also prepared. Figure 1.2 shows a uniform surface morphology in the case pure unsaturated polyester (UPE), while the incorporation of P-MMT shows increased surface roughness and poor dispersion due to the agglomeration of MMT within the UPE matrix. LQASMMT displayed little exfoliation because MMT modified with QAS are hydrophilic in nature; BQASMMT, on the other hand, had better exfoliation with increased interplanar d-spacing, due to the presence of benzyl substituted surfactant. The composite was also observed to become more hydrophobic, making it to have better adhesion to the polymer matrix.

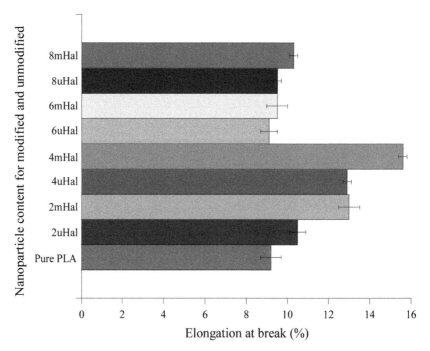

FIGURE 1.1 **(See color insert.)** % Elongation-at-break of unmodified and modified Hal nanotube-reinforced PLA nanocomposites. Adapted from Ref. [32].

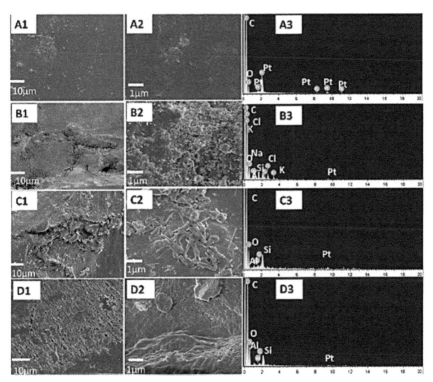

FIGURE 1.2 SEM and EDX analysis of: (a) UPE resin, (b) P-MMT, (c) LQASMMT, and (d) BQASMMT composites. 1 and 2 represent the different magnification, and 3 represents the EDX data of the corresponding clays.

Source: Reprinted with permission from Ref. [30]. © 2016 Elsevier.

1.2.2 THERMAL PROPERTY

Energy generation and storage, at one stage or the other, require materials with high heat or flame resistance as the case may be in practice. As it was explained previously, a single material may not be able to meet all the requirements; hence, a composite can mitigate this drawback. The incorporation of the nanoparticle can further improve the thermal properties and other relevant properties of composites. In the case of thermal properties, improvement does not always necessarily mean increasing the thermal property. Improvement depends on the intended area of application because increasing the melt temperature (T_m), for instance, from between 100°C to 150°C, may not be regarded as an improvement if the T_m needed to be brought to a lower (<100°C) temperature. The thermal properties of individual components in the case of polymer-based nanocomposites must be taken into account as the processing conditions (temperature and time), and morphology is greatly affected by this factor. When speaking of thermal properties, the followings are issues that come to mind:

a. **Thermal Conductivity:** This is the ability of the nanocomposite materials to transfer heat from one part to another; mostly from higher temperature region to lower. Materials with low thermal conductivity and/or temperature resistance, find significant applications in the electrical and electronic industries, where they are largely used for insulation [39, 40].
b. **Thermal Transition Temperature:** This is the temperature required to initiate a change in the molecular structure of polymer, enabling free movement of the molecules, therefore, causing solid-like polymer to behave just like viscous liquids [41].
c. **Thermal Stability:** The temperature ranges, for which a material can find useful application, defines its thermal stability, after this temperature, degradation steps in for the material [42].
d. **Thermal Expansion:** Most polymer-based nanocomposites are known to possess high thermal expansion coefficients due to the weak bond that exists within their molecules. Hence, such materials tend to have an unstable dimension in an area where the temperature varies per time as a result of climatic or operational conditions. When the temperature is lowered, they contract, and when it is raised, they expand [43, 44].

The thermal conductivity of most polymers ranges between 0.1 and 0.5 $Wm^{-1}K^{-1}$, which is not good enough, where materials with high heat conduction are required [45]. Thermal conductivity can be increased by adding fillers, such as boron nitride, graphene, aluminum nitride, aluminum oxide, silicon nitride, graphite, CNT, diamond, metal particle, etc. that have heat conduction ability [46]. Whenever electrical insulation and high thermal properties are required, aluminum nitride, aluminum oxide, boron nitride, etc. are mostly utilized.

The thermal conductivity of most thermoplastics is shown in Table 1.2. Most thermosets have higher thermal conductivity than their thermoplastics counterparts. In a report by Chen et al., [46], different thermosets and their thermal conductivities were presented, showing the structure of the individual monomers and hardeners used. In a composite, the matrix

Preparation and Properties of Nanocomposites

thermal conductivity has an enormous impact on the thermal conductivity of the final nanocomposite. Epoxy resin is amorphous, similar to most thermosets, and it is one of the most used thermosets, having low thermal conductivity, ranging between 0.17 and 0.21 $Wm^{-1}K^{-1}$ [45].

TABLE 1.2 General Thermoplastic Polymers and Their Thermal Conductivity [45]

Thermoplastic Polymer	Thermal Conductivity at Room Temperature ($Wm^{-1}K^{-1}$)
High density polyethylene (HDPE)	0.33–0.53
Ultrahigh molecular weight polyethylene (UHMWPE)	0.41–0.51
Commercial thermotropic liquid crystalline polymers (LCP)	0.30–0.40
Polyoxymethylene (Homo) (POM)	0.30–0.37
Low density polyethylene (LDPE)	0.30–0.34
Poly(ethylene vinyl acetate) (EVA)	0.35
Polyphenylene sulfide (PPS)	0.30
Poly(butyleneterephthalate) (PBT)	0.25–0.29
Polytetrafluoroethylene (PTFE)	0.27
Polyamide-6,6 (PA66)	0.24–0.33
Polyamide-6 (PA6)	0.22–0.33
Polyetheretherketone (PEEK)	0.25
Polysulfone (PSU)	0.22
Polymethylmethacrylate (PMMA)	0.16–0.25
Polycarbonate (PC)	0.19–0.21
Urethanebase TPE (TPU)	0.19
Poly(acrylonitrile-butadiene-styrene) copolymer (ABS)	0.15–0.20
Polyvinyl chloride (PVC)	0.13–0.29
Polyvinylidenedifluoride (PVDF)	0.19
Styrene/polybutadienecopolymer (SB)	0.17–0.18
Styrene-acrylonitrilecopolymer (SAN)	0.15–0.17
Poly(ethyleneterephthalate) (PET)	0.15
Polystyrene (PS)	0.10–0.15
Polyvinylidenechloride (PVDC)	0.13
Polyisobutylene (PIB)	0.12–0.20
Polypropylene (PP)	0.11–0.17
Polyimide, thermoplastic (PI)	0.11

1.3 EFFECTS OF NANOPARTICLES (NPS) IN COMPOSITES

The incorporation of NPs as strengthening agents, to composites, results in an enormous assurance, to not only to improve the obtainable properties of these composites, but in addition, adds several other properties that will exploit and broaden their areas application. NPs have good properties which have made it become the major attractive materials in many fields of applications, such as: in health, engineering, environment, etc. Researches on NPs have vast potentials; viz.: the ability to amend weakly soluble and weakly absorption of particles, which can make, otherwise engineering and biological resources, into useful and capable materials that have better performances.

NPs can be described as objects with at least one of its dimensions, ranging in size between 1 to 100 nm owing to their large surface area-to-volume ratio. The performances and properties of nanocomposites are majorly controlled by the particle-particle interactions and nanoparticle-matrix interactions [47]. There are two forces that exist in nanocomposites; they are the repulsive forces of particle-particle interaction (like charges) and electrostatic forces of nanoparticle-matrix interactions (unlike charges). However, the particle-particle interaction weakens the composites performances, whereas the nanoparticle-matrix interaction produces composites with enhanced performances. The attraction and repulsion forces between the particles, due to van der Waals and electrostatic forces, affect the particle-particle interaction in the composite [48, 49]. The reduction in the repulsive forces between the particles can be achieved by decreasing the particle size in the nanoscale [48]. The NPs have a large surface area-to-volume ratio; hence, more particles are exposed to the particles of the matrix. The large surface area-to-volume ratio enhances more particles collision (reactions), which can be attributed to the reason why the nanocomposites exhibit improved performance. There are different types of NPs, such as the organic type (polymers), inorganic/oxide NPs, carbon-based, and metallic-based NPs. These NPs, when added to the composite, have enormously beneficial effects, such as mechanical, electrical, and health benefits. This section of the chapter deals with the different types of nanoparticle-based composites; their mechanical, electrical and health beneficial effects and the factors that influence their behaviors (improvement of composites properties).

1.3.1 TYPES OF NANOPARTICLE-BASED COMPOSITES

There are different types of NPs-based composites, which are used in various fields of application. Polymeric particles (organic NPs) are the dendrimers, which are mostly used in the pharmaceutical industries. Dendrimers are an attractive class of polymers with controlled structure, covalent bonds, and are distinct NPs [50]. They are used to deliver drugs efficiently. Polyamidoamines (PAMAM) complex (a dendrimer) with Ketoprofen, have been reported to improve the drug permeation through the skin as penetration enhancer [51]. The inorganic non-metallic NPs are synthetic amorphous silica and the oxide NPs. The carbon-based NPs are CNT and carbon black, while the metal-based NPs comprise: NPs of gold, silver, iron, platinum, zirconium, palladium, cadmium, etc. The diagrammatic representation of the different types of NPs, with examples, is represented in Figure 1.3.

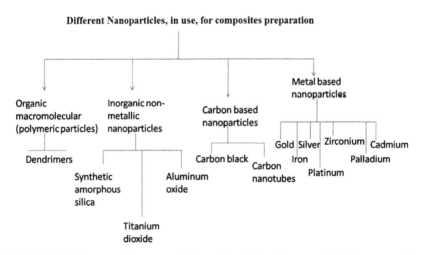

FIGURE 1.3 Diagrammatic representation of the different nanoparticles for composites production.

1.3.2 BENEFICIAL EFFECTS OF INCORPORATION OF NANOPARTICLES (NPS) TO COMPOSITES

The inclusion of NPs in composites has a drastic effect on the mechanical, electrical, and health properties. It has been reported that the incorporation of nanoparticle at a controlled amount can contribute to the properties of the resulting composites [6, 17]. Conventional materials may not meet the required property, but the modification of the composites can significantly improve the properties; modifications, e.g., surface treatment [1], incorporation of compatibilizer [20, 45], and nanoparticle inclusion [46].

1.3.2.1 MECHANICAL BENEFICIAL EFFECT OF NANOPARTICLES (NPS)

The beneficial effect of the incorporation of NPs to improve mechanical properties, such as: tribological, tensile modulus, tensile strength, hardness, tear resistance, rebound resilience, abrasion and flex resistance have been widely explored [5, 6, 53]. There are many reports from various researchers that affirmed that the incorporation of NPs could improve the mechanical properties of the composites. According to the work by Chang and Friedrich [54], the tribological properties of short fiber-reinforced polyphenylene sulfide (SFPS) were considerably improved when NPs was added. They affirmed that the tribological advantages gained by SFPS might perhaps, be accredited to the rolling capacity of NPs. The NPs behaved as spacers who reduce the adhesion forces between the contact surface of the steel and the composites materials and reduce the stress absorption that could have reacted on the individual fibers. In a separate research, Cho and Bahadur [55] stated that the inclusion of 2 vol. % nano-CuO can usually boost the wear resistance of SFPS. According to the study, the useful effects of NPs were greatly appreciated in the growth of thin and homogeneous transfer films. Guo et al., [56], used surface adapted nano-SiO_2 filler to augment the tribo-properties of epoxy composites packed with short carbon fibers. It was reported that the enhancement mechanisms of NPs can be easily recognized by the improved strength of the matrix and enhanced properties of the transfer film. The following mechanical properties: tensile strength, hardness, flexes resistance; abrasion resistance, tensile modulus, etc. were also improved by the addition of NPs to the composites materials. Rosso et al., [57], reported the improvement in mechanical properties, such as tensile modulus and fracture toughness, with the inclusion of 5 vol% silica NPs. Borba et al., [58], reported increases in the composite mechanical properties, viz: tensile strength, flex resistance, rebound resilience, hardness and abrasion resistance when 2 wt% of montmorillonite clay- NPs was incorporated to 5 wt% of styrene-butadiene-styrene triblock copolymer/curaua fiber composite with MA-g-SEBS, as compatibilizer. The inclusion of NPs can also increase the thermo-mechanical properties of the composite. In the work of Bayani et al., [25], it was observed that the thermal stability of composite was improved when nano-TiO_2 was incorporated; the improvement was due to the heat transfer capacity of particles and the impeded diffusion of unstable decomposed products [25]. In the area of improvement on the toughness of epoxy adhesive with silica NPs, Kinloch et al., [59], established that the inclusion of silica NPs to a rubber-toughened epoxy, resulted in incredibly significant improvement in the toughness, glass transition temperature, and single-overlap shear strength.

1.3.2.2 HEALTH BENEFICIAL EFFECTS OF NANOPARTICLES (NPS)

Silver in its lump state (not in NPs form), is a metal that is well-known for its wide range of action against bacteria, definite viruses, and fungi. However, its properties are considerably improved when they are in the form of NPs (silver NPs) [60–62]. Silver NPs have an attraction with molecular groups that have sulfur and phosphorus, which are contained in bacteria membranes and inside bacteria cells. These NPs allowed free movement of silver ions, which can damage trans-membrane electron movement and stop DNA duplication [61, 63, 64]. Quite a number of studies have verified the fact that silver NPs are non-hazardous to animals and humans' cells [65, 66]. Silver NPs are used as antimicrobials and are added in biomaterials and food casing and in the refrigerator, in covering for water filters, air purifiers, washing machines and to decrease the surface microbial biofilm formation [61, 63, 64].

1.3.2.3 ELECTRICAL BENEFICIAL EFFECTS OF NANOPARTICLES (NPS)

The large surface area-to-volume ratio properties of NPs will allow a large portion of NPs to react with the particles of the composite. The incorporation of NPs will help to tailor any void that can cause the electrons to scatter when an electric charge passes through the composite, hence, improving the electrical conductivity of the composite. However, in order to have clear and electrically conductive nanocomposites materials, the followings: silver NPs, oxide of indium tin, CNTs, carbon nanofibers (CNFs), etc., are good candidates' NPs that can induce electrical conductivity in composites.

1.3.3 FACTORS INFLUENCES THE EFFECTS OF NANOPARTICLES (NPS) IN COMPOSITES

The successful reinforcement for the desired properties (mechanical, electrical, magnetic, optical or thermal) of nanocomposites, depend on NPs adequate dispersion, amount loaded, types of NPs and the use of coupling agent. The benefit of adding NPs in composites as reinforcing materials has not been completely realized, owing to the dispersion problem. The dispersion problems encountered in NPs are attributed to their well-built contacts between themselves. Uddin and Sun [67] stated that uneven dispersion of NPs was the cause of the wide differences in the experimental outcome of some researches on polymeric nanocomposites. According to them [67], some researchers reported improvement in the matrix properties [67–69], while others recorded undesirable effects [70–73], owing to the NPs addition. It was further stated that the subject on the dispersion of NPs requires further researches in the: techniques for uniform dispersion of high NPs loading, dispersion of different types of high NPs for their versatile use and how dispersion varies with the particles size. The smaller the nanoparticle sizes (at least one of its dimensions $\leq$ 100 nm), the easier agglomerates are formed, hence, making it difficult to disperse in the matrix. The desired properties of nanocomposites are achieved when the particles are evenly dispersed and within a certain range of weight percentage (of nanoparticle) loading. However, research has shown that the properties of nanocomposites tend to reduce after reaching the maximum, at nanoparticle loading of between 3 to 5 wt% [74–77]. These properties can include tensile and flexural strength, impact strength, flame retardance, water absorption, and morphology.

1.4 METHODS OF PREPARATION OF NANOCOMPOSITES

Scientists and engineers have, over time, combined organic and inorganic composites, at the nanoscale, to create smart materials with outstanding properties and applications. Looking at the natural world, researchers have been able to develop a new approach that comes up with materials that can have, at the same time, numerous advantages in many fields of applications [78]. The right choice of preparation techniques is needed in order to obtain NMs with appropriate properties [79]. Each particular preparative technique that is chosen to produce nanocomposites has great consequences on the types of material that may be produced [80]. In this section, methods of preparation of different nanocomposites are briefly discussed.

1.4.1 DIFFERENT TYPES OF NANOCOMPOSITES AND THEIR METHODS OF PREPARATIONS

There are three different types of nanocomposites. They include ceramics matrix nanocomposites (CMNC), metallic nanocomposites (MMNC) and polymer matrix nanocomposites (PMNC). In spite of their nano-dimensions, the majority of the processing methods of the aforementioned types of nanocomposites are relatively similar as in micro composites. This is also correct, even for CNTs-reinforced composites [81]. A lot of experimental techniques have been published for the preparation of ceramic matrix nanocomposites [81–84]. However, among the reported techniques for the preparation of CMNC, the most common methods used are: the conventional powder method, polymer precursor route, spray pyrolysis and the vapor method (which includes CVD and PVD). The most frequent methods for the processing of MMNC are vapor techniques, electrodeposition, and the chemical methods (which include colloidal and sol-gel) processes, while the methods used to process polymer nanocomposites include: layered materials and those of CNTs [81].

The main effective chemical processes that permit the incorporation of metal NPs into the organic phase is the sol-gel process. The phrase "sol-gel" is related to two reaction steps, viz: sol and gel. The sol is a colloidal suspension of solid particles in a liquid phase, and gel is the unified network formed between phases [85]. However, the exploitation of the sol-gel method to prepare useful materials is extremely old; as it has been in use for over 17,000 years [86]. The expansion of the sol-gel method as it applies to ceramic, continues until today due to the work of Yamane [87] and Yoldas [88]. They established that watchful dehydration of gels can generate monoliths that are attributed to increased research on the sol-gel process, which are still relevant until this day [89, 90]. The sol-gel production route is made up of two main reactions, viz: hydrolysis and condensation. Both reaction steps are multi-steps processes, which happen in sequence. The hydrolysis reaction is a cleaving of organic chain bonding to metal and followed by the substitution with hydroxyl (OH) groups through nucleophilic count. Condensation is dependent on oxygen, metal, oxygen bond formation. By means of clarity, condensation reaction releases small molecules, such as water or alcohol, as products of the reactions [85, 91, 92].

1.5 AREAS OF APPLICATION OF NANOCOMPOSITES IN THE ENERGY INDUSTRIES

NMs exhibit unique physical, chemical and biological phenomena and characteristics at the nanoscale level when compared to those of individual atoms and molecules or bulk matter. In this regard, they have a diversity of applications in various engineering and technological fields [93]. The energy industry is one field, where nanoscale science and engineering continue to be successfully applied.

The area of scientific research interest is in the application of nanoscience to the energy industry, which includes ways of developing clean, affordable and renewable energy sources and reduction of energy consumption and the related pollution on the environment. Thus, the applications of NMs and nanocomposites in the energy industry encompass the following areas:

1.5.1 SOLAR ENERGY HARVESTING

Globally, there is an ever-increasing demand on renewable energy sources as a substitute for non-renewable energy due to the obvious depletion of such energy. In addition to this challenge, non-renewable energy sources that use fossil fuels, such as coal, petroleum, and natural gas, also contribute greatly to environmental pollution and global warming. These challenges make the exploration of alternative environmentally-friendly renewable energy sources, e.g., solar energy, an attractive option [94, 95]. Solar energy is generated from the radiant light and heat from the sun. This energy is clean, free, abundantly available, and renewable, and it has a minimal environmental impact. An immense amount of power, $\sim 1.8 \times 10^{11}$

Preparation and Properties of Nanocomposites 21

MW, is intercepted from the sun by the earth [96]. This power can efficiently and beneficially be converted into electricity and heat through the application of nanoscience and nanotechnology. Harvesting and modification of solar energy can be done in different ways, viz.:

i. Use of photovoltaic (PV) technology to directly convert light into the electrical current;
ii. Use of solar collectors for solar thermal systems;
iii. Use of artificial photosynthesis to produce either carbohydrates or hydrogen, via water splitting;
iv. Use of 'passive solar' technologies to maximize solar lighting and heating;
v. Use of biomass technology, whereby plants use solar radiation to drive chemical transformations and create complex carbohydrates. These carbohydrates can then be used to produce electricity, steam, or biofuels.

These energy-related processes and their applications are part of the solar economy concept [90]. Nanotechnology related to solar collectors, photocatalysis, solar PV, dye-sensitized solar cells (DSSC) and fuel cells, are used to enhance the efficient conversion of solar energy to more useful forms of energy, e.g., electricity and heat [98].

Nanotechnology is a powerful tool that can be employed in supporting the efficient use, sustainable energy conversion, storage, and conservation of solar energy. Technologies, such as solar collectors, photo-catalysis, solar PV, DSSC and fuel cells, are means through which nanotechnology is used to enhance the conversion of solar energy to more useful forms of energy [98].

1.5.1.1 SOLAR COLLECTORS

Solar collectors are green heat exchangers that absorb solar radiant light, transfer this energy to a collector fluid by passing into contact with it and causing the internal energy of the absorbing body to increase. The properties of the collector fluid greatly influence the performance of the solar collector. The energy absorbed can be converted to thermal energy via the solar thermal applications, or be converted to electrical energy via PV applications [99]. Solar collectors can be used in residential and small commercial applications, such as water heating systems in homes, solar space heating, solar desalination, solar drying devices, electricity production, and small solar power plants [99].

In order to improve the efficiency of a solar collector, the absorbing collector fluid medium can be replaced with a nanofluid, i.e., a high thermal conductivity fluid containing suspended solid NPs [98]. NPs in the fluid has increased surface area-to-volume ratio, which enhances the solar energy collection and efficiency by exposing more conducting surfaces to the sunlight [100]. Furthermore, nanofluids have high absorption capacity when compared to ordinary base fluids, and they are stable under moderate temperature, improve the radiative properties of the base fluids, reduce sedimentation, fouling and clogging of pumps and pipes due to their nano-size. These characteristics can enhance, very greatly, the efficiency of solar collectors [99].

1.5.1.2 PHOTOVOLTAIC (PV) CELLS

The PV technology converts radiant energy from solar light into electrical energy. This occurs when light falls on a semiconductor material, thereby causing electron excitation and strongly enhancing conductivity [94]. Two types of PV technology that are available in the market include: (a) crystalline silicon-based PV cells and (b) thin-film technologies made from various semiconducting materials that include amorphous silicon, cadmium telluride and copper indium gallium dieseline [94].

Conventional PV cells, have the following advantages: environmentally-friendly, produce no noise, have no moving parts, no emissions, do not require the use of fuels and water, have minimal maintenance requirements, long lifetime (up to 30 years), they generate electricity wherever there is light (solar or artificial), they operate even in cloudy conditions, their modular energy can be customized for any application from a watt to a multi-megawatt power plant. Some of their disadvantages include: PV cells have no output at night when there is no light, lower output in an unfavorable weather, use of toxic materials in some solar cell types, high initial costs that overshadow the low maintenance costs, large area requirements for large scale applications. PV generates direct current (DC), and therefore, special DC appliances or inverters are needed. For off-grid applications, energy storage is needed [101].

PV technology has been established over many years. However, research efforts have continued to reduce the cost per unit of power by using less costly materials and improving efficiency. Study areas include: replacing the costly silicon that was conventionally used in PV technology by the less costly organic nanofilms, quantum dots (QDs) and dye-sensitized cells, among other areas [102].

The use of nanocrystal QDs, which are NPs made of direct bandgap semiconductors in PV technology, leads to thin-film SCs that are based on a conductive transparent oxide (CTO) substrate, with a nanocrystal coating. QDs are efficient light emitters. They emit multiple electrons per solar proton, with different absorption and emission spectra, depending on the particle size. In this way, they raise the theoretical efficiency of the PV cells [97].

1.5.1.3 DYE-SENSITIZED SOLAR CELLS (DSSC)

DSSC are semiconductor PV devices that enable optical absorption and charge separation of solar radiation by linking a light-absorbing dye sensitizer to a photoanode, which is usually a wide bandgap semiconductor of nano-crystalline morphology [100]. In DSSC, light absorption takes place largely in dye molecules attached to the surface of NPs of a wide bandgap semiconductor, usually TiO_2. Upon excitation with visible light, the dye injects electrons from its excited state into the TiO_2 conduction band, leading to a charge-separated state, as expressed in the chemical reaction equation below:

$$Proton + TiO_2 + dye \rightarrow TiO^-_2 + dye^+$$

The energy in the form of voltage or current can then be tapped from this reaction. Having a nanoporous TiO_2 with a large surface area per unit volume and weight leads to increased light absorption. The optimization of the dye/TiO_2 layer is heavily dependent on a nanoscale fabrication and properties, a large TiO_2-layer surface area to incorporate enough dye and a structure that is thin and open enough to enable efficient charge transport with minimum losses [103].

Wang and Wu [104] reported the potential application of ZnO NMs as transparent photo-anodes in DSSCs due to the material's coupled piezoelectric, semiconductor, and photoexcitation properties. Some drawback of DSSCs built with one-dimensional nanostructure is the insufficient internal surface area and deficient dye loading. The integration of ZnO or TiO$_2$ nanowires with 3D optical fibers or planar waveguides has been shown to address these drawbacks. In addition, this approach also enables the construction of DSSCs with increased flexibility, adaptability, and efficiency.

1.5.1.4 PHOTO-CATALYSIS

Photo-catalysis process is the acceleration of a photo-reaction mechanism in the presence of a catalyst, accompanied by the absorption of solar energy by an adsorbed substrate. The resulting photo-generated catalysis activity (PCA) depends on the ability of the catalyst to create electron-hole pairs, which can then undergo redox reactions with other catalytic species [105]. Materials with photo-catalytic properties are used to convert solar energy into chemical energy to further oxidize or minimize materials and generate useful components together with hydrogen and hydrocarbons. Furthermore, pollutants and bacteria are removed from the wall surfaces [106].

A widely used material in photo-catalysis is titania (TiO$_2$) nanoparticle. This material is chemically stable, non-toxic, durable, and inexpensive. It is transparent to the visible light and has strong oxidizing abilities to decompose organic pollutants, e.g., super-hydrophilicity [107, 108]. The photocatalytic properties of TiO$_2$ result from the formation of photo-generated charge carriers (hole and electron), which occur upon the absorption of ultraviolet (UV) light, corresponding to the bandgap. TiO$_2$ two-dimensional structured nanosheets have a flat surface, high aspect ratio, low turbidity, and excellent adhesion to substrates. They can thus, be effectively applied in the self-cleaning glass. This unique structure might provide potential advantages for purification, separation, and storage [108]. With the discovery of this phenomenon, the application range of TiO$_2$ coating can be applied in various areas, including sterilization, air purification, water purification, defogger, self-cleansing, and H$_2$ generation. [98].

1.5.2 APPLICATION OF OIL RECOVERY ENHANCEMENT IN THE PETROLEUM INDUSTRY

Increased production from oil reservoirs for the petroleum industries depend on various techniques and parameters. Nanotechnology can be used to introduce the mechanisms that can enhance oil recovery from oil reservoirs. These mechanisms include viscosity reduction, interfacial tension (IFT) reduction, wettability alteration, and selective plugging [109]. Various studies relating to the use of nanotechnology for the enhancement of oil recovery have been reported by various researchers. Table 1.3 summarizes the different NPs and their potential mechanisms for enhancing oil recovery.

1.5.3 ENHANCING BATTERIES EFFICIENCY

A battery is an electrochemical power source that instantly converts chemical energy into electrical energy. It has voltaic cells arranged in series, composed of a cathode, electrolyte, and anode. The two electrodes; the cathode as a positive terminal and anode as a negative terminal, are immersed in an electrolyte, through which electrode systems with different potentials are formed and therefore, enabling tapping of electrical energy at the two electrodes [93].

For over three decades, huge investments have been made in developing LIBs, and they are now a part of our daily life as power sources for electric vehicles (EV) and portable electronic devices. For instance, since 1991, the Sonny Company has commercially produced LIBs for miniaturization of electronic devices. Subsequently, further progress has been noted in the development of new electrode and electrolyte materials and the associated increase in the energy density of commercial batteries.

The use of nanosized and nanostructured materials in batteries promises new possibilities in terms of increased energy density, exceptionally high rate of charge and discharge ability, and recyclability. The self-monitoring analysis and reporting technology (SMART) devices require fast charging and long operation time. This can be achieved through the use of NMs that enable the realization of charging speeds that are faster than those achieved from conventional bulk materials, without any significant degradation in storage capacity [111].

There has been a considerable amount of research on rechargeable lithium batteries. One reason for this is that the Li-ion chemistry in these batteries yields an increase of 100–150% of the storage capacity of energy per unit weight and volume in comparison to the aqueous batteries. However, there are some associated disadvantages to these batteries. These include low energy and power density, large volume changes on reaction, safety challenges, and cost. These shortcomings are being addressed

TABLE 1.3 Summary of the Nanoparticles and Their Potential Effect on Enhancing Oil Recovery, After Negin, Ali [110] and Bera and Belhaj [109]

Nanoparticle	Desired Effect for Oil Recovery Enhancement	Remark
Aluminum Oxide (Al_2O_3) Copper(II) oxide, CuO Iron oxide, (Fe_2O_3/Fe_3O_4) Nickel Oxide (Ni_2O_3)	Viscosity reduction	Nanoparticles effect viscosity reduction in heavy oils leading to improved oil recovery
SiO2 Nanoparticles Hydrophilic and lipophilic polysilicon (HLP) Polyacrylamide micro-gel Nano-spheres Polymer-coated nanoparticles Ferro-fluid	Interfacial tension (IFT) reduction	Nanoparticles influence surfactant flooding. This increases the capillary number by reducing interfacial tension of oil-water systems thereby enhancing oil recovery
Tin Oxide (SnO_2) SiO_2 nanoparticles Alumina coated nanoparticles Hydrophobic silicone oxide Hydrophobic silicon oxide Spherical fumed silica nanoparticles Neutrally wet polysilicon (NWP) nanoparticles Lipophobic and Hydrophilic Polysilicon nanoparticles Polymer-coated nanoparticles	Wettability alteration	Wettability of rock surface is changed, affecting capillary pressure and relative permeability leading to favorable oil recovery efficiency
Nono-sized colloidal dispersion gels (CDG) Polymer nanoparticles	Sweep and displacement efficiency	Used as agents to improve oil mobility ratio thereby enhancing oil recovery

through the application of nanotechnology to the field of rechargeable batteries. Researches in the use of NMs for both the electrodes and the non-aqueous electrolyte nano-batteries are ongoing activities [97].

Goriparti et al., [112] reported that certain physicochemical properties of nanostructures can (and will) enable increased lithium storage, increased Li-ion flux at the electrode/electrolyte interface, reduced diffusion length for both Li-ions and electrons and minimal anode volume change during the charging/discharging process. Combined together, these improvements can readily, guaranty increased energy density and high power density devices. For example, graphene nanomaterial, with its good electrical properties, can be suitably used as graphene/metal anodes. The use of CNTs has shown promising experimental results for the improvement of battery performance. However, the production cost is a hindrance to their application as battery anode activating material.

A battery category with good potential for high energy capacity application is the one that incorporates carbonaceous and titanium oxides as anode materials. The former makes use of well-developed soft carbon technology. However, hard carbon technology presents a good alternative, particularly for a high capacity application, e.g., in the EV sector. Titanium oxide anode technology is yet to be well-developed for use in batteries. They exhibit poor energy density, high reversible capacity, and high power density, just as they have safety challenges. Nonetheless, they have the potential for application in high power batteries in hybrid electric vehicles (HEVs) and systems that require high power.

1.5.4 HYDROGEN ENERGY PRODUCTION

The continued global increase in demand for energy has led to the harnessing of alternative renewable energy sources. Hydrogen generated from renewable energy sources is a potential alternative fuel, which can lead to less dependence on fossil fuels. Hydrogen is a clean and environmentally-friendly fuel that does not emit any greenhouse gas [97, 100, 102]. Furthermore, H_2 combustion with oxygen produces only water vapor and heat, in line with the following chemical reaction:

$$H_2 + \frac{1}{2}O_2 \rightarrow H_2O + heat \quad \Delta H = 120 kJg^{-1}$$

In addition, hydrogen is the lightest element and offers an optimal energy-to-weight ratio of any fuel. Being one of the atoms in water, hydrogen gas (H$_2$) is also the most abundant element. It can act both as an alkali metal by giving away the sole electron and as a halogen by accepting an electron. Hydrogen is an important industrial commodity, since it is the key component in ammonia synthesis, petroleum refining, and petrochemical processing.

Hydrogen gas can be generated by decomposing water, by using low-cost primary sources of energy, such as solar, wind and tidal energy or biomass decomposition. In this regard, hydrogen acts as an energy carrier, transporting energy from the generation site to another point. However, the low density of hydrogen gas necessitated the development of safe and cost-effective methods to store hydrogen. The gas is also highly reactive and inflammable. These issues hinder the use of H$_2$ on a commercial scale [113].

Fuel cell technology considered in the framework of hydrogen energy, electro-chemically converts hydrogen and oxygen as an oxidizing agent into the water and in the process produces electricity and heat. This is an environmentally-friendly and non-carbon dioxide emitting process. Fuel cells ensure efficiency and safe storage of hydrogen and thus, address the challenges of the commercial use of hydrogen gas. However, there are drawbacks in fuel cell technology, which include: high cost, durability, and operability challenges. It is believed that these drawbacks can be addressed by nanotechnology.

Hydrogen energy is used extensively in space vehicles, due to its excellent energy-to-weight ratio when compared to any other fuel. Furthermore, fuel cells with nanostructured materials are effectively used to enhance the conversion of hydrogen energy into electricity [100]. Water splitting, by photo-catalysis, also called artificial photosynthesis, is a research niche area that aims at the production of affordable hydrogen. By using a variety of semiconductor nanoparticulate catalyst systems, based on CdS, SiC, CuInSe$_2$, or TiO$_2$ are nanotechnology approaches being studied for this purpose [97].

1.5.5 WIND POWER PRODUCTION

Conversion of wind energy into a useful form of energy is another alternative to non-renewable fossil fuels. Wind power is abundantly available, renewable, widely distributed, clean, and produces no greenhouse gas emissions during the generation and transmission/operation. Wind electricity is predicted to increase considerably in the next decade, with the fastest developments taking place in China and India. Wind power can substantially contribute towards the increasing global electricity demand [114].

In the early development of windmills, wood and canvas were used to make windmill blades, due to their reasonable price, availability, and ease of manufacture. Later, small blades were also made from light-weight metals, e.g., aluminum and its alloys. However, these materials require repeated maintenance and limit the blade shape to a flat plate. These blade shapes have low aerodynamic efficiency, needed to capture wind energy. Modern wind blades can now be manufactured by using polymer composites, made of additives and polymer matrices, thermoplastics, thermosets, and elastomers. They are considered as relatively inexpensive materials. Furthermore, CNTs can be added to the polymer resin, thereby resulting in CNT/polymer nanocomposites. The CNT/polymer nanocomposites have a wide potential for their application due to the exceptional mechanical, electrical, and thermal characteristics of CNTs [114].

In wind turbines, the energy produced is proportional to the square of its blade length. The use of longer blades on wind turbines is, therefore, advantageous in energy production. Nanocomposite materials have remarkable strength-to-weight and stiffness-to-weight ratios. They can thus, be used to manufacture, not only long but also strong blades. Nanotechnology can also be used to increase wind turbine efficiency. This is achieved by applying nano-lubricants and low-friction coatings on the blades. These coatings reduce energy losses caused by tribological issues, such as micro-pitting, wear scuffing, and spalling in gearboxes [100]. Further research on NMs for the minimization of environmental factors and increased strength for composite blades of the windmill is ongoing. Apart from CNTs, other multifunctional NMs, such as CNFs, graphene, and TiO_2, have attracted great interest in this area of research [115].

1.5.6 ENERGY STORAGE IN SUPERCAPACITORS

Supercapacitors and ultra-capacitors are also referred to as electrochemical capacitors (ECs). Like the batteries, they are used to store electrical energy, but using a different mechanism. Whereas batteries store electrical energy chemically, ultra-capacitors store it physically through the separation of the positive and negative charges. The disadvantages of classical capacitors are three-fold:

a. The costly premium performance electrodes due to the miniaturization;
b. The long cycle life requirements of approximately 10^5 cycles; and
c. Low efficiency.

The surface area of each electrode that makes up the capacitor is the dominant factor for power density and maximum power output. The use of nanostructured materials dramatically increases this surface area. For example, it can be as high as 1000 m^2/g of carbon [97].

Supercapacitors have a remarkably high power density and fast charge than conventional batteries. However, they have a very low energy density, besides being heavy and bulky [116]. The maximum power density of supercapacitors is proportional to the reciprocal of the internal resistance. Various factors, referred to as equivalent series resistance (ESR), contribute to the internal resistance. These include the electronic resistance of the electrode carbon material and the interface resistance between the electrode and the current collector. Nanotubes can be used to escalate the power density of supercapacitors. This is connected to the nanoscale tubular morphology of these materials, which offers a unique combination of low electrical resistivity and high porosity in a readily accessible structure. Studies on the use of thin-film electrodes with multi-walled aligned nanotubes have shown a significant increase in the specific power density of supercapacitors [98].

The use of MnO_2 nanowire/CNT paper composite material in supercapacitors has shown improved electrochemical performance and capacitance of the composite electrode. This is attributed to the high conductivity, flexibility, and active substrate behavior of CNT paper on one hand and the facilitation of the contact between the electrolyte and the other active components of the superconductor. Other nano-composites that enhance electrode capacitance of superconductors include: Globular-shaped reduced graphene oxide–Co_3O_4 nanocomposite and graphene oxide–multi-walled carbon nanotube (MWCNT) composites [116].

1.5.7 OTHER ENERGY RELATED SYSTEMS AND APPLICATIONS

Other applications of nanotechnology in the energy industry, published online by the National Nanotechnology Initiative of the United States include: efficient lighting systems to reduce illumination energy consumption, lighter and stronger vehicle chassis materials for the transportation sector, advanced electronics for low energy consumption, nano-engineered low-friction lubricants for improved efficiency in machine gears, pumps and fans, light-responsive smart coatings for glass, to complement alternative heating/cooling schemes and high-light-intensity, fast-recharging lanterns for emergency works [117].

Many other alternative energy sources are being explored, studied, and implemented worldwide, in a bid to reduce the energy dependence on fossil fuels and minimize the associated greenhouse effects on earth. These alternative energy systems include vibration energy, waste heat harvesting, recycling, biomass, catalysts, and micro-organisms, among others. The use of nanotechnology, directly and indirectly, to produce high-efficiency energies and reduce greenhouse gases, by employing these alternative energy systems, is a priority [115].

1.6 ENVIRONMENTAL, HEALTH, AND SAFETY CONCERNS

Nanoscience and nanotechnology offer great opportunities to better the life of humankind. Without a doubt, science is poised to address the challenges of environmental pollution and the need for increased and sufficient clean energy at an affordable cost. However, it is important that all these are done without introducing any new risk, for example, inducing new forms of pollution or endangering the lives of people, animals, and plants [118].

The growth in nanotechnology applications in the energy and other fields, such as electronics, medicine, defense, and food entails some public environmental health and safety concerns. There is, therefore, need for scientists, researchers, engineers, and policymakers to work collectively in creating better and safer NMs. This will help to reduce (and possibly eliminate) any negative effects of nanotechnology and its products [119].

1.7 FUTURE TRENDS AND CHALLENGES

Nanotechnology research and development have been growing rapidly for the past two and half decades, and most likely, this trend will continuously increase worldwide for the next few decades. Breakthroughs in nanotechnology will result in major business changes. Obsolescence of the present technologies and the capital shifts from traditional business to novel technologies are likely to pose some challenges. Nanoscience research and nanotechnology are key enablers of a variety of renewable and sustainable energy. Some key areas where nanotechnology is likely to have a critical impact in the near future, include: generating terawatt-scale solar energy at a lower cost when compared to that of fossil fuel-based energy generation and economically facilitating the conversion of solar energy into environmentally-friendly chemical fuels [120]. Nanotechnology has the potential to lead to the development of novel hybrid electronic/ionic conduction, storage mechanisms, electrolytes, and new electrochemical compounds. These will greatly improve battery power, energy density, battery lifetime, and efficiency of diffusion and displacement reactions in the batteries [121].

CNTs with their potential to absorb and store huge quantities of hydrogen are likely to contribute greatly to the development of hydrogen-fueled vehicles. This development is being championed by the world's four major automobile companies, namely: General Motors, Ford, Toyota, and Daimler-Benz AG [122]. The use of this clean, environmentally-friendly fuel that does not emit any greenhouse gas will reduce carbon emission from vehicles and reduce ozone layer effects. Water-splitting by nanophotocatalysis is a very promising, clean, sustainable, attractive, and economical way of directly generating hydrogen from a primary renewable energy source [97].

Generally, nanotechnology is critical to the development of materials which would result in energy, fuel, material and economic savings as well as the development of remarkable materials with properties at the subatomic level, which are yet to be discovered. Nanotechnology is likely to lead to the replacement of the existing chips by super chips, plastic semiconductors, stronger and lighter jet fighters, super fuel cells, and super batteries. Sustained future researches are required in order to produce novel NPs with excellent heat conductive and mechanical properties. Increasing the aspect ratio of fillers can increase the thermal conductivity with low nanoparticle content in the composite. Hence, further improvements in smart nanocomposite materials that can be achieved and adopted for various applications in the energy industries are obviously imperative.

1.8 CONCLUSION

This chapter has presented the different properties of nanocomposites, used for energy applications, methods of preparation, and areas of application. NPs have widened the area of application of polymeric materials; most especially in the energy sector. The current demand has influenced the types of materials used in the production of renewable energy for sustainable development. Energy generation and storage are largely influenced by the choice of material(s) used. Carbon cryogels and titania (TiO_2), in their various forms, have shown prominence for energy applications. Utilization of NPs for electrodes or electro-catalysts can increase the surface area and result in electrodes with increased reaction rates, therefore improving the properties of the nanocomposite materials.

The thermal conductivity of nanocomposites can be increased by adding fillers with heat conduction ability, such as boron nitride, graphene, aluminum nitride, aluminum oxide, silicon nitride, graphite, CNT, diamond and metal particle into the polymer matrix. The thermal conductivity of most polymers ranges between 0.1 and 0.5 $Wm^{-1}K^{-1}$. Various factors have been advanced to contribute to the properties of nanoparticle-based composites. These factors include particle size, loading, and method of dispersion. The smaller surface area of nanoparticle increased the contact area between the particle and the matrix. It was equally reported that nanocomposites with nanoparticle loading above 5 wt% tend to show a reduction in properties. Furthermore, if the NPs are not well dispersed within the composites, it forms agglomeration, which has shown a tendency to reduce the properties of the nanocomposites. The future of the energy sector of any country depends on nanotechnology. Based on the characteristics of nanocomposite materials and their area of applications, a variety of preparation routes can be applied to prepare the nanocomposite materials with large surface area to volume ratio. Each preparation route determines the properties of the nanocomposites and where it can be applied. The future of the materials used in the energy industries depends a lot on the level of advancement in the field of nanotechnology.

ACKNOWLEDGMENTS

The financial support from the CSIR-BIS South Africa is greatly appreciated by ID Ibrahim, AA Eze, and OS Adekomaya. C Kambole gratefully acknowledges the financial support from Copperbelt University, Kitwe, and Zambia.

KEYWORDS

- energy application
- nanocomposites
- nanoparticle
- renewable energy
- thermal properties

REFERENCES

1. Banos, R., Manzano-Agugliaro, F., Montoya, F., Gil, C., Alcayde, A., & Gómez, J., (2011). Optimization methods applied to renewable and sustainable energy: A review. *Renewable and Sustainable Energy Reviews, 15,* 1753–1766.
2. Martiskainen, M., & Coburn, J., (2011). The role of information and communication technologies (ICTs) in household energy consumption-prospects for the UK. *Energy Efficiency, 4,* 209–221.
3. Lee, T. Y., & Chen, C. L., (2009). Wind-photovoltaic capacity coordination for a time-of-use rate industrial user. *IET Renewable Power Generation, 3,* 152–167.
4. Zhou, W., Lou, C., Li, Z., Lu, L., & Yang, H., (2010). Current status of research on optimum sizing of stand-alone hybrid solar–wind power generation systems. *Applied Energy, 87,* 380–389.
5. Alexandre, M., & Dubois, P., (2000). Polymer-layered silicate nanocomposites: Preparation, properties and uses of a new class of materials. *Materials Science and Engineering: R: Reports, 28,* 1–63.
6. Ibrahim, I. D., Jamiru, T., Sadiku, R. E., Kupolati, W. K., & Agwuncha, S. C., (2016). Dependency of the mechanical properties of sisal fiber reinforced recycled polypropylene composites on fiber surface treatment, fiber content and nanoclay. *Journal of Polymers and the Environment,* 1–8.
7. Spitalsky, Z., Tasis, D., Papagelis, K., & Galiotis, C., (2010). Carbon nanotube–polymer composites: chemistry, processing, mechanical and electrical properties. *Progress in Polymer Science, 35,* 357–401.
8. Odegard, G., Clancy, T., & Gates, T., (2005). Modeling of the mechanical properties of nanoparticle/polymer composites. *Polymer, 46,* 553–562.
9. Bonnet, P., Sireude, D., Garnier, B., & Chauvet, O., (2007). Thermal properties and percolation in carbon nanotube-polymer composites. *Applied Physics Letters, 91,* 1910.

10. Huang, X., & Jiang, P., (2015). Core–shell structured high-k polymer nanocomposites for energy storage and dielectric applications. *Advanced Materials, 27*, 546–554.
11. Hu, X., Li, G., & Yu, J. C., (2009). Design, fabrication, and modification of nanostructured semiconductor materials for environmental and energy applications. *Langmuir, 26*, 3031–3039.
12. Arico, A. S., Bruce, P., Scrosati, B., Tarascon, J. M., & Van Schalkwijk, W., (2005). Nanostructured materials for advanced energy conversion and storage devices. *Nature Materials, 4*, 366–377.
13. Kamat, P. V., (2007). Meeting the clean energy demand: nanostructure architectures for solar energy conversion, *The Journal of Physical Chemistry C, 111*, 2834–2860.
14. Orilall, M. C., & Wiesner, U., (2011). Block copolymer based composition and morphology control in nanostructured hybrid materials for energy conversion and storage: solar cells, batteries, and fuel cells. *Chemical Society Reviews, 40*, 520–535.
15. Proctor, C. M., Kuik, M., & Nguyen, T. Q., (2013). Charge carrier recombination in organic solar cells. *Progress in Polymer Science, 38*, 1941–1960.
16. McCulloch, I., Heeney, M., Bailey, C., Genevicius, K., MacDonald, I., Shkunov, M., Sparrowe, D., Tierney, S., Wagner, R., & Zhang, W., (2006). Liquid-crystalline semiconducting polymers with high charge-carrier mobility. *Nature Materials, 5*, 328–333.
17. Barroso, M., Pendlebury, S. R., Cowan, A. J., & Durrant, J. R., (2013). Charge carrier trapping, recombination and transfer in hematite (α-Fe 2 O 3) water splitting photoanodes. *Chemical Science, 4*, 2724–2734.
18. Rothenberger, G., Moser, J., Graetzel, M., Serpone, N., & Sharma, D. K., (1985). Charge carrier trapping and recombination dynamics in small semiconductor particles. *Journal of the American Chemical Society, 107*, 8054–8059.
19. Liu, Q., Guo, Y., Chen, Z., Zhang, Z., & Fang, X., (2016). Constructing a novel ternary Fe (III)/graphene/gC 3 N 4 composite photocatalyst with enhanced visible-light driven photocatalytic activity via interfacial charge transfer effect. *Applied Catalysis B: Environmental, 183*, 231–241.
20. Ndjawa, G. O. N., Graham, K. R., Mollinger, S., Wu, D. M., Hanifi, D., Prasanna, R., Rose, B. D., Dey, S., Yu, L., & Brédas, J. L., (2017). Open-circuit voltage in organic solar cells: The impacts of donor semicrystallinity and coexistence of multiple interfacial charge-transfer bands. *Advanced Energy Materials, 7*(12), 1601995. https://doi.org/10.1002/aenm.201601995 (Accessed on 17 October 2019).
21. Laquai, F., Andrienko, D., Deibel, C., & Neher, D., (2017). Charge carrier generation, recombination, and extraction in polymer–fullerene bulk heterojunction organic solar cells. In: *Elementary Processes in Organic Photovoltaics* (pp. 267–291). Springer.
22. Candelaria, S. L., Chen, R., Jeong, Y. H., & Cao, G., (2012). Highly porous chemically modified carbon cryogels and their coherent nanocomposites for energy applications. *Energy & Environmental Science, 5*, 5619–5637.
23. Ibrahim, I. D., Jamiru, T., Sadiku, E. R., Kupolati, W. K., Agwuncha, S. C., & Ekundayo, G., (2016). Mechanical properties of sisal fiber-reinforced polymer composites: A review. *Composite Interfaces, 23*, 15–36.
24. Ibrahim, I. D., Jamiru, T., Sadiku, E. R., Kupolati, W. K., & Agwuncha, S. C., (2016). Impact of surface modification and nanoparticle on sisal fiber reinforced polypropylene

nanocomposites. *Journal of Nanotechnology*. http://dx.doi.org/10.1155/2016/4235975 (Accessed on 25 July 2019).
25. Bayani, M., Ehsani, M., Khonakdar, H. A., Seyfi, J., & HosseinAbadi-Ghaeni, M. H., (2016). An investigation of TiO_2 nanoparticles effect on morphology, thermal and mechanical properties of epoxy/silica composites. *Journal of Vinyl and Additive Technology*. doi: 10.1002/vnl.21558.
26. Jacob, M., Thomas, S., & Varughese, K. T., (2004). Mechanical properties of sisal/oil palm hybrid fiber reinforced natural rubber composites. *Composites Science and Technology, 64*, 955–965.
27. Ismail, H., Nizam, J., & Khalil, H. A., (2001). The effect of a compatibilizer on the mechanical properties and mass swell of white rice husk ash filled natural rubber/linear low density polyethylene blends. *Polymer Testing, 20*, 125–133.
28. Ibrahim, I., Mohamed, F., & Lavernia, E., (1991). Particulate reinforced metal matrix composites—a review. *Journal of Materials Science, 26*, 1137–1156.
29. Habraken, W., Wolke, J., & Jansen, J., (2007). Ceramic composites as matrices and scaffolds for drug delivery in tissue engineering. *Advanced Drug Delivery Reviews, 59*, 234–248.
30. Shah, K. J., Shukla, A. D., Shah, D. O., & Imae, T., (2016). Effect of organic modifiers on dispersion of organoclay in polymer nanocomposites to improve mechanical properties. *Polymer, 97*, 525–532.
31. Pasbakhsh, P., Ismail, H., Fauzi, M. A., & Bakar, A. A., (2010). EPDM/modified halloysite nanocomposites. *Applied Clay Science, 48*, 405–413.
32. Krishnaiah, P., Ratnam, C. T., & Manickam, S., (2017). Development of silane grafted halloysite nanotube reinforced polylactide nanocomposites for the enhancement of mechanical, thermal and dynamic-mechanical properties. *Applied Clay Science, 135*, 583–595.
33. Zare, Y., (2016). Study of nanoparticles aggregation/agglomeration in polymer particulate nanocomposites by mechanical properties. *Composites Part A: Applied Science and Manufacturing, 84*, 158–164.
34. Horny, N., Kanake, Y., Chirtoc, M., & Tighzert, L., (2016). Optimization of thermal and mechanical properties of bio-polymer based nanocomposites. *Polymer Degradation and Stability, 127*, 105–112.
35. Mohammadi, M., Ziaie, F., Majdabadi, A., Akhavan, A., & Shafaei, M., (2017). Improvement of mechanical and thermal properties of high energy electron beam irradiated HDPE/hydroxyapatite nano-composite. *Radiation Physics and Chemistry, 130*, 229–235.
36. Yang, S., Fan, H., Jiao, Y., Cai, Z., Zhang, P., & Li, Y., (2017). Improvement in mechanical properties of NBR/LiClO 4/POSS nanocomposites by constructing a novel network structure. *Composites Science and Technology, 138*, 161–168.
37. Wang, X., Wang, L., Su, Q., & Zheng, J., (2013). Use of unmodified SiO_2 as nanofiller to improve mechanical properties of polymer-based nanocomposites. *Composites Science and Technology, 89*, 52–60.
38. Ma, H. L., Zhang, Y., Hu, Q. H., He, S., Li, X., Zhai, M., & Yu, Z. Z., (2013). Enhanced mechanical properties of poly (vinyl alcohol) nanocomposites with glucose-reduced graphene oxide. *Materials Letters, 102*, 15–18.

39. Kemaloglu, S., Ozkoc, G., & Aytac, A., (2010). Properties of thermally conductive micro and nano size boron nitride reinforced silicon rubber composites. *Thermochimica Acta, 499*, 40–47.
40. Zhou, W., Wang, C., Ai, T., Wu, K., Zhao, F., & Gu, H., (2009). A novel fiber-reinforced polyethylene composite with added silicon nitride particles for enhanced thermal conductivity. *Composites Part A: Applied Science and Manufacturing, 40*, 830–836.
41. Chrissafis, K., & Bikiaris, D., (2011). Can nanoparticles really enhance thermal stability of polymers? Part I: An overview on thermal decomposition of addition polymers. *Thermochimica Acta, 523*, 1–24.
42. Bogoeva-Gaceva, G., Raka, L., & Dimzoski, B., (2008). Thermal stability of polypropylene/organo-clay nanocomposites produced in a single-step mixing procedure. *Adv. Comp. Lett., 17*, 161–164.
43. Golebiewski, J., & Galeski, A., (2007). Thermal stability of nanoclay polypropylene composites by simultaneous DSC and TGA. *Composites Science and Technology, 67*, 3442–3447.
44. Liao, C. Z., & Tjong, S. C., (2011). Effects of carbon nanofibers on the fracture, mechanical, and thermal properties of PP/SEBS-g-MA blends. *Polymer Engineering & Science, 51*, 948–958.
45. Huang, X., Jiang, P., & Tanaka, T., (2011). A review of dielectric polymer composites with high thermal conductivity. *IEEE Electrical Insulation Magazine, 27*, 8–16.
46. Chen, H., Ginzburg, V. V., Yang, J., Yang, Y., Liu, W., Huang, Y., Du, L., & Chen, B., (2016). Thermal conductivity of polymer-based composites: Fundamentals and applications. *Progress in Polymer Science, 59*, 41–85.
47. Pukánszky, B., & Fekete, E., (1999). Adhesion and surface modification. In: *Mineral Fillers in Thermoplastics I* (pp. 109–153). Springer.
48. Islam, M. S., Masoodi, R., & Rostami, H., (2013). The effect of nanoparticles percentage on mechanical behavior of silica-epoxy nanocomposites, *Journal of Nanoscience.* http://dx.doi.org/10.1155/2013/275037 (Accessed on 25 July 2019).
49. Russel, W. B., Saville, D. A., & Schowalter, W. R., (1989). *Colloidal Dispersions.* Cambridge University Press.
50. Baig, T., Nayak, J., Dwivedi, V., Singh, A., Srivastava, A., & Tripathi, P. K., (2015). A review about dendrimers: synthesis, types, characterization and applications. *International Journal of Advances in Pharmacy, Biology and Chemistry*, 44–59.
51. Tripathy, S., & Das, M. K., (2013). Dendrimers and their applications as novel drug delivery carriers. *Journal of Applied Pharmaceutical Science, 3*(09) 142–149.
52. Ibrahim, I. D., Jamiru, T., Sadiku, R. E., Kupolati, W. K., Agwuncha, S. C., & Ekundayo, G., (2015). The use of polypropylene in bamboo fiber composites and their mechanical properties–A review. *Journal of Reinforced Plastics and Composites, 34*, 1347–1356.
53. Ray, S. S., & Okamoto, M., (2003). Polymer/layered silicate nanocomposites: A review from preparation to processing. *Progress in Polymer Science, 28*, 1539–1641.
54. Chang, L., Zhang, Z., Ye, L., & Friedrich, K., (2007). Tribological properties of epoxy nanocomposites: III. Characteristics of Transfer Films, *Wear, 262*, 699–706.
55. Cho, M., & Bahadur, S., (2005). Study of the tribological synergistic effects in nano CuO-filled and fiber-reinforced polyphenylene sulfide composites. *Wear, 258*, 835–845.

56. Guo, Q. B., Rong, M. Z., Jia, G. L., Lau, K. T., & Zhang, M. Q., (2009). Sliding wear performance of nano-SiO 2/short carbon fiber/epoxy hybrid composites. *Wear, 266*, 658–665.
57. Rosso, P., Ye, L., Friedrich, K., & Sprenger, S., (2006). A toughened epoxy resin by silica nanoparticle reinforcement. *Journal of Applied Polymer Science, 100*, 1849–1855.
58. Borba, P. M., Tedesco, A., & Lenz, D. M., (2014). Effect of reinforcement nanoparticles addition on mechanical properties of SBS/Curauá fiber composites. *Materials Research, 17*, 412–419.
59. Kinloch, A., Mohammed, R., Taylor, A., Eger, C., Sprenger, S., & Egan, D., (2005). The effect of silica nano particles and rubber particles on the toughness of multiphase thermosetting epoxy polymers. *Journal of Materials Science, 40*, 5083–5086.
60. Morones, J. R., Elechiguerra, J. L., Camacho, A., Holt, K., Kouri, J. B., Ramírez, J. T., & Yacaman, M. J., (2005). The bactericidal effect of silver nanoparticles. *Nanotechnology, 16*, 2346.
61. Kalishwaralal, K., BarathManiKanth, S., Pandian, S. R. K., Deepak, V., & Gurunathan, S., (2010). Silver nanoparticles impede the biofilm formation by Pseudomonas aeruginosa and Staphylococcus epidermidis. *Colloids and Surfaces B: Biointerfaces, 79*, 340–344.
62. Neves, P. B. A. D., Agnelli, J. A. M., Kurachi, C., & Souza, C. W. O. D., (2014). Addition of silver nanoparticles to composite resin: effect on physical and bactericidal properties *in vitro*. *Brazilian Dental Journal, 25*, 141–145.
63. Li, Q., Mahendra, S., Lyon, D. Y., Brunet, L., Liga, M. V., Li, D., & Alvarez, P. J., (2008). Antimicrobial nanomaterials for water disinfection and microbial control: Potential applications and implications. *Water Research, 42*, 4591–4602.
64. Allaker, R., (2010). The use of nanoparticles to control oral biofilm formation. *Journal of Dental Research, 89*, 1175–1186.
65. Sayes, C. M., Fortner, J. D., Guo, W., Lyon, D., Boyd, A. M., Ausman, K. D., Tao, Y. J., Sitharaman, B., Wilson, L. J., & Hughes, J. B., (2004). The differential cytotoxicity of water-soluble fullerenes. *Nano Letters, 4*, 1881–1887.
66. Sondi, I., & Salopek-Sondi, B., (2004). Silver nanoparticles as antimicrobial agent: A case study on *E. coli* as a model for gram-negative bacteria. *Journal of Colloid and Interface Science, 275*, 177–182.
67. Uddin, M., & Sun, C., (2009). Effect of nanoparticle dispersion on mechanical behavior of polymer nanocomposites. in: *Proceedings of the 50th Materials Conference* (pp. 4–7). American Institute of Aeronautics and Astronautics.
68. Yasmin, A., Abot, J. L., & Daniel, I. M., (2003). Processing of clay/epoxy nanocomposites by shear mixing. *Scripta Materialia, 49*, 81–86.
69. Adebahr, T., Roscher, C., & Adam, J., (2001). Reinforcing nanoparticles in reactive resins. *European Coatings Journal*, 144–149.
70. Kinloch, A., Lee, J., Taylor, A., Sprenger, S., Eger, C., & Egan, D., (2003). Toughening structural adhesives via nano-and micro-phase inclusions. *The Journal of Adhesion, 79*, 867–873.
71. Yong, V., & Hahn, H. T., (2004). Processing and properties of SiC/vinyl ester nanocomposites. *Nanotechnology, 15*, 1338.

72. Haggenmueller, R., Du, F., Fischer, J. E., & Winey, K. I., (2006). Interfacial in situ polymerization of single wall carbon nanotube/nylon 6, 6 nanocomposites. *Polymer, 47*, 2381–2388.
73. Zilg, C., Thomann, R., Finter, J., & Mülhaupt, R., (2000). The influence of silicate modification and compatibilizers on mechanical properties and morphology of anhydride-cured epoxy nanocomposites. *Macromolecular Materials and Engineering, 280*, 41–46.
74. Guo, Z., Liang, X., Pereira, T., Scaffaro, R., & Hahn, H. T., (2007). CuO nanoparticle filled vinyl-ester resin nanocomposites: Fabrication, characterization and property analysis. *Composites Science and Technology, 67*, 2036–2044.
75. Rodgers, R. M., Mahfuz, H., Rangari, V. K., Chisholm, N., & Jeelani, S., (2005). Infusion of SiC nanoparticles into SC-15 epoxy: An investigation of thermal and mechanical response. *Macromolecular Materials and Engineering, 290*, 423–429.
76. Subramaniyan, A. K., & Sun, C., (2006). Enhancing compressive strength of unidirectional polymeric composites using nanoclay. *Composites Part A: Applied Science and Manufacturing, 37*, 2257–2268.
77. Choi, Y. K., Sugimoto, K. I., Song, S. M., Gotoh, Y., Ohkoshi, Y., & Endo, M., (2005). Mechanical and physical properties of epoxy composites reinforced by vapor grown carbon nanofibers. *Carbon, 43*, 2199–2208.
78. Bensaude-Vincent, B., Arribart, H., Bouligand, Y., & Sanchez, C., (2002). Chemists and the school of nature. *New Journal of Chemistry, 26*, 1–5.
79. Ruiz-Hitzky, E., Aranda, P., Darder, M., & Ogawa, M., (2011). Hybrid and biohybrid silicate based materials: Molecular vs. block-assembling bottom–up processes. *Chemical Society Reviews, 40*, 801–828.
80. Wang, D., Zhu, J., Yao, Q., & Wilkie, C. A., (2002). A comparison of various methods for the preparation of polystyrene and poly (methyl methacrylate) clay nanocomposites. *Chemistry of Materials, 14*, 3837–3843.
81. Camargo, P. H. C., Satyanarayana, K. G., & Wypych, F., (2009). Nanocomposites: Synthesis, structure, properties and new application opportunities. *Materials Research, 12*, 1–39.
82. Thostenson, E. T., Ren, Z., & Chou, T. W., (2001). Advances in the science and technology of carbon nanotubes and their composites: A review. *Composites Science and Technology, 61*, 1899–1912.
83. Nakahira, A., & Niihara, K., (1992). Strctural ceramics-ceramic nanocomposites by sintering method: roles of nano-size particles. *Journal of the Ceramic Society of Japan, 100*, 448–453.
84. Stearns, L. C., Zhao, J., & Harmer, M. P., (1992). Processing and microstructure development in Al_2O_3-SiC 'nanocomposites.' *Journal of the European Ceramic Society, 10*, 473–477.
85. Maneeratana, V., (2007). *Alkoxide-Based Precursors for Direct Electrospinning of Alumina Fibers*. PhD Thesis, ISBN: 9780549456681. University of Florida, USA.
86. Wright, J. D., & Sommerdijk, N., (2001). *Sol-Gel Materials: Chemistry and Applications* (Vol. 4, pp. 1–14). Gordon and Breach Science Publishers, Amsterdam, the Netherlands.

87. Yamane, M., Aso, S., & Sakaino, T., (1978). Preparation of a gel from metal alkoxide and its properties as a precursor of oxide glass. *Journal of Materials Science, 13*, 865–870.
88. Yoldas, B. E., (1975). Alumina gels that form porous transparent Al_2O_3. *Journal of Materials Science, 10*, 1856–1860.
89. West, J., & Hench, L., (1990). The sol-gel process, *Chem. Rev., 90*, 33–72.
90. Bahloul, W., Bounor-Legaré, V., David, L., & Cassagnau, P., (2010). Morphology and viscoelasticity of PP/TiO2 nanocomposites prepared by in situ sol–gel method. *Journal of Polymer Science Part B: Polymer Physics, 48*, 1213–1222.
91. Brinker, C. J., & Scherer, G. W., (2013). *Sol-Gel Science: The Physics and Chemistry of Sol-Gel Processing*. Academic Press.
92. Schubert, U., Huesing, N., & Lorenz, A., (1995). Hybrid inorganic-organic materials by sol-gel processing of organofunctional metal alkoxides. *Chemistry of Materials, 7*, 2010–2027.
93. Bashir, S., & Liu, J., (2015). Chapter 1 - nanomaterials and their application. In: *Advanced Nanomaterials and Their Applications in Renewable Energy* (pp. 1–50). Elsevier, Amsterdam.
94. Sagadevan, S., (2013). Recent trends on nanostructures based solar energy applications: A review. In: *Rev. Adv. Mater. Sci., 44*–61.
95. Han, N., & Ho, J. C., (2014). 3 - one-dimensional nanomaterials for energy applications A2 - Tjong, sie-chin. In: *Nanocrystalline Materials* (2nd edn., pp. 75–120). Elsevier, Oxford.
96. Sukhatme, S., & Nayak, K., (2008). *Solar Energy Principles of Thermal Collection and Storage* (3rd edn.). Tata McGraw Hill Education Private Limited. ISBN (13), 78-0-07-026064-1.
97. Serrano, E., Rus, G., & García-Martínez, J., (2009). Nanotechnology for sustainable energy. In: *Renewable and Sustainable Energy Reviews*, 2373–2384.
98. Abdin, Z., Alim, M. A., Saidur, R., Islam, M. R., Rashmi, W., Mekhilef, S., & Wadi, A., (2013). Solar energy harvesting with the application of nanotechnology. In: *Renewable and Sustainable Energy Reviews*, 837–852.
99. Hussein, A. K., (2016). Applications of nanotechnology to improve the performance of solar collectors–recent advances and overview. In: *Renewable and Sustainable Energy Reviews*, 767–792.
100. Hussein, A. K., (2015). Applications of nanotechnology in renewable energies-A comprehensive overview and understanding. In: *Renewable and Sustainable Energy Reviews*, 460–476.
101. Dhankhar, M., Pal Singh, O., & Singh, V. N., (2014). Physical principles of losses in thin film solar cells and efficiency enhancement methods. In: *Renewable and Sustainable Energy Reviews*, 214–223.
102. Tegart, G., (2009). Energy and nanotechnologies: Priority areas for Australia's future. In: *Technological Forecasting and Social Change*, 1240–1246.
103. Zäch, M., Hägglund, C., Chakarov, D., & Kasemo, B., (2006). Nanoscience and nanotechnology for advanced energy systems. In: *Current Opinion in Solid State and Materials Science*, 132–143.
104. Wang, Z. L., & Wu, W., (2012). Nanotechnology-enabled energy harvesting for self-powered micro-/nanosystems. In: *Angewandte Chemie International Edition*, 11700–11721.

105. Xiong, J., Han, C., Li, Z., & Dou, S., (2015). Effects of nanostructure on clean energy: Big solutions gained from small features. In: *Science Bulletin*, 2083–2090.
106. Zhang, L., Ding, Y., Povey, M., & York, D., (2008). ZnO nanofluids–A potential antibacterial agent. In: *Progress in Natural Science*, 939–944.
107. Colmenares, J. C., Luque, R., Campelo, J. M., Colmenares, F., Karpiński, Z., & Romero, A. A., (2009). Nanostructured photocatalysts and their applications in the photocatalytic transformation of lignocellulosic biomass: An overview. In: *Materials*, 2228–2258.
108. Nakata, K., & Fujishima, A., (2012). TiO$_2$ photocatalysis: Design and applications. In: *Journal of Photochemistry and Photobiology C: Photochemistry Reviews*, 169–189.
109. Bera, A., & Belhaj, H., (2016). Application of nanotechnology by means of nanoparticles and nanodispersions in oil recovery-A comprehensive review. In: *Journal of Natural Gas Science and Engineering*, 1284–1309.
110. Negin, C., Ali, S., & Xie, Q., (2016). Application of nanotechnology for enhancing oil recovery- a review. *Petroleum*, 2(4) 324–333.
111. Myung, S. T., Amine, K., & Sun, Y. K., (2015). Nanostructured cathode materials for rechargeable lithium batteries. *Journal of Power Sources*, 219–236.
112. Goriparti, S., Miele, E. F., De Angelis, E., Di Fabrizio, R., Proietti, Z., & Capiglia, C., (2014). Review on recent progress of nanostructured anode materials for Li-ion batteries. In: *Journal of Power Sources*, 421–443.
113. Shashikala, K., (2012). Chapter-hydrogen storage materials A2. In: Banerjee, S., & Tyagi, A. K., (eds.), *Functional Materials, Elsevier* (pp. 607–637), London.
114. Ma, P. C., & Zhang, Y., (2014). Perspectives of carbon nanotubes/polymer nanocomposites for wind blade materials. *Renewable and Sustainable Energy Reviews*, 30, 651–660.
115. Asmatulu, R., & Khan, W. S., (2013). Chapter 10-nanotechnology safety in the energy industry. In: *Nanotechnology Safety* (pp. 127–139). Elsevier, Amsterdam.
116. Liu, H. K., (2013). An overview-functional nanomaterials for lithium rechargeable batteries, supercapacitors, hydrogen storage, and fuel cells. In: *Materials Research Bulletin*, 4968–4973.
117. United States National Nanotechnology Initiative, Nanotechnology benefits and applications. https://www.nano.gov/you/nanotechnology-benefits. Online (Accessed on 25 July 2019).
118. Pautrat, J. L., (2011). Nanosciences: Evolution or revolution? In: *Comptes. Rendus. Physique*, 605–613.
119. Khan, W. S., & Asmatulu, R., (2013). Chapter 1-nanotechnology emerging trends, markets, and concerns. In: *Nanotechnology Safety* (1–16). Elsevier, Amsterdam.
120. Brinker, C. J., & Ginger, D., (2011). Chapter 6-nanotechnology for sustainability: Energy conversion, storage, and conservation. In: *Nanotechnology Research Directions for Societal Needs in 2020* (pp. 261–303). Springer.
121. Brinker, C. J., & Ginger, D., (2011). Nanotechnology for sustainability: Energy conversion, storage, and conservation. In: *Nanotechnology Research Directions for Societal Needs in 2020* (pp. 261–303). Springer.
122. Shi, D., Guo, Z., & Bedford, N., (2015). 3-carbon nanotubes. In: *Nanomaterials and Devices* (pp. 49–82). William Andrew Publishing, Oxford.

CHAPTER 2

Recent Progress in Nanocrystalline Oxide Thin Films

PALLABI GOGOI,[1] D. PAMU,[2] and L. ROBINDRO SINGH[1]

[1]*Department of Nanotechnology, North Eastern Hill University, Shillong, Meghalaya, India*

[2]*Department of Physics, Indian Institute of Technology Guwahati, Guwahati, Assam, India*

ABSTRACT

Recently, the worldwide telecommunication industry is developing towards slighter and lighter technology. Thin film technology has been playing an important role in the field of miniaturization of various microwave devices and integration of various optical devices. Scientists and researchers have been giving intensive and continuous effort in the development of thin films. The dielectric thin films with good optical properties are used in the integrated optical device as antireflection coating, dielectric mirror, etc. Fabrication of thin-film capacitors using high dielectric constant materials can provide higher charge storage densities when they exhibit lower dielectric loss. Ferroelectric, piezoelectric oxide thin films of low leakage current, low hysteresis, and high stability are used in sensor and tunable microwave device applications. In this chapter, the recent progress in the fabrication of oxide thin films and their applications in different fields are discussed.

2.1 INTRODUCTION

Thin films are homogeneous solid material which can be extended infinitely in two dimensions but restricted in the third dimensions; i.e., thickness. The thin film is fabricated by depositing materials on any substrate atom-by-atom process. The typical thickness of the thin films is in between 1 nm–10 μm.

The science of the thin film has become an important area of researcher for the scientific community because of its various applications in recent technology. The constant need of miniaturization of technology provides importance and inspiration for the development of ever smaller and lighter alternatives of existing bulk material. Thin films have distinct advantages over bulk material because of the nanoscale size-dependent properties of the material. Thin films are very sensitive to their structures. Their properties change with variation of thickness, uniformity, deposition techniques, etc. The substrate temperature, deposition rate, environmental conditions, purity of the material to be deposited, residual gas are some of the important factor affecting the physical, electrical, optical and other properties of the film [1–3].

Number of researchers and scientific communities have been developing different new deposition techniques and successfully deposited good quality thin film of different materials. Among them, oxide thin films are showing higher impact in the technological field because of their excellent dielectric, electric, and optical properties in thin-film form. Dielectric oxide films exhibit high dielectric constant, low loss even in a microwave frequency range, which makes these film very useful in energy storage and telecommunication industries [4–6]. The ferroelectric and piezoelectric oxide films are leading in sensor manufacturing [7–10]. In this chapter, the characterization, deposition techniques of oxide thin films are presented. Furthermore, the recent advances of the oxide film in the application filed are discussed.

2.2 CHARACTERIZATION OF OXIDE THIN FILMS

2.2.1 STRUCTURAL PROPERTIES

When we say about structural properties of a thin film, it includes mainly two basic properties: a) crystallographic structure of the film, b) morphology of the film. The measurements methodologies use to find out them are briefly discussed below.

1. **Crystallographic Structure:** X-ray diffraction (XRD) is the most versatile and non-destructive technique revealing detailed information about the crystallographic structure of materials. Every

crystalline substance gives particular XRD patterns. It is based on the scattering of X-rays by crystal.

X-ray with a wavelength similar to the distance between the crystal planes can be reflected such that the angle of reflection is equal to the angle of incidence. This is described by Bragg's law:

$$2d \sin \theta = n\lambda \quad (1)$$

where λ is the wavelength, d is the average spacing between layer or raw of atom.

Figure 2.1 shows the diffracted X-ray from a crystal plane. The refinement of the XRD patterns can give the value of lattice parameters, size, shape, and internal stress of the crystalline film.

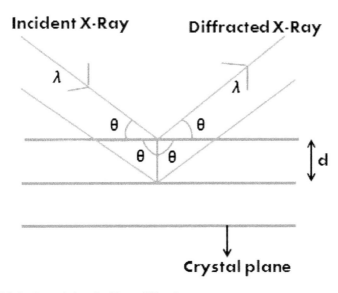

FIGURE 2.1 Bragg's law for X-ray diffraction.

2. **Morphology of the Film:** Scanning electron microscopy (SEM) and atomic force microscopy (AFM) are used to characterize the morphology, chemical composition and crystallographic information of the films. SEM can produce images of a sample by scanning the surface with a focused beam of electrons. The energy of electrons is dissipated as different signals depend on the interactions between the electrons and samples. SEM can achieve

resolution better than 10 nm. For very small topographic details on the surface of a sample, field emission scanning electron microscopy (FESEM) is used. In FESEM external electromagnetic field is used for emission of electron. AFM is the alternate method to characterize the topography of the deposited film. It can provide a 3D surface profile of the sample on a nanoscale. The surface roughness of the thin film can also measure using AFM. Energy-dispersive X-ray spectroscopy is used for the elemental composition analysis of the thin films.

2.2.2 OPTICAL PROPERTIES

One of the most easy and popular ways to find out the optical properties of thin films is the enveloping technique [11]. All the optical constants such as refractive index, absorption coefficient, optical bandgap, packing density, porosity ratio can be determined from transmission spectra. Further, the thickness of the films can also find out from this method. The typical transmission spectrum of a film deposited onto a transparent substrate is shown in Figure 2.2.

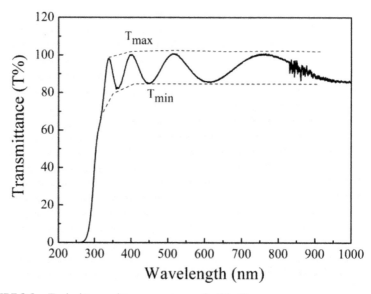

FIGURE 2.2 Typical transmittance spectrum of a thin film.

The refractive index (n) of the film is derived by employing the envelop method based on the following expressions [11],

$$n = [N + (N^2 - n_s^2)^{0.5}]^{0.5} \tag{2}$$

where,

$$N = 2n_s^2 \left(\frac{T_{max} - T_{min}}{T_{max} T_{min}}\right) + \frac{n_s^2 + 1}{2} \tag{3}$$

n_s is the refractive index of the substrate used, T_{max} and T_{min} are the corresponding transmittance maximum and minimum at a certain wavelength λ.

The thickness of the films (d) can be calculated by the following expression,

$$d = \frac{\lambda_1 \lambda_2}{2(\lambda_1 n_2 - \lambda_2 n_1)} \tag{4}$$

where, n_1 and n_2 are the refractive indices of two adjacent maximum and minimum at wavelengths λ_1 and λ_2.

The optical packing density (p) of the films is calculated using the relation [12],

$$p = \left(\frac{n_f^2 - 1}{n_f^2 + 2}\right)\left(\frac{n_b^2 + 2}{n_b^2 - 1}\right) \tag{5}$$

where, n_b is the bulk refractive index of the bulk sample, and n_f is the observed film refractive index.

The porosity ratio (the volume of pores per volume of film) of the films obtained using the following expression [13],

$$P = 1 - \frac{n_f^2 - 1}{n_b^2 - 1} \tag{6}$$

where n_f and n_b are the refractive indices for the film and bulk sample, respectively.

The optical bandgap (E_g) of the films can be calculated using the Tauc relation [14],

$$(\alpha h\nu)^n = B(h\nu - E_g) \tag{7}$$

where $h\nu$ is the incident photon energy, n determines the type of electronic transition between the valence band and conduction band, α is the absorption coefficient, B is a measure of crystalline order. Extrapolating the linear portion of $(\alpha h\nu)^n$ verses $(h\nu)$ curve, the E_g of the film would be obtained.

2.2.3 MICROWAVE AND BROADBAND DIELECTRIC PROPERTIES

The split post dielectric resonator (SPDR) is the most accurate method for measuring dielectric properties such as dielectric constant and loss tangent of the substrate and thin-film at spot frequency in the range of 1–20 GHz [15]. The schematic sketch of the SPDR is shown in Figure 2.3. The thin film used for measurement is placed in between two low loss dielectric resonators. The film is in contact with only one of the resonator. There is a small air gap between the film and another resonator. SPDR typically operated with $TE_{01\delta}$ mode so that the electric field remains continuous across the dielectric interface.

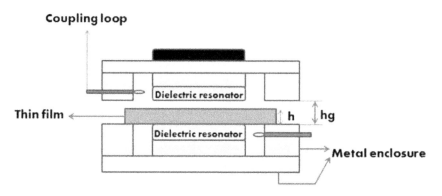

FIGURE 2.3 (See color insert.) Typical cross-sectional view of SPDR fixture.

To cancel out the effect of the substrate in the resonance frequency, one has to measure the resonance frequency and quality factor (f_{01}, Q_{01}) of the empty resonator and resonator with the bare substrate (f_0, Q_0). After the film deposition, the resonance frequency and quality factor of the (f_s, Q_s) substrate coated with the film have to be measured again. The real part of the permittivity of the sample is found on the basis of the measurements of the resonant frequencies and thickness of the sample as an iterative solution using the following expression [16],

$$\varepsilon' = \frac{1+(f_0 - f_s)}{hf_0 K_s(\varepsilon', h)} \quad (8)$$

Here, h is the sample thickness; f_0 is the resonance frequency of the empty SPDR, and f_s is the resonance frequency of the SPDR with the film. K_s is a function of the sample's dielectric constant ε' and h of the film.

The loss tangent is evaluated by using the equation (9),

$$\tan \delta = \frac{Q^{-1} - Q_{Dr}^{-1} - Q_c^{-1}}{P_{es}} \quad (9)$$

where, Q is related to the unloaded quality factor of SPDR with the film, Q_{Dr} is the dielectric losses in the empty resonators, and Q_c is related to the metal losses for the resonant cavity containing the thin film.

The LCR meters are generally used for measurement of the capacitance and dissipation factor of thin-film in the radio frequency (RF) region by fabricating parallel plate capacitor of a thin-film. The parallel plate capacitor method involves sandwiching a thin sheet of the material between two electrodes to form a capacitor. The relative permittivity is calculated using the following relation [17],

$$C = \frac{\varepsilon_r \varepsilon_o A}{d} \quad (10)$$

where, C is the capacitance of the material and ε_r, and ε_o are the relative permittivity of the material and free space, respectively. A is the area of the electrode, and d is the thickness of the film.

2.3 DEPOSITION TECHNIQUES OF OXIDES THIN FILMS

Researchers have been developing newer methods and techniques for the fabrication of thin films. Among them, most commonly used techniques are: physical vapor deposition (PVD), sputtering, chemical vapor deposition (CVD), sol-gel technique, etc. But in addition to these varieties of techniques such as inkjet printing, screen-printing, microcontact printing, nanopatterning, electrochemical deposition, drop coating are used to form useful devices [18–23].

For oxide thin film fabrication-sputtering technique is very popular because of its better adhesion of the material to the substrate, low cost, surface uniformity, control of film thickness. The sputtering is a momentum transfer process between the positive ions and material surface (cathode) atoms as a result of which ejection of atoms takes place. Sir W. R. Grove discovered the sputtering phenomenon in 1852. The measure of ejection of the rate of atoms is known as sputtering yield (Y), and it can be expressed as:

$$Y = \frac{3}{4\pi^2} \alpha \frac{4m_1 m_2}{(m_1 + m_2)^2} \frac{E}{U_S} \quad (11)$$

where, α is a dimensionless parameter. It depends on the mass ratio and ion energy. m_1 and m_2 (amu) are the masses of the target atom and projectile. E is the projectile energy, and U_s is the binding energy of the surface atom.

There are several sputtering techniques, among them direct current (DC) sputtering, RF sputtering, magnetron sputtering and reactive sputtering are commonly used for thin film deposition. In DC sputtering technique, plasma is created in the sputtering chamber by the application of DC voltage between the electrodes. Here Ar gas is used as a sputtering gas. But the low deposition rate and high thermal load on the substrate are the main drawbacks of this technique. Moreover, only conducting material can be used for deposition using DC sputtering. In DC sputtering, when the energetic Ar ions hit the target; their electric charge is neutralized, and they become to Ar atoms. For the insulating target, instead of the Ar neutralization process, a positive charge layer will form on the target surface. It causes the repulsion of bombarding ions, and eventually, the sputtering process stops. To overcome this problem, the RF sputtering technique is used for deposition of insulating/dielectric thin film. A frequency of 13.56 MHz is applied to the cathode via impedance matching network to create plasma. The magnetron sputtering is used to get better confinement of plasma and to improve the ionization efficiency of the atom, which results in a high deposition rate onto the substrate. In this technique, a magnetic field is produced parallel to the cathode surface to trap the emitted electrons. This technique can be used in both DC and RF sputtering. In reactive magnetron sputtering along with inert gas (Ar) some reactive gases are intentionally added to react with sputtering material. This process significantly affects the deposition rate and stoichiometry of the film.

Kannan [24] deposited zinc oxide thin film by DC sputtering on a glass substrate. They obtained a 40 nm thick crystalline ZnO film by maintaining 300°C substrate temperatures during film deposition. Crystallographic orientation was observed by Li [25] while depositing $Pb_{0.6}Sr_{0.4}TiO_3$

thin film using RF sputtering on LaNiO$_3$ buffered silicon substrate. As-deposited nanocrystalline was observed under a pure Ar atmosphere. Jain [26] fabricated perfect (100) orientation of Pb$_{0.6}$Sr$_{0.4}$TiO$_3$/LaNiO$_3$ thin film deposited on SiO$_2$/Si substrate by sputtering. The thicknesses of the films were 250 nm. To study the effect of deposition environment, many researchers have tested different deposition conditions by changing the substrate temperature, sputtering gas and varying the sputtering gas ratio which are Ar and O$_2$ while using RF sputtering for thin film deposition. They observed a remarkable difference in the properties of thin films. A list of oxide thin film deposited by RF sputtering under different deposition conditions is given in Table 2.1.

From the tabulating data, it is clear that the properties of the films change with the deposition environment while using the RF sputtering method. The crystallinity, density of the film, surface uniformity, and thicknesses of the films are greatly influenced by deposition conditions, and all these parameters are related to film quality. Therefore, optimization of deposition conditions depending upon the use of the film is very important. Furthermore, choosing a substrate for deposition of the thin film is also important. To grow good quality thin film, the lattice parameters of the substrate and film material should be matched. The increase of lattice mismatch enhances the internal stress of the film, which degrades the film quality.

Use of Sol-gel technique for deposition of thin-film is promising because of better control of the reaction with homogeneity and doping content. Also, the required reaction temperature is very low in Sol-gel technique. In this method, the sol-gel solution of the required sample is deposited onto the surface of the substrate by dip coating, spin coating, roll-to-roll coating, droplet-assisted deposition, and electro assisted deposition process. It is one of the most cost-efficient methods that provide an opportunity to deposit thin films even on a large area. Jain [47] prepared Ba$_{0.5}$Sr$_{0.5}$TiO$_3$: MgO heterostructured thin film by Sol-gel method and obtained high structural quality films with low dielectric loss and low leakage current. The film thickness was 520 nm, and LaAlO$_3$ substrate was used for film growth. The composite film was annealed at 1100°C in the air for 6 h. The Ba$_{0.5}$Sr$_{0.5}$TiO$_3$/MgTiO$_3$ composite thin films were also successfully fabricated by sol-gel process onto Pt(111)/Ti/SiO$_2$/Si(100) substrate [48]. The Ag-doped (Ba$_{0.6}$Sr$_{0.4}$) TiO$_3$ thin films prepared by Sol-gel method exhibited crystallinity after annealed at 700C for 2 h. The film thickness was 280 nm [49]. Surendran [50], prepared Ni and Zn doped MgTiO$_3$ thin

TABLE 2.1 List of Oxide Thin Films Deposited by RF Sputtering Under Different Deposition Conditions

Material Name	Substrate Used	Substrate Temperature	Sputtering Gas Ratio (Ar/O$_2$)	Thickness	References
(K$_{0.5}$Na$_{0.5}$)NbO$_3$	Quartz, Pt/TiO$_2$/SiO$_2$/Si	400°C	0:100, 25:75, 50:50, 75:25, 100:0	468–498 nm	[27]
(K$_{0.5}$Na$_{0.5}$)NbO$_3$ +0.5wt.% Dy$_2$O$_3$	Quartz, Pt/TiO$_2$/SiO$_2$/Si	400°C	0:100, 25:75, 50:50, 75:25, 100:0	410–430 nm	[28]
KNN + 1 wt.% Gd$_2$O$_3$	Quartz, Pt/TiO$_2$/SiO$_2$/Si	400°C	0:100, 25:75, 50:50, 75:25, 100:0	270–280 nm	[29]
Mg$_2$TiO$_4$	Si(100), Quartz, Pt/TiO$_2$/SiO$_2$/Si	Room temperature (RT)	0:100, 25:75, 50:50, 75:25, 100:0	------	[30]
Mg$_2$TiO$_4$	Si(100), Quartz	RT	0:100, 25:75, 50:50, 75:25, 100:0	200–300 nm	[31]
MgTiO$_3$	n-type Si(100)	200°C, 300°C, 400°C	100:0, 90:10, 80:20, 70:30	------	[32]
MgTiO$_3$	Si(1 0 0)	100–400°C	100:0	6 μm	[33]
Mg$_4$Ta$_2$O$_9$	p-type Si(10 0)	250°C, 350°C, 450°C	100:0	4.82 μm	[34]
(Mg$_{0.95}$Ni$_{0.05}$)TiO$_3$	SiO$_2$, Pt/TiO$_2$/SiO$_2$/Si	RT	0:100, 25:75, 50:50, 75:25, 100:0	------	[35]
Mg(Zr$_{0.05}$Ti$_{0.95}$)O$_3$	SiO$_2$, Pt/TiO$_2$/SiO$_2$/Si	RT	0:100, 25:75, 50:50, 75:25, 100:0	350± 10 nm	[36]
Co doped MgTiO$_3$	Pt/TiO$_2$/SiO$_2$/Si	300°C	0:100, 25:75, 50:50, 75:25, 100:0	280 ± 10 nm	[37]
(Mg$_{0.95}$Co$_{0.05}$)TiO$_3$	SiO$_2$	300°C	0:100, 25:75, 50:50, 75:25, 100:0	300 ± 10 nm	[38]
Ba$_5$Nb$_4$O$_{15}$	Pt/Ti/SiO$_2$/Si, SiO$_2$	RT	0:100, 25:75, 50:50, 75:25, 100:0	429–568 nm	[39]
BaWO$_4$	SiO2	RT, 200°C, 400°C, 600°C, 800°C	100:0	450± 10 nm	[40]
Ba$_5$Nb$_4$O$_{15}$–BaWO$_4$	(100) oriented Pt/Ti/SiO$_2$/Si, SiO$_2$	RT	0:100, 25:75, 50:50, 75:25, 100:0	630 nm	[41]
Al$_2$O$_3$	Indium tin oxide (ITO) covered glass slides	RT	1:0, 10:1, 100:1	160 nm	[42]

TABLE 2.1 (Continued)

Material Name	Substrate Used	Substrate Temperature	Sputtering Gas Ratio (Ar/O_2)	Thickness	References
$Na_{0.5}K_{0.5}NbO_3$	$LaAlO_3$ substrates	650°C	3:1	1.0 μm	[43]
(K, Na)NbO_3	(001)$SrRuO_3$/Pt/MgO	580–650°C	20:1	0.5–2.5 μm	[44]
$(K_{0.50}Na_{0.50})NbO_3$	$SrRuO_3$-buffered $SrTiO_3$	650°C	4:1	300 nm	[45]
$K_{0.48}Na_{0.52}NbO_3$	Pt/Ti/SiO_2/Si	600°C	0:1	300 nm	[46]

film on platinum-coated silicon substrates [(111) Pt/TiO$_2$/SiO$_2$/(100)] Si using Sol-gel technique. The 500 nm thick films were annealed at 650°C for 1 h to get crystallinity. It has been observed that getting the required thickness of the film by Sol-gel techniques is very complex. It requires several coating-baking processes, which may affect the film quality.

Wu [51] used the aerosol deposition method to prepare MgTiO$_3$ thick film. This technique is effective for the fabrication of densified thick film. In the aerosol deposition method, fine or ultra-fine sample powder is mixed with gas to form an aerosol, which will be sprayed through a nozzle to form a film coat on substrate. The energy conservation mechanism in this method is still not sufficiently clear [52]. In this method, the nanocrystalline film can be directly formed from an initial bulk powder on almost any substrate material without sintering. Zenga [53] prepared MgTiO$_3$ thin film on Si and SrTiO$_3$ substrate by atmospheric pressure metalorganic CVD techniques. SrTiO$_3$ and MgTiO$_3$ thin film grown by the vapor-solid reaction on (100) and (110) TiO$_2$ single crystals has been studied by Lotnyk [54]. Interestingly, they found different orientation relationship depending on the substrate orientation and reaction temperature.

Pulse laser deposition (PLD) is also an effective method used for fabricating oxide thin film [55]. This method can provide accurate stoichiometric film by using a very small target. PLD can grow crystalline film much lower substrate temperature than other film growth technique within very less time. BaTiO$_3$, SrRuO$_3$, DyScO$_3$, (k, Na)NbO$_3$, ZnO, BiFeO$_3$, Indium tin oxide (ITO), TiO$_2$ are examples of some of the oxide films successfully deposited by PLD method. Furthermore, the Pechini method [56], electron beam evaporation method [57] are also used for fabrication of oxide thin film.

2.4 APPLICATIONS OF OXIDE THIN FILMS

Recently, oxide nanocrystalline thin films are gaining attention due to the wide range of device applications. The extensive use of thin-film in a diverse field of electronics, optics, space science, defense, and other industries have attracted many researchers for higher studies of the thin film. Active devices and passive components, piezoelectric devices, micro-miniaturization of power supply, rectification and amplification, sensor elements, storage of solar energy and its conservation to other forms, magnetic memories, dielectric mirror, antireflection coating are some of the important uses of the thin film. The applications of the films depend on the properties of the film. Accordingly, researchers are trying to investigate new thin-film materials and to improve the properties of already identified materials.

$MgTiO_3$ thin films possess very good physical and electrical properties such as moderate dielectric constant, low dielectric loss, and high-temperature stability. These properties make $MgTiO_3$ film a suitable material in integrated electronic such as a storage capacitor in dynamic random access memory (DRAM). The successful fabrications of $MgTiO_3$ metal-insulator-metal (MIM) capacitors have achieved a major role in analog circuit applications and DRAM. Earlier, SiO_2 and Si_3N_4 were used in the MIM capacitor. But their low dielectric constant (SiO_2~3.9 and Si_3N_4~7) limited their applications [58]. After that, the Al_2O_3 MIM capacitor has been introduced. But the large leakage current and poor voltage linearity restricted its applications [59]. The leakage current of $MgTiO_3$ thin film is very low ~ 1.051×10^{-9} A/cm^2 at 5 V and capacity density is ~1.2 nF/μm^2 along with the dielectric constant = 16.3 and dielectric loss = 0.0021 at 1 MHz [58]. Though the reported properties are reliable for MIM applications, continuous research has been carried out to improve these properties by using different dopant and additives. Remarkable enhancement in the properties of $MgTiO_3$ thin film has been observed after Ni, Zr, Co, Zn doping [35–37, 50]. Santhosh [37] observed a progressive improvement in the dielectric properties of $(Mg_{0.95}Co_{0.05}TiO_3)$ thin film deposited by RF sputtering. They obtained a maximum dielectric constant 17.3 and loss tangent ~1.1×10^{-3} at a spot frequency of 10 GHz, which leads the usability of $MgTiO_3$ films in higher frequency device applications. Further, these films exhibited good optical properties (refractive index ~2.08, bandgap ~4.16 eV), which are suitable for antireflection coating [38].

The perovskite sodium potassium niobate (($K_{0.5}Na_{0.5}NbO_3$ (KNN)) thin films are also showing significant influence in various applications such as sensors, non-volatile memories, varactors, actuators, electro-optic devices, high-frequency tunable devices due to their outstanding performance in dielectric, ferroelectric and piezoelectric properties [59–65]. But the difficulties in obtaining stoichiometric KNN thin film due to the volatile nature of alkali element causes degradation of electrical properties. Different deposition techniques are introduced to rectify this problem [66–69]. Various dopants like La, Ta, Sb are used to enhance the electrical properties of KNN thin films [70–72]. Mn, Co, and Ti-doped KNN thin films significantly decreased the leakage current with enhanced dielectric and ferroelectric properties [73–75]. The dielectric loss was found to decrease in V, Ba, and Y doped KNN films [9, 76, 78]. Mahesh [28, 29] reported KNN + 0.5 wt.% Dy_2O_3, and KNN + 1.0 wt.% Gd_2O_3 films deposited by RF reactive magnetron sputtering and observed strong optical nonlinearity with minimum leakage current on both the films which make the films very suitable for nonlinear photonics and sensor applications. Tsujiura [79] successfully fabricated piezoelectric MEMS energy harvester of (K, Na) NbO_3 thin films. They deposited the thin film on microfabricated stainless steel cantilever. The power generation of the thin film energy harvester was very high. They obtained a large output power, i.e., 1.6 µW under vibration at 393 Hz.

The (Mg-Zn-Ti) oxide thin film deposited on equilateral prism is a very good candidate for humidity sensor [80]. The humidity sensitivity of the film is ~ 2, which is quite useful for optoelectronic humidity sensor. The ZrO_2/TiO_2 multilayered thin films prepared by Sol-gel process also fulfill the characteristics of humidity sensor such as high sensitivity, short response and recovery time, low hysteresis, and excellent stability [81]. The Ag-doped $(Ba_{0.6}Sr_{0.4})$ TiO_3 thin films are investigated for tunable microwave applications [49]. A tunability of 41.2% was observed for the film along with dielectric constant 269.3 and quality factor 48.1. The obtained properties of the thin film are very reliable for the tunable resonator and tunable filter applications. $BaWO_4$ and $Ba_5Nb_4O_{15}$ thin films are also showing presence in electric and smart window applications [38, 39]. $Ba_5Nb_4O_{15}$- $BaWO_4$ composite thin film was reported by Anil [40]. They observed the electrochromic response of the film while depositing by RF sputtering technique, which can be useful for smart window applications.

After the first fabrication of memristor by Stanly Willium in 2008, it has been showing potential in several areas of integrated circuit design and computing. These memristors are mainly used in non-volatile random access memory technology. The ZnO based thin films show good memristive response. Ayana [82] prepared sol-gel derived multilayer undoped and Al-doped ZnO film for memristive applications. They fabricated Pt/ZnO/Pt/SiO_2 and Pt/ZnO (Al) /Pt/SiO_2 memristive cells. The observed current-voltage exhibited desired switching characteristics. Jeong [83] introduced Graphene oxide (G-O) thin film for non-volatile memory applications. They prepared the G-O thin film by spin casting. The oxide thin films are also used in supercapacitor fabrications. Due to high energy storage and efficiency, the supercapacitors are widely used in power applications. The cobalt oxide thin films are studied by Kandalkar [84] for the mentioned purpose. The supercapacitor properties of the films are studied by galvanostatic charge-discharge and cyclic voltammetry (CV) method. The cobalt oxide thin film deposited onto a copper substrate exhibited a maximum specific capacitor of 127.58 F/g with good energy efficiency and stable nature, which is useful in energy storage applications. The heat mirror of hafnium oxide was built for energy-efficient window application by Al-Kuhaili [85]. They deposited the thin films by hafnium oxide by electron beam evaporation process. The obtained amorphous, transparent films exhibited transmittance of 72.4% in visible range and 67.0% reflectance in the near-infrared region.

The systematic investigation on oxide thin films reveals that the oxide thin films have great potential in the advancement of recent technologies. Getting as-prepared crystallinity of some films are still a challenge for the researchers. Further, the comparative studies of bulk and a thin film of different materials are very rare. Attempts should be made to low the cost of production of a thin film with the use of environment-friendly materials and techniques.

KEYWORDS

- atomic force microscopy
- field emission scanning electron microscopy
- radio frequency
- scanning electron microscopy
- split post dielectric resonator
- x-ray diffraction

REFERENCES

1. Chopra, K. L., (1969). *Thin Film Phenomena* (pp. 23–43), McGraw Hill, New York.
2. Goswami, A., (1995). *Thin Film Fundamental* (pp. 21–31). New Age International Pvt. Limited, New Delhi.

3. Vankar, V. D., (1985). In: Chopra, K. L., & Malhotra, L. K., (eds.), *Thin Film Technology and Applications* (pp. 14–26). Tata McGraw hill, New Delhi.
4. Pamu, D., Sudheendran, K., Krishna, M. G., James, K. C., & Bhatnagar, A. K., (2009). Ambient temperature stabilization of crystalline zirconia thin films deposited by direct current magnetron sputtering. *Thin Solid Films, 517*, 1587–1591.
5. Ferraris, M., Verne, E., Appendino, P., Moisescu, C., Krejewski, A., Ravaglioli, A., & Piancastelli, A., (2000). Coatings on zirconia for medical applications. *Biomaterials, 21*, 765–773.
6. Pamu, D., Sudheendran, K., Krishna, M. G., Raju, K. C. J., & Bhatnagar, A. K., (2007). Microwave dielectric behavior of nanocrystalline titanium dioxide thin films. *Vacuum, 81*, 686–694.
7. Blomqvist, M., Khartsev, S., & Grishin, A., (2006). Electro-optic effect ferroelectric $Na_{0.5}K_{0.5}NbO_3$ thin films on oxide substrates. *Integr. Ferroelectr., 80*, 97–106.
8. Abazari, M., Choi, T., Cheong, S. W., & Safari, A., (2010). Nanoscale characterization and local piezoelectric properties of lead-free KNN-LT-LS thin films. *J. Phys D: Appl. Phys., 43*, 25405 (1–5).
9. Abazari, M., & Safari, A., (2009). Effects of doping on ferroelectric properties and leakage current behavior of KNN-LT-LS thin films on $SrTiO_3$ substrate. *J. Appl. Phys., 105*, 94101–94105.
10. Abazari, M., Akdoğan, E. K., & Safari, A., (2008). Effect of manganese doping on remnant polarization and leakage current in $(K_{0.44}Na_{0.52}Li_{0.04})(Nb_{0.84}Ta_{0.10}Sb_{0.06})O_3$ epitaxial thin films on $SrTiO_3$. *Appl. Phys. Lett., 92*, 212903 (1–3).
11. Li, T., Wang, G., Remiens, D., & Dong, X., (2013). Characteristics of highly (001) oriented (K, Na)NbO_3 films grown on $LaNiO_3$ bottom electrodes by RF magnetron sputtering. *Ceram. Int., 39*(2), 1359–1363.
12. Kugler, V. M., Soderlind, F., Music, D., Helmersson, U., Andreasson, J., Lindback, T., Music, D., Helmersson, U., Andreasson, J., & Lindback, T., (2003). Low temperature growth and characterization of (Na, K)NbOx thin films. *J. Cryst. Growth, 254*, 400–404.
13. Ye, Q., Liu, P. Y., Tang, Z. F., & Zhai, L., (2007). Hydrophilic properties of nano-TiO_2 thin films deposited by RF magnetron sputtering. *Vacuum, 81*, 627–631.
14. Tauc, J. C., (1972). In: Abeles, F., (ed.), *Optical Properties of Solids* (p. 277). North-Holland Publishing, Amsterdam.
15. Sebastian, M. T., Ubic, R., & Jantunen, H., (2015). Low loss dielectric ceramic materials and their applications. *International Materials Reviews, 60*, 392–412.
16. Krupka, J., Gabelich, S., Derzakowski, K., & Pierce, B. M., (1999). Composition of split post dielectric resonator and ferrite disc resonator techniques for microwave permittivity measurements of polycrystalline ytterium iron garnet, *Measur. Sci. Tech., 10*, 1004–1008.
17. Kingery, W. D., & Berg, M., (1955). Study of the initial stages of sintering solids by viscous flow, evaporation-condensation, and self-diffusion. *J. Appl. Phys., 26*, 1205–1212.
18. Leaver, K. D., & Chapman, B. N., (1971). *Thin Film* (pp. 1–5). Wykeham Publishers, London.
19. Heavens, O. S., (1969). *Thin Film Physics* (pp. 14–49). Mc-Graw Hill, New York.

20. Maissel, L. I., & Glang, R., (1970). *Handbook of Thin Film Technology* (pp. 1, 2). Mc-Graw Hill, New York.
21. Coutts, T. J., (1978). *Active and Passive Thin Film Devices* (pp. 1–20). Academic press, London.
22. Ohring, M., (1992). *The Materials Science of Thin Film* (p. 292). Academic press, San Diego.
23. Brodie, I., & Muray, J. J., (1992). *The Physics of Micro/Nano-Fabrication* (p. 15). Phenum, New York.
24. Kannan, P. K., Saraswathi, R., & Rayappan, J. B. B., (2010). A highly sensitive humidity sensor based on DC reactive magnetron sputtered zinc oxide thin film. *Sensors and Actuators A, 164*, 8–14.
25. Li, K., Remiens, D., Costecalde, J., Sama, N., Du, G., Li, T., Dong, X., & Wang, G., (2013). Crystallographic orientation dependence of dielectric response in lead strontium titanate thin films. *J. Crystal Growth, 377*, 143–146.
26. Jain, M., Majumder, S. B., Yuzyuk, Yu, I., Katiyar, R. S., Bhalla, S., Miranda, F. A., & Keuls, F. W. V., (2003). Dielectric properties and leakage current characteristics of sol-gel derived $(Ba_{0.5}Sr_{0.5})TiO_3:MgTiO_3$ thin film composites. *Ferroelectric Letters, 30*, 99–107.
27. Mahesh, P., & Pamu, D., (2014). Effect of deposition temperature on structural, mechanical, optical and dielectric properties of RF sputtered nanocrystalline $(K_xNa_{1-x})NbO_3$ thin films. *Thin Solid Films, 562*, 471–477.
28. Mahesh, P., Thota, S., & Pamu, D., (2014). Dielectric and AC-conductivity studies of Dy_2O_3 doped $K_{0.5}Na_{0.5}NbO_3$ ceramics. *AIP Advances, 4*, 087113 (1–11).
29. Mahesh, P., Bharti, G. P., Khare, A., & Pamu, D., (2016). Optical and dielectric studies on radio frequency sputtered Gd_2O_3 doped $K_{0.5}Na_{0.5}NbO_3$ thin films for nonlinear photonic and microwave tunable device applications. *Journal of Alloys and Compounds, 682*, 634–642.
30. Bhuyan, R. K., Kumar, T. S., Perumal, A., & Pamu, D., (2013). Effect of annealing and atmosphere on the structure and optical properties of Mg_2TiO_4 thin films, obtained by the radio frequency magnetron sputteringmethod. *Journal of Experimental Nanoscience, 8*, 371–381.
31. Bhuyan, R. K., Kumar, T. S., Perumal, A., Ravi, S., & Pamu, D., (2012). Optical properties of ambient temperature grown nanocrystalline Mg_2TiO_4 thin films deposited by RF magnetron sputtering. *Surface and Coatings Technology, 221*, 196–200.
32. Huang, C. L., & Chen, Y. B., (2006). Structure and electrical characteristics of RF magnetron sputtered $MgTiO_3$. *Surface & Coatings Technology, 200*, 3319–3325.
33. Huang, C. L., & Chen, Y. B., (2005). The effect of deposition temperature and RF power on the electrical and physical properties of the $MgTiO_3$ thin films. *Journal of Crystal Growth, 285*, 586–594.
34. Huang, C. L., & Chen, J. Y., (2009). The effect of RF power and deposition temperature on the structure and electrical properties of $Mg_4Ta_2O_9$ thin films prepared by RF magnetron sputtering. *Journal of Crystal Growth, 311*, 627–633.
35. Gogoi, P., Srinivas, P., Sharma, P., & Pamu, D., (2016). Optical dielectric characterization and impedance spectroscopy of Ni-substituted $MgTiO_3$ thin films. *Journal of Electronic Materials, 45*(2), 899–909.

36. Gogoi, P., Kumar, T. S., Sharma, P., & Pamu, D., (2015). Structural, optical, dielectric and electrical studies on RF sputtered nanocrystalline Zr doped MgTiO$_3$ thin films. *Journal of Alloys and Compounds*, *619*, 527–537.
37. Kumar, T. S., Gogoi, P., Thota, S., Raju, K. C. J., & Pamu, D., (2014). Structural and dielectric studies of Co doped MgTiO$_3$ thin films fabricated by RF - magnetron sputtering. *AIP Advances*, *4*, 67142 (1–14).
38. Kumar, T. S., Gogoi, P., Bhasaiah, S., Raju, K. C. J., & Pamu, D., (2015). Structural optical and microwave dielectric studies of Co doped MgTiO$_3$ thin films fabricated by RF magnetron sputtering. *Mater. Res. Express*, *2*, 056403.
39. Kumar, C. A., & Pamu,, D., (2015). Dielectric optical and electric studies on nanocrystallineBa$_5$Nb$_4$O$_{15}$thin films deposited by RF magnetron sputtering. *Applied Surface Science*, *340*, 56–63.
40. Kumar, C. A., Kumar, T. S., & Pamu, D., (2015). Irreversible thermochromic response of RF sputtered nanocrystalline BaWO$_4$ films for smart window applications. *AIP Advances*, *5*, 107232–107239.
41. Kumar, C. A., & Pamu, D., (2016). Dielectric and optical characterization of RF sputtered Ba$_5$Nb$_4$O$_{15}$–BaWO$_4$ composite films for electronic and smart window applications. *Journal of Electronic Materials*, *45*, 3101–3112.
42. Voigt, M., & Sokolowski, M., (2004). Electrical properties of thin RF sputtered aluminum oxide films. *Materials Science and Engineering B*, *109*, 99–103.
43. Blomqvist, M., Koh, J. H., Khartsev, S., & Grishin, A., (2002). High-performance epitaxial Na$_{0.5}$K$_{0.5}$NbO$_3$ thin films by magnetron sputtering. *Appl. Phys. Lett.*, *81*(2), 337–339.
44. Kanno, I., Mino, T., Kuwajima, S., Suzuki, T., Kotera, H., & Wasa, K., (2007). Piezoelectric properties of (K, Na)NbO$_3$ thin films deposited on (001)SrRuO$_3$/Pt/MgO substrates. *IEEE Trans, Ultrason. Ferroelectr. Freq. Control*, *54*(12), 2562–2566.
45. Wu, J., & Wang, J., (2009). Phase transitions and electrical behavior of lead-free K$_{0.50}$Na$_{0.50}$NbO$_3$ thin film. *J. Appl. Phys.*, *106*(6), 066101–066103.
46. Kim, J. S., Lee, H. J., Lee, S. Y., Kim, I. W., & Lee, S. D., (2010). Frequency and temperature dependence of dielectric and electrical properties of radio-frequency sputtered lead-free K0.48Na0.52NbO3 thin films. *Thin Solid Films*, *518*(22) 6390–6393.
47. Jain, M., Majumder, S. B., & Katiyar, R. S., (2002). Dielectric properties of sol–gel-derived MgO:Ba0.5Sr0.5TiO3 thin-film composites. *Appl. Phys. Lett.*, *81*, 3212–3214.
48. Bian, Y., & Zhai, J., (2014). Low dielectric loss Ba$_{0.6}$Sr$_{0.4}$TiO$_3$/MgTiO$_3$ composite thin films prepared by a sol–gel process. *Journal of Physics and Chemistry of Solids*, *75*, 759–764.
49. Kim, K. T., Kim, C., Senior, D. E., Kim, D., & Yoon, Y. K., (2014). Microwave characteristics of sol-gel based Ag-doped (Ba$_{0.6}$Sr$_{0.4}$)TiO$_3$ thin films. *Thin Solid Films*, *565*, 172–178.
50. Surendran, K. P., Wu, A., Vilarinho, P. V., & Ferreira, V. M., (2010). Ni and Zn doped MgTiO3 thin films: Structure, microstructure and dielectric characteristics. *J. Appl. Phys.*, *107*, 114112–114118.
51. Wu, Y. C., Wang, S. F., & Teng, L. G., (2012). Microstructures and dielectric properties of MgTiO$_3$ thick film prepared using aerosol deposition method. *Ferroelectrics*, *435*, 137–147.

52. Akedo, J., Nakano, S., Park, J., Baba, S., & Ashida, K., (2008). The aerosol deposition method- for production of high performance micro device with low cost and low energy consumption. *Synthesiology, 1*, 130–138.
53. Zenga, J., Wang, H., Song, S., Zhang, Q., Cheng, J., Shang, S., Wang, M., Wang, Z., & Lin, C., (1997). Preparation and characterization of MgTiO$_3$ thin films byatmospheric pressure metalorganic chemical vapor deposition. *Journal of Crystal Growth, 178*, 355–359.
54. Lotnyk, A., Senz, S., & Hesse, D., (2007). Orientation relationships of SrTiO$_3$ and MgTiO$_3$ thin films grown by vapor-solid reactions on (100) and (110) TiO$_2$ (Rutile) single crystals. *J. Phys. Chem. C., 111*, 6372–6379.
55. Mahesh, P., Bharti, G. P., Khare, A., & Pamu, D., (2016). Optical and dielectric studies on radio frequency sputtered Gd2O3 doped K0.5Na0.5NbO3 thin films for nonlinear photonic and microwave tunable device applications. *Journal of Alloys and Compounds, 682*, 634–642.
56. Zhu, J., Zhou, J., Xu, L. S., Shen, J., Yang, X. Y., & Chen, W., (2008). Preparation of Ca((Mg$_{1/3}$Nb$_{2/3}$)$_{2/3}$Ti$_{1/3}$)O$_3$ microwave dielectric thin films by the pechini method, synthesis and reactivity in inorganic. *Metal-Organic, and Nano-Metal Chemistry, 38*, 168–172.
57. Lee, C. H., & Kim, S., (2003). The characteristics of magnesium titanate thin film as buffer layer by electron beam evaporation. *Integrated Ferroelectrics, 57*, 1265–1270.
58. Huang, C. L., Wang, S. Y., Chen, Y. B., Li, B. J., & Lin, Y. H., (2012). Investigation of the electrical properties of metaloxidemetal structures formed from RF magnetron sputtering deposited MgTiO$_3$ films. *Current Applied Physics, 12*, 935–939.
59. Chen, S. B., Lai, C. H., Chin, A., Hsieh, J. C., & Liu, J., (2002). High-density MIM capacitors using Al$_2$O$_3$ and AlTiOx dielectrics. *IEEE Electron Device Lett., 23*, 185–187.
60. Cho, C. R., Park, S. H., Moon, B. M., Sundqvist, J., Hårsta, A., & Grishin, A., (2002). Na$_{0.5}$K$_{0.5}$NbO$_3$ thin films for MFIS_FET type non-volatile memory applications. *Integr. Ferroelectr., 49*(1), 21–30.
61. Zhen, Y., Wang, M., Wang, S., & Xue, Q., (2014). Humidity sensitive properties of amorphous (K, Na)NbO$_3$ lead free thin films. *Ceram. Int., 40*(7), 10263–10267.
62. Kanno, I., Ichida, T., Adachi, K., Kotera, H., Shibata, K., & Mishima, T., (2012). Power generation performance of lead free (K, Na)NbO$_3$ piezoelectric thin film energy harvester. *Sens. Actuators A: Phys., 179*, 132–136.
63. Cho, C. R., Grishin, A., Andrèasson, J., Lindbäck, T., Abadei, S., & Gevorgian, S., (2000). Ferroelectric Na0.5K0.5NbO$_3$ films for voltage tunable microwave devices. *MRS Proc., 656* DD3.2.
64. Zhu, B., Zhang, Z., Ma, T., Yang, X., Li, Y., Shung, K. K., & Zhou, Q., (2015). (100)-Textured KNN-based thick film with enhanced piezoelectric property for intravascular ultrasound imaging. *Appl. Phys. Lett., 106*(17), 173504 (1–4).
65. Cho, C. R., Koh, J. H., Grishin, A., Abadei, S., & Gevorgian, S., (2000). Na$_{0.5}$K$_{0.5}$NbO$_3$ÕSiO$_2$ÕSi thin film varactor. *Appl. Phys. Lett., 76*(13), 1761–1763.
66. Blomqvist, M., Khartsev, S., Grishin, A., & Petraru, A., (2010). RF sputtered Na$_{0.5}$K$_{0.5}$NbO$_3$ films on oxide substrates as optical wave guiding material. *Integr. Ferroelectr., 54*(1), 631–640.

67. Söderlind, F., Käll, P. O., & Helmersson, U., (2005). Sol–gel synthesis and characterization of $Na_{0.5}K_{0.5}NbO_3$ thin films. *J. Cryst. Growth, 281*(2–4), 468–474.
68. Ahn, C. W., Lee, S. Y., Lee, H. J., Ullah, A., Bae, J. S., Jeong, E. D., Choi, J. S., Park, B. H., & Kim, I. W., (2009). The effect of K and Na excess on the ferroelectric and piezoelectric properties of $K_{0.5}Na_{0.5}NbO_3$ thin films. *J. Phys. D: Appl. Phys., 42*(21), 215304, 215305.
69. Goh, P. C., Yao, K., & Chen, Z., (2011). Lithium diffusion in (Li, K, Na)NbO_3 piezoeletric thin films and the resulting approach for enhanced performance properties. *Appl. Phys. Lett., 99*(9), 092902, 092903.
70. Goh, P. C., Yao, K., & Chen, Z., (2010). Lead-free piezoelectric $(K_{0.5}Na_{0.5})NbO_3$ thin films derived from chemical solution modified with stabilizing agents. *Appl. Phys. Lett., 97*(10), 102901–102903.
71. Akmal, M. H. M., Warikh, A. R. M., Azlan, U. A. A., Azam, M. A., & Ismail, S., (2016). Effect of amphoteric dopant on the dielectric and structural properties of yttrium doped potassium sodium niobate thin film. *Mater. Lett., 170*, 10–14.
72. Ahn, C. W., Seog, H. J., Ullah, A., Lee, S. Y., Kim, J. W., Kim, S. S., Park, M., No, K., & Kim, I. W., (2012). Effect of Ta content on the phase transition and piezoelectric properties of lead-free $(K_{0.48}Na_{0.48}Li_{0.04})(Nb_{0.995-x}Mn_{0.005}Ta_x)O_3$ thin film. *J. Appl. Phys., 111*(2), 024110–024116.
73. Lee, S. Y., Ahn, C. W., Kim, J. S., Ullah, A., Lee, H. J., Hwang, H. I., Choi, J. S., Park, B. H., & Kim, I. W., (2011). Enhanced piezoelectric properties of Ta substituted-$(K_{0.5}Na_{0.5})NbO_3$ films: A candidate for lead-free piezoelectric thin films. *J. Alloy. Compd., 509*(20), L194–L198.
74. Wang, L., Ren, W., Ma, W., Liu, M., Shi, P., & Xiaoqing, W., (2015). Improved electrical properties for Mn-doped lead-free piezoelectric potassium sodium niobateceramics. *AIP Advnc., 5*, 97120 (1–10).
75. Zhang, Y., Li, W., Cao, W., Feng, Y., Qiao, Y., Zhang, T., & Fe, W., (2017). Mn doping to enhance energy storage performance of lead-free 0.7NBT-0.3ST thin films with weak oxygen vacancies. *Appl. Phys. Lett., 110*, 243901–243905.
76. Wang, L., Zuo, R., Liu, L., Su, H., Shi, M., Chu, X., Wang, X., & Li, L., (2011). Preparation and characterization of sol–gel derived (Li, Ta, Sb) modified (K, Na)NbO_3 lead-free ferroelectric thin films. *Mater. Chem. Phys., 130*(1/2), 165–169.
77. Li, N., Li, W. L., Wang, L. D., Zhang, S. Q., Ye, J. W., & Fei, W. D., (2011). Effect of V doping on phase composition and electrical properties of K0.4Na0.6Nb1−xVxO3 thin films. *J. Alloy. Compd., 509*(31), 8028–8031.
78. Akmal, M. H. M., Warikh, A. R. M., Azlan, U. A. A., Azam, M. A., & Ismail, S., (2016). Effect of amphoteric dopant on the dielectric and structural properties of yttrium doped potassium sodium niobate thin film. *Mater. Lett., 170*, 10–14.
79. Tsujiura, Y., Suwa, E., Kurokawa, F., Hida, H., Suenaga, K., Shibata, K., & Kanno, I., (2009). Lead-free piezoelectric MEMS energy harvesters of (K, Na)NbO_3 thin films on stainless steel cantilevers. *Japn. J. Applied Physics, 52*, 09KD13–5.
80. Yadav, B. C., Yadav, R. C., & Dwivedi, P. K., (2010). Sol–gel processed (Mg–Zn–Ti) oxide nanocomposite film deposited on prism base as an opto-electronic humidity sensor. *Sensors and Actuators B, 148*, 413–419.
81. Biju, K. P., & Jain, M. K., (2008). Sol–gel derived $TiO_2:ZrO_2$ multilayer thin films for humidity sensing application. *Sensors and Actuators B, 128*, 407–413.

82. Ayana, D. G., Prusakova, V., Collini, C., Nardi, M. V., Tatti, R., Bortolotti, M., et al., (2016). Sol-gel synthesis and characterization of undoped and Al-doped ZnO thin films formemristive application. *AIP Advnc., 6*, 111306, 111307.
83. Jeong, H. Y., Kim, J. Y., Kim, J. W., Hwang, J. O., Kim, J. E., Lee, J. Y., et al., (2010). Graphene oxide thin films for flexible nonvolatile memory applications. *Nano Lett., 10*, 4381–4386.
84. Kandalkar, S. G., Gunjakar, J. L., & Lokhande, C. D., (2008). Preparation of cobalt oxide thin films and its use in supercapacitor application. *Applied Surface Science, 254*, 5540–5544.
85. Al-Kuhaili, M. F., (2004). Optical properties of hafnium oxide thin films and their application in energy-efficient windows. *Optical Materials, 27*, 383–387.

CHAPTER 3

Application of Magnetic Fluid in the Energy Sector

KINNARI PAREKH[1] and R. V. UPADHYAY[1,2]

[1]Dr. K. C. Patel R&D Center, Charotar University of Science and Technology, Changa – 388421, Anand District, Gujarat, India, E-mail: kinnariparekh.rnd@charusat.ac.in

[2]P D Patel Institute of Applied Sciences, Charotar University of Science and Technology, Changa – 388421, Anand District, Gujarat, India

ABSTRACT

In the present chapter, the review of energy applications of magnetic fluid is presented. Few studies mentioning the use of magnetic fluid for voltage generator, energy harvester and energy storage is reviewed; however, the study is yet at the preliminary level and needs focused research with the novel design and synthesis of special magnetic fluid for the purpose. The use of magnetic fluid for energy transfer devices, especially for heat transfer using convection pipe study and automatic cooling devices (ACD) shows potential. A special type of magnetic fluid, known as temperature-sensitive magnetic fluid (TSMF) can be used to achieve enhanced heat transfer. Results published in the literature are reviewed and presented in this article. However, the work is still, to the author's knowledge, is the tip of an iceberg. The motive to write this article is to explore the potential benefits of magnetic fluid in the energy sector as well as to spread awareness in the field of heat transfer.

3.1 INTRODUCTION

The energy sector is highly influenced by the need of magnetic materials as they play a vital role in improving the functioning of devices especially in electric power generation and conversion (soft magnets), in electric motors and transportation (hard magnets), in refrigeration (magnetocaloric materials), etc. Indeed, magnetic materials are essential components of motors, generators, transformers, and actuators, which furnish the continuous and unstoppable need of energy supply at its optimum properties. Energy storage is also very important application in magneto-electronic devices as far as the designing of magnetic materials is concerned. In all these application areas, the magnetic materials used are in solid form. Replacing liquid magnet over solid magnet offers additional benefits such as to conform the shape, easily injected to the hard-to-access locations, negligible friction, etc. [1].

Magnetic fluid, also known as 'Ferrofluid' (colloidal suspension of magnetic nanoparticles (NPs) in liquid medium), is one such nanotechnology-based liquid magnetic material which is extensively used in microfluidic devices [1]. Magnetic fluid research was first initiated in the 1960s by S. Pappell at NASA to regulate and control the liquid fuel in space. Soon after that the field was extensively developed by Ronald E. Rosensweig by proposing various Ferrohydrodynamic (FHD) (a branch related to the influence of strong magnetic field on the fluid motion) applications of such fluid with the initial objective of using heat into useful work without the use of any additional mechanical parts [1]. The idea is to exploit tremendous magnetic response due to ferri/ferromagnetic NPs (~10 nm) along with the fluid characteristics of carrier, mainly influenced by a gradient magnetic field and a temperature gradient. Subsequently, substantial magnetic forces will be induced in the fluid, which leads to the motion of the fluid. Most of the applications of magnetic fluid use the concept of remote positioning or control of fluid flow using the magnetic force. Figure 3.1 shows the pictorial view of magnetic fluid compositions.

3.2 APPLICATIONS OF MAGNETIC FLUID

The magnetism induced in magnetic fluids makes them attractive for varieties of applications where the combination of liquid state and magnetic properties is essential. NASA initially used them as rotating shaft seals in satellites, and they are now serving the same purpose in bearings, vacuum devices, and centrifuges to computer hard disk drives. Therefore, magnetic fluids are novel and technologically advanced materials. Such fluids are widely used as a lubricant, airtight seals of rotary shafts, dampers,

Application of Magnetic Fluid in the Energy Sector

coolant, smart switches, level sensors, energy conversion devices, etc. [1–8]. These fluids are also used in targeted drug delivery, hyperthermia, protein separation, MRI agents, biosensors, etc. [9–13]. A large variety of new magnetic fluid based devices are upcoming at a spectacular rate. In this respect, magnetic fluid will also find its place in the energy sector in various domains like energy harvesting, energy storage, energy collector, energy transfer, energy conversion, and conservation, etc. (Figure 3.2). In each domain, the principal mechanism underlying the usage of magnetic fluid is different. In the following section, the application of magnetic fluid in various energy sectors are discussed in detail.

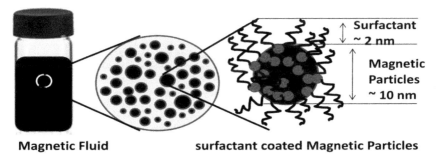

FIGURE 3.1 (See color insert.) Components of magnetic fluid–a schematic representation.

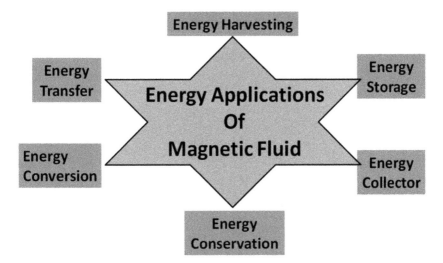

FIGURE 3.2 Applications of magnetic fluid in energy sector.

3.2.1 ENERGY HARVESTER

In 2003 onwards, a research has been initiated in the area of energy harvester because of its unique opportunities to match with an uninterrupted power supply for minimal power consumption electronics. This offers an added advantage of its use in a hazardous environment too [14–18]. The principal is to extract the mechanical energy from the system and convert to an electric potential in response to mechanical stimuli. This is achieved by amalgamating the active materials with electromechanical coupling mechanisms. Among piezoelectric, electrostatic, and electromagnetic transduction as the energy harvesting mechanisms, most of the literature is available for the investigation of electromagnetic transduction [19–23]. Piezoelectric energy harvesters provide higher output voltages and are more efficient for relatively high vibration frequencies. Whereas the low-frequency vibrators utilize electromagnetic energy harvesters [24], since, they generate a low output voltage. Different configurations with the same basic principle have been proposed; external vibration causes the motion of a solid magnet placed inside a stationary coil. The change in the magnetic flux produces a current in the coil, which is then feed to an electric load. When magnetic fluids (a colloidal suspension of nano-scale permanent magnetic particles) is used in place of a magnet, being a liquid magnet it has additional advantage such as to conform the shape, easily injected to the hard-to-access locations along with the scalable electromagnetic energy harvesters [25].

Bibo et al., [26] have reported the excitation of surface waves (both horizontal and rotational) with large amplitude when the frequency of seismic excitations resonates with one of the model frequencies of magnetized magnetic fluid. The magnetic flux will be created due to the alignment of magnetic particles inside the sloshing fluid and induces electromagnetic force in a coil wound around the container generating an electric current. The use of magnetic fluid has an added advantage of tunability of model frequencies to very low values. Thus, effective energy harvester with low-frequency excitation is made possible. Moreover, it is possible to get a response from infinitely closely spaced frequencies, which is generally not possible with the solid magnets. This additional feature of magnetic fluid based harvester improves the performance of the device for random as well as non-stationary excitations.

Chae et al., [27] have prepared a linear vibration energy harvester using magnetic fluid lubricant and stacked rectangular permanent magnets. The external lateral vibration causes the back and forth motion of the magnet assembly on the channel. But the motion has less friction and wear due to the use of magnetic fluid. This will considerably improve the reliability of the device over the long run. Aluminum housing wounded by copper wire was used to generate electric power. A device for proof-of-concept harvester was fabricated and tested it with a vibration exciter for different input frequencies and accelerations. Maximum 493 μW power was generated using 5 μL magnetic fluid lubrication. This was 4.37% higher than the power generated without magnetic fluid. The improvement in reliability with long span due to magnetic fluid lubrication was verified. The device with magnetic fluid showed 1.02% decrease in output power after 93,600 cycles whereas 59.73% decrement in output power was observed for a device without magnetic fluid. Similarly, Robert Gherca et al., [25] have observed 25% higher efficiency with magnetic fluids in harvesting generator in a horizontal and vertical position. Alzami et al., [28] have reported a nonlinear analytical model, which governs the electro-magneto-hydrodynamics of an electromagnetic magnetic fluid-based vibratory energy harvester, and they compare their theory with experimental findings. Kim et al., [29] have used the oscillating magnetic field for the power generation through contact-electrification of magnetic fluid droplets. The important advantage of their device is that it eliminates the need to vibrate the device for actuating the magnetic fluid droplet for the power generation; instead, the shape of the magnetic fluid inside the device can be modulated by using an external magnetic field.

Wearable harvester that can be attached on a shoe to offer continuous power supply for wearable sensors and device is capable of storing 0.014 mW and 0.149 mW average power during walking and running conditions of a human being [30]. When a non-resonant, broadband electromagnetic vibration-energy harvester using magnetic fluid liquid bearing is used to place on the back of a human, walking at various speeds, the output power was found to increase with the increase in walking speed from 0.44 m/s (walking) to 3.56 m/s (running), and it reaches to 18.1 μW at 3.56 m/s [31].

Markus Zahn has proposed magnetic field-based micro/nanoelectromechanical systems (MEMS/NEMS) devices that use magnetic NPs (10 nm diameter), with and without a carrier liquid [32], and named it as a nanogenerator. In this nanogenerator, rotation of single-particle next to the representative coil at an angular speed of 1000 revolutions per second delivered 3 x 10^{-9} W/m² time average powers. Similarly, a voltage generator using magnetic fluid is reported by [33]. In their report, they present the magnetic flux change due to the bubble of low boiling point immiscible liquid in a column filled with magnetic fluid. When the system is heated up to the boiling temperature of the immiscible fluid, bubbles are generated which enters into the magnetic fluid column and induces an electromotive force around 1 mV signal. The maximum signal produced until now using this concept is in millivolt. A large output power generator is yet very challenging using magnetic fluid.

3.2.2 ENERGY STORAGE

Most commonly used facile sources of energy are Li-ion batteries, fuel cell, supercapacitors, and hydrogen production cell. Lithium-ion batteries that possess high charge and discharge ability, high energy density, and high power capacity are very efficient, cost-effective, and attractive energy source for power application and electric vehicles (EV). These batteries are considered to be one of the most promising power sources for popular mobile devices, like mobile phones, notebooks, etc. Magnetic NPs such as Fe_3O_4 can also have a potential in this area [34] and used as an anode for lithium ions batteries [34–38], but, there are some drawbacks with Fe_3O_4 to use it as anode. For example, it undergoes volume change during the process of lithiation and de-lithiation, it possesses low conductivity that results in lower capacity and power cycle ability, which lowers the efficiency of the lithium-ion batteries. To overcome these problems, Fe_3O_4 has been used as hybrid material with various other materials (such as 3D grapheme foam [38] that enhances the properties of Fe_3O_4 and makes it suitable for the best performance in the batteries.

The difference between the availability of energy source and needs of energy has led us to think for the advancement in energy storage devices. The thermal energy storage in the form of latent heat appears a key feature of energy management with the emphasis on efficient storage and retrieval of the waste heat and solar energy in industry and buildings [39]. The most proficient way to store thermal energy is the storage of latent heat. This requires a medium with large thermal conductivity, large heat capacity, and enhanced heat transfer coefficient. The magnetic or nonmagnetic nanofluid finds its application in this area by suitably tuning its thermo-physical properties. Such nanofluid will improve the operational efficiencies and reliability of systems that is based on the conversion of solar thermal energy.

Organic phase change materials (PCM) have several potential applications in a latent heat storage system with high storage capacity owing to their characteristic properties like high heat of fusion, self-nucleating ability, and non-toxic/non-reactive nature. PCM can store or release sizable quantity of thermal energy during solidification or melting. Such PCM will show enhanced thermal conductivity if the NPs can be dispersed in it. One such nanofluid based PCMs was prepared by Liu et al., [40] using TiO_2 NPs in saturated $BaCl_2$ aqueous solution, and this sample showed large thermal conductivity over the base material. Similarly, the heating and cooling rate of PCMs can be improved by using copper NPs as additives [41]. 1 wt% of copper NPs composites have shown a reduction of 30.3 and 28.2% in the heating and cooling time, respectively. This has not affected the latent heat and phase- change temperature even after 100 thermal cycles. Similarly, magnetic fluid has also showed a large enhancement in thermal conductivity [42–55], which can be further increased using an external magnetic field. The development of this area is very encouraging to get the good efficiency of the device.

3.2.3 ENERGY CONSERVATION

In this particular area, the nanofluid is expected to be work as a good absorbent material of energy in any form and to store it. To achieve this goal, the thermal and physical properties of nanofluids have to be modified. Any one renewable energy sources such as solar energy can be used as a source of heat. Presently the direct absorption solar collector is well-established technology, but its efficiency is limited by the absorption capacity of working fluid. The conventional fluid used for this purpose in solar collectors has very poor thermal absorption property. The use of nanofluid in place of working fluid for solar collector will serve a purpose. The nanofluid comprising of NPs such as CNTs, graphite, and silver in solar collector has been studied very recently [56]. And up to 5% efficiency improvement was observed with such nanofluids. Both the experiment as well as the numerical models showed an initial increase in efficiency with increasing volume fraction, but above critical volume fraction, the efficiency does not showed any increase.

Solar flat plate collectors have 70% higher efficiency than solar direct energy conversion systems, which has an efficiency of around 17% [57]. These collectors are useful as domestic appliances, space heater, industry applications where low temperature is required, etc. Currently, a large number of solar collectors using the principle of concentrating solar power (CSP) are available in the market as a source of heat for a conventional power plant. Omid Mahian et al., [58] has presented review article on enhanced efficiency and performance of the solar thermal system, solar water heater, thermal energy storage, solar cells (SCs) and solar stills, however, a very limited number of research work in the area of solar collectors augmented with nanofluids is reported. The technology combining solar thermal configuration systems with the nanofluids is a topic of recent research which is aimed to develop an innovative approach of nanofluid based solar collectors. Tyagi et al., [59] have theoretically investigated low-temperature Al_2O_3 water-based nanofluids with direct absorption solar collectors (DASC) with varying particle volume fraction (0.1% to 5%) to study its influence on the collector's efficiency. Significant increase in the collector's efficiency (8%) was observed not only varying the particles volume fraction (from 0.8% to 1.6%), but also the transmissivity of glass cover and collector height. A marginal increase in efficiency is observed for the NPs with different size.

Use of magnetic fluids as absorbent media for the solar energy [60] is yet to be exploited. The research shows that petroleum-based magnetic fluids showed enhancement of the spectral absorption coefficient in the range 500–600 nm, which promises the potential use of magnetic fluids as absorbent media for solar energy [61, 62]. Circulating magnetic fluid in transparent pipes, with a diameter of 10 mm, shows remarkably high overheating of magnetic fluid, which is useful for the thermal conversion of solar energy [63].

3.2.4 ENERGY CONVERSION DEVICES

A direct conversion of thermal energy to the energy of fluid motion can be calculated using the change of fluid magnetization with temperature change. This effect is more prominent in the temperature range near to the Curie temperature (T_C). In order to understand that how heat energy can be converted into the mechanical energy, one has to consider the FHD Bernoulli's equation for a steady-state or time-invariant flow as [1]:

$$p^* + \frac{1}{2}\rho v^2 + \rho gh - \mu_0 \bar{M}H = const \quad (1)$$

Here, p^* is pressure, v is fluid velocity, g is acceleration due to gravity, h is a rise in fluid height in the presence of the field, μ_0 permeability in a vacuum, H is applied a magnetic field. The fluid properties are denoted by ρ, fluid density, and $\bar{M}$, the area under the magnetization curve.

Concept of fluid motion will be visualized by considering a situation as depicted in Figure 3.3. A relatively cooler magnetic fluid is entered into the solenoid due to the magnetic attraction, and thus it starts the motion of the fluid. At station 2, the fluid warms up because of the externally supplied heat, where the fluid magnetization starts decreases with increase in temperature. As the temperature reaches the T_c, the fluid magnetization reduces to zero. Such non-magnetic fluid gets push from the cooler fluid, which is entered into the solenoid due to its magnetic nature. The hot fluid will dissipate the heat while moving and again becomes magnetic when it gets cooled. This type of fluid is known as temperature-sensitive magnetic fluid (TSMF). The TSMF generally uses low-T_c magnetic NPs such as Mn–Zn ferrite. A typical curve of saturation magnetization as a function of temperature for bulk Mn-Zn ferrite and magnetite is compared and shown in Figure 3.4 [64].

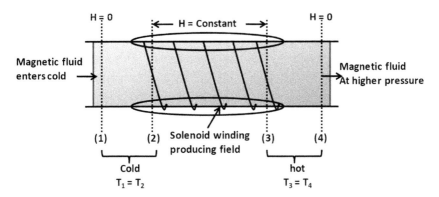

FIGURE 3.3 The magnetocaloric pump using direct conversion of heat energy to flow, a magnetic fluid with the temperature-dependent magnetic moment is heated in the presence of magnetic field. Between station 1 and 2 as well as 3 and 4 magnetic fields is zero, while between 2 and 3, a uniform magnetic field is applied and heat supplied through an

external source. Magnetic body force elevates the fluid pressure and drives the fluid into the motion. There are no moving mechanical parts. [1].

FHD Bernoulli's equation between the station 1 and 2, under the assumptions of (i) negligible change in the gravitational potential energy and (ii) a constant kinetic energy in a tube of uniform cross-section, is written as;

$$p_1^* = p_2^* - \mu_0 \left(\bar{M}H\right)_2 \tag{2}$$

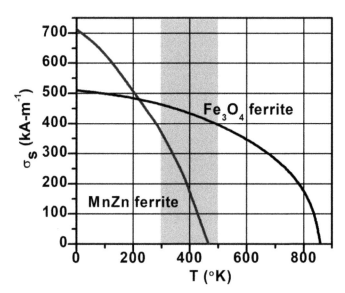

FIGURE 3.4 Saturation magnetization of Fe_3O_4 and mixed ferrite (Mn-Zn ferrite) as a function of temperature.

Station 1 is a field-free region of space, H = 0. Similarly, FHD Bernoulli's equation between station 3 and 4 gives,

$$p_3^* - \mu_0 \left(\bar{M}H\right)_3 = p_4^* \tag{3}$$

FHD Bernoulli equation is not applicable inside the solenoid, because in that region the assumption of an isothermal flow field (inherent in its deviation) does not hold. In the presence of a magnetic field, the equation of motion for a magnetic fluid is given as,

Application of Magnetic Fluid in the Energy Sector 75

$$\rho \frac{Dv}{Dt} = -\nabla p^* + \mu_0 M \nabla H + \eta \nabla^2 v + \rho g \qquad (4)$$

where, D/Dt is substantial differentiation, η is the viscosity of the fluid, and M is fluid magnetization. However, inside the solenoid the field is uniform, so ∇H = 0, also neglecting the acceleration (Dv/Dt) of the fluid, friction force and gravity in Eq. (4), fluid flow between station 2 and 3 reduces to,

$$p_2^* = p_3^* \qquad (5)$$

Now, the change in pressure between station 1 and 4 can be computed as:

$$\Delta p^* = p_4^* - p_1^* = \mu_0 H \left[\bar{M}(T_1) - \bar{M}(T_4) \right] = \mu_0 H \Delta \bar{M} \qquad (6)$$

From Eq. (6), it is seen that the pressure difference created between station 1 and 4 is directly proportional to the difference of magnetization at two temperatures, i.e., $\Delta p \, \alpha \Delta \bar{M}$. In other words, it can be said that increasing pressure difference accelerates the fluid flow. Equation (6) is also known as Kelvin body force acting on a particle per unit volume [63–67]. Using this principle, an automatic energy transport device (AETD) or automatic cooling device (ACD) can be designed [68–71]. Flow and energy transport in thermo-magnetic convection is governed by the equations as the equation of continuity:

$$\nabla \cdot v = 0 \qquad (7)$$

Equation of motion as given by equation (4) and equation of energy conservation [72]:

$$\rho c_p (v \cdot \nabla) T + \mu_0 T \frac{\partial M}{\partial T} (v \cdot \nabla) H = \nabla \cdot (\lambda \nabla T) \qquad (8)$$

where, T is the absolute temperature, ∂M/∂T is a pyromagnetic co-efficient of the magnetic nanofluid, c_p is specific heat capacity, and is thermal conductivity. The process is scientifically interesting in the coupling of thermodynamics with magnetic.

Additional cycle analysis and a design configuration have been studied using this no-moving-part converter in application to topping cycles for nuclear reactors and high reliability, light-weight space power supplies [73]. A laboratory proof of principle demonstration has been given [74], and their effort was reported in the development of magnetic fluids based on the liquid metal carrier to increase the heat release rate in the process [75]. The concept was also suggested for removal of heat from nuclear fusion reactors where the high-intensity magnetic field is available as part of the process [76]. In more modest use of the principle, there would seem to be much opportunity for applications in heat pipes, self-actuated pumps in thermal loops, and other devices.

3.2.5 ENERGY TRANSPORT IN COOLING DEVICE BY MAGNETIC FLUID

The need of rapidly improving the device performance has initiated the miniaturization and quantum devices with a spectacular pace. The trends toward faster speeds (in the multi-GHz range) and smaller features (to <100 nm) for microelectronic devices, more power output for engines, and brighter beams for optical devices are continuously demanding increasing heat rejection rate. This creates a challenge for high-tech industries like microelectronics, transportation, manufacturing, metrology, etc. Liquid cooling in such cases is a welcome solution over the conventional cooling methods such as the use of fins and micro-channel.

An electro-mechanical devices development is restricted by their operating temperature, and so a heat transport (cooling) device with better functioning is required for the thermal engineering. In practice, the power density (heat flux) of Central Processing Units (CPUs) is estimated to reach 1000 kW/m^2 because of the high-density integration of transistors [77]. This has attracted researchers towards long-distance energy (heat) transport device (typically the heat pipe) with high ability of transporting thermal energy [78]. A prototype heat pipe device [79] was built, which is able to transport heat as far as around 10 m distance. Due to the potential ability of achieving higher heat transfer of the TSMF, the research in the long-distance heat transport device (heat pipe) using TSMF has emerged recently. As discussed above, the flow of TSMF can be initiated by the gradient in thermal energy superimposed with a magnetic field gradient which transfers the heat from the source to sink. The magnetically-driven heat transport (cooling) device thus transfers heat for a long distance without any mechanical pump. In a series of research works [80–84], the authors have developed a device with proposing an innovative energy carrier of a binary TSMF. The binary TSMF is a mixture of kerosene-based

TSMF with a low boiling point organic liquid such as n-Hexane. The binary TSMF uses the boiling gas- bubble kinematics and its latent heat to increases the magnetic driving pressure together with heat transfer.

The original concept of energy conversion using TSMF was explained by Resler and Rosensweig [85], where they proposed a new thermal engine (power cycle). In their proposed power cycle, the working fluid of TSMF was thought to be operated by applying an appropriate magnetic field, and with which the cycle transfers the heat from the heat source to the radiator without any mechanical pumps and generates mechanical power output. The cycle simulation suggests a possible magnetohydrodynamic (MHD) power generation system. Namely, in the cycle, the thermal energy is converted to the kinetic energy of TSMF, and then it is converted to the mechanical power output using MHD power generator. A magnetic fluid when exposed to an external magnetic field and a gradient temperature is present then it would show variable susceptibility owing to a non-uniform magnetic body force. This thermo-magnetic convection leads to the heat transfer. This type of heat transfer can be useful when conventional convection heat transfer is inadequate; e.g., in miniature microscale devices or under reduced gravity conditions.

Several experimental and theoretical research related to the energy conversion in the application of a power device or heat transport device, have been reported [86–97]. As for design of heat transport (cooling) device, not for the power cycle application, it is thought that the boiling heat transfer (latent heat) is a key factor to enhance the heat transfer of TSMF. Nevertheless, till recent years most of the researches have focused on the heat transport by a single-liquid phase flow of TSMF [86–94], and a very few research was reported in the literature under the condition of a gas-liquid boiling two-phase flow [80–84, 95–97]. Among those, Kamiyama et al., [96] investigated the boiling two-phase flow under magnetic pressure. In their experimental research, TSMF was heated by a laser beam power device, and the boiling gas-bubbles distribution under a non-uniform magnetic field was visualized. They showed that the boiling gas-bubble enhances the magnetic pressure in an effective manner which drives the motion of TSMF. However, heat transfer characteristic was not reported may be because of their focus on to develop a power (generating) cycle.

The smart energy conversion using the TSMF enables the advantages like: (i) no external electric energy consumption to drive the energy carrier, (ii) A large amount of energy (heat) transport with phase change, (iii) A long-distance of energy (heat) transport, (iv) Simple configuration which enables the device to miniaturize, (v) No directional limit for heat transport, in terms of horizontal or vertical position, etc.

3.2.6 ENERGY TRANSFER

Special magnetic fluids, with tunable temperature-responsive magnetization, thermal conductivity, specific heat capacity, and viscosity, are required as multifunctional 'smart materials' to remove heat. A magnetocaloric pump is constituted by the combined effect of an external magnetic field and the temperature gradient created in the fluid. Rosensweig [98] has the first time reported the development of the conceptual embodiment and the viability study to use magnetic fluid for getting room temperature (RT) refrigeration with a permanent magnet. They found the technical problem of getting low yield stress at lower particle concentration when it is magnetized. The miniature model of AETD was developed and tested using a TSMF as a coolant by Lian et al., [99, 100]. The performance of automatic energy transport systems has been studied for different loop structures, viz; the single magnet pump square loop, the circular loop, and the two-magnet pump loop. Their findings show that the two magnet pump AETD is a preferred choice for transporting greater heat loads as compared to the other two. Electronic cooling using an AETD using thermo-magnetic effect was also tested by Xuan et al., [101]. The generation of waste heat in the system was used to drive the fluid. They report that a stronger thermo-magnetic convection enhances the heat dissipation rate as heat load increases, which indicates a self-regulating characteristic of such devices.

A water-based Mn–Zn ferrite magnetic fluid in a self-pumping magnetic cooling device shows cooling (ΔT) by ~20°C and ~28°C with 0.3 T magnetic field where the initial temperature of the heat load was 64°C and 87°C, respectively [102]. A decrease in fluid magnetization with an increase in temperature increases the velocity of fluid flow; consequently, the heat transfer rate increases. Yamaguchi et al., [103] have reported experimentally a long-distance (5 m) heat transport using the binary TSMF of transferring the thermal energy of 35.8W when the device (circulation loop) is horizontally placed.

The other example of AETD is power/distribution transformer. The heat transfer in an oil-cooled distribution transformer is because of the density variation of oil (follows the Archimedes law). But this is relatively weak, and hence a large temperature gradient is observed across the oil reservoir. The inefficiency in heat dissipation creates a localized region of intense temperature in the space between the core and windings and between the windings, this is known as the "hot spot." The creation of hot spot resulted into the degradation of the winding insulation as well as the other conductive components of a transformer yielding a continuous sparking. This will decompose the insulating oil by forming an oil-carbon and reduces the flashpoint of the oil. Over a long period of time, this leads to a failure of a transformer.

The normal lifespan of a transformer upon loading is also determined by the hot spot temperature (HST). ANSI (American Standards Institute) has specified the permitted value of HST at rated load based on arise in the winding resistive temperature [104]. To prevent or to reduce the hot spot winding temperature, a device is assisted by fins or external pumping devices. However, the fins not only increase the dimension but also increase the weight of a transformer. Similarly, the oil pump consumes power, requires regular maintenance, and makes a system complex and massive. Therefore, one has to think for modifying the cooling properties of transformer oil so as to improve the rate of heat dissipation.

In the last decade, many researchers [105–117] have checked the potential ability of magnetic fluid for the reduction in "hot spot" at the transformer core and its winding. The main principle is to choose a TSMF whose T_c is close to the temperature of a transformer (393 K) so that maximum convection due to magnetic buoyancy can be achieved [105]. The winding of a transformer is a source of generating a magnetic field; also, the maximum temperature is also generated at the winding; hence, the magnetic fluid experiences a drag towards the winding. A typical high voltage transformer generates about 2T magnetic field while the maximum temperature tolerance of 383°K. The transformer is submerged in oil tank; hence the surrounding is at ambient temperature with zero magnetic fields. This creates a magnetic field gradient (∇H) and the temperature gradient (∇T) in the same direction. Thus, the surrounding fluid feels attractive force at the core. At the winding, the fluid temperature rises because of the heat generated by it. The hot fluid losses its magnetization, as well as its density, reduces as compared to the surrounding fluid creating a gradient in fluid magnetization (∇M) and density gradient ($\nabla \rho$). The weak magnetic fluid pushes by the cooler magnetic fluid at the core due to the presence of a strong magnetic field. Thus, the additional flow is created in the transformer due to the thermo-magnetic convection of magnetic fluid. The resultant convective flow transfers the heat, and the fluid regains its magnetization when it gets cooled and completing the regenerative cycle [118]. It is expected that when a T_C of the magnetic fluid is close to the transformer temperature, then maximum heat will transfer, and the heat conduction will be optimum [105]. This result shows the potential benefits of TSMF in other applications, such as the passive cooling of electrical circuits, machinery, and processes [119]. Thus TSMF technology offers promising enhancement of the conventional oil flow rate [105, 108, 111, 118, 120, 121].

The loading of a transformer is mainly restricted by the temperature of its winding. The temperature of the winding increases just by increasing the load on the transformer and when the transformer is continuously working at its nameplate rating or goes beyond maximum allowed rating, then an intense temperature zone is created at the winding, known as 'hot spot' temperature. This maximum hotspot temperature is decided by the transformer regulatory standards. The standards are set based on the effect of increasing hotspot temperature on oil degradation and the deterioration in insulating strength of oil and surrounding of windings. The failure of the insulating system leads to the failure of a transformer functioning and its life span. Any of several incidences like transformer overloading, moisture, poor quality of the oil or insulating paper, extreme rise in transformer temperature, etc., are responsible for the deterioration of the insulation of a transformer. The life span of a properly sized, designed and maintained transformer is about 20–30 years when operated at the nameplate load. By increasing the loading on the transformer beyond its nameplate rating for an extended period of time reduces the life of a transformer drastically.

The theoretical calculations showed the 10–30 K temperature reduction of the winding for 2D [122] and 3D [116] model transformer when submerged in magnetic fluid instead of oil and operated under nominal conditions. Experimentally V. Segal [108] has reported 3.6 K reductions in winding temperature of a transformer filled with magnetic fluid containing Fe_3O_4 particles. Stoian et al., [112] had observed 3.4 K reduction in the coil core temperature powered by a 50 Hz AC power when magnetic fluid was used. The magnetite magnetic fluid used in the above experiments has a T_C of 858 K [1], which is quite higher than the temperature of the transformer. This may be a reason to observe the disparity in the reduction of HST in experiment and theory. To reach to the theoretical value of temperature reduction, one needs to use a magnetic fluid having T_C closer to the temperature of a transformer along with the high pyromagnetic coefficients from ambient temperature to the operating temperature of a transformer. Such low T_C magnetic fluid was prepared by J. Patel et al., [121, 123] and filled in a 3 KVA single-phase prototype transformer. About 22°C temperature reduction in the transformer core was obtained when specially synthesized TSMF was used. They have also studied the temperature rise inside a winding temperature of a transformer when it is operated for under load, normal load, overload and planned overload conditions to test the feasibility of TSMF filled transformer under actual load variation when working in field. The performance of TSMF is superior over transformer oil under the entire situation. Moreover, the normal life expectancy of the transformer calculated using the ASTM standards, found to increase by nine times under overload conditions. Whereas, the expected normal life doubles under planned overload condition. This study showed the potential benefit of using TSMF in distribution transformer with the expectation of delivering more power than its nameplate rating and without using any external accessories. It also shows the improvement in the normal life of a transformer. The proper use of TSMF in transformer offers the advantage of the reduction in the basic investment cost and maintenance cost of the transformer.

3.3 CONCLUSION

This review focuses on the energy applications of magnetic fluid as an energy harvester, energy storage, energy conservation, energy conversion and energy transfer using the principle of controlling fluid magnetization by the magnetic field and/or thermal energy gradient. Researchers in the field have generated few microvolts to millivolt signal using the magnetic fluid. A large output power generator using magnetic fluid is a challenging task. Similarly, a very preliminary work is reported using magnetic fluids as absorbent media for the solar energy. A proper combination of magnetic fluid properties will have the capacity to absorb and store a sufficient amount of energy. Using the TSMF with the magnetic field gradient and temperature gradient, the energy conversion devices were developed conceptually. Practically, the researchers have built heat transfer pipes, prototype devices for automatic cooling and the prototype model transformer to study the feasibility and efficiency of TSMF in such devices. The results of their experiments are very promising. But to bring this research a step ahead to make it commercialized needs careful investigations. In summary, the use of magnetic fluid for the energy sector is upcoming slowly, but it has a tremendous potential to be uncovered.

ACKNOWLEDGMENTS

Authors would like to thank Charotar University of Science and Technology, Changa for providing facilities to carry out most of the experimental work. We also acknowledge the funding agencies, Department of Science and Technology (DST), Government of India, New Delhi and GUJCOST, Gandhinagar for the financial support under various research schemes.

KEYWORDS

- **coolant**
- **electrical transformer**
- **energy harvester**
- **heat transfer devices**
- **magnetic fluid**
- **magnetic properties**

REFERENCES

1. Rosensweig, R. E., (1997). *Ferrohydrodynamics*. Dover Publications, New York.
2. Popplewell, J., (1984). Technological applications of ferrofluids. *Physics in Technology, 15*(3), 150–156.
3. Bailey, R. L., (1984). Lesser known applications of ferrofluids. *J. Magn. Mater., 39*(1/2), 178–182.
4. Raj, K., & Yamamura, A., (1982). *Ferrofluid Seal Apparatus* (Vol. 340, p. 233). US Patent US4.
5. Trivedi, K., Kothari, A., Parekh, K., & Upadhyay, R. V., (2018). Effect of particle concentration on lubricating properties of magnetic fluid. *J. Nanofluids., 7*(3), 420–427.

6. Trivedi, K., Parckh, K., & Upadhyay, R. V., (2017). Nanolubricant: Magnetic nanoparticle based. *Mater. Res. Express.*, *4*(11), 114003.
7. Bhatt, R. K., Trivedi, P. M., Sutariya, G. M., Upadhyay, R. V., & Mehta, R. V., (1993). Ferrofluid based inclination sensor. *Indian J. Pure Appl. Phys.*, *31*, 113–115.
8. Shah, K., Upadhyay, R. V., & Aswal, V. K., (2014). Nano-MRF: A material for damping application. *Solid State Phenomena*, *209*, 35–38.
9. Pankhurst, Q. A., Connolly, J., Jones, S. K., & Dobson, J., (2003). Applications of magnetic nanoparticles in biomedicine. *J. Phys. D: Appl. Phys.*, *36*(13), R167–R181.
10. Ganguly, R., Gaind, A. P., Sen, S., & Puri, I. K., (2005). Analyzing ferrofluid transport for magnetic drug targeting. *J. magn. Magn. Mater.*, *289*, 331–334.
11. Tartaj, P., Del Puerto Morales, M., Veintemillas-Verdaguer, S., González-Carreño, T., & Serna, C. J., (2003). The preparation of magnetic nanoparticles for applications in biomedicine. *J. Phys. D: Appl. Phys.*, *36*(13), R182–R197.
12. Jordan, A., Scholz, R., Wust, P., Fähling, H., & Felix, R., (1999). Magnetic fluid hyperthermia (MFH): Cancer treatment with, A. C., magnetic field induced excitation of biocompatible superparamagnetic nanoparticles. *J. Magn. Magn. Mater.*, *201*(1–3), 413–419.
13. Mehta, R. V., (2017). Synthesis of magnetic nanoparticles and their dispersions with special reference to applications in biomedicine and biotechnology. *Materials Science and Engineering: C.*, *79*, 901–916.
14. Roundy, S., Wright, P. K., & Rabaey, J., (2003). A study of low level vibrations as a power source for wireless sensor nodes. *Computer Communications*, *26*(11), 1131–1144.
15. Priya, S., (2005). Modeling of electric energy harvesting using piezoelectric windmill. *Appl. Phys. Lett.*, *87*(18), 184101.
16. Elvin, N. G., Lajnef, N., & Elvin, A. A., (2006). Feasibility of structural monitoring with vibration powered sensors. *Smart Mater. Structures*, *15*(4), 977.
17. Jia, D., Liu, J., & Zhou, Y., (2009). Harvesting human kinematical energy based on liquid metal magnetohydrodynamics. *Phys. Lett. A.*, *373*(15), 1305–1309.
18. Lallart, M., Pruvost, S., & Guyomar, D., (2011). Electrostatic energy harvesting enhancement using variable equivalent permittivity, *Phys. Lett. A.*, *377*, 3921–3924.
19. Shearwood, C., & Yates, R. B., (1997). Development of an electromagnetic microgenerator. *Electronics Lett.*, *33*(22), 1883, 1884.
20. Amirtharajah, R., & Chandrakasan, A. P., (1998). Self-powered signal processing using vibration-based power generation. *IEEE J. Solid-State Circuits*, *33*(5), 687–695.
21. Williams, C. B., Shearwood, C., Harradine, M. A., Mellor, P. H., Birch, T. S., & Yates, R. B., (2001). Development of an electromagnetic micro-generator. *IEEE Proceedings-Circuits, Devices and Systems*, *148*(6), 337–342.
22. Serre, C., Pérez-Rodríguez, A., Fondevilla, N., Martincic, E., Martínez, S., Morante, J. R., Montserrat, J., & Esteve, J., (2008). Design and implementation of mechanical resonators for optimized inertial electromagnetic microgenerators. *Microsystem Technologies*, *14*(4/5), 653–658.
23. Mann, B. P., & Sims, N. D., (2009). Energy harvesting from the nonlinear oscillations of magnetic levitation. *J. Sound Vibration*, *319*(1/2), 515–530.
24. Rahimi, A., Zorlu, Ö., Muhtaroğlu, A., & Külah, H., (2011). A compact electromagnetic vibration harvesting system with high performance interface electronics. *Procedia Engineering*, *25*, 215–218.

25. Gherca, R., & Olaru R., (2011). "Harvesting vibration energy by electromagnetic induction." *Annals of the University of Craiova, Electrical Engineering Series, 35*, pp. 7–12.
26. Bibo, A., Masana, R., King, A., Li, G., & Daqaq, M. F., (2012). Electromagnetic ferrofluid-based energy harvester. *Phys. Lett. A., 376*(32), 2163–2166.
27. Chae, S. H., Ju, S., Choi, Y., Chi, Y. E., & Ji, C. H., (2017). Electromagnetic linear vibration energy harvester using sliding permanent magnet Array and ferrofluid as a lubricant. *Micromachines, 8*(10), 288.
28. Alazmi, S., Xu, Y., & Daqaq, M. F., (2016). Harvesting energy from the sloshing motion of ferrofluids in an externally excited container: Analytical modeling and experimental validation. *Phys. Fluids., 28*(7), 077101.
29. Kim, D., & Yun, K. S., (2015). Energy harvester using contact-electrification of magnetic fluid droplets under oscillating magnetic field. *J. Phys.: Conference Series, 660*(1), 012108.
30. Wu, S., Luk, P. C., Li, C., Zhao, X., Jiao, Z., & Shang, Y., (2017). An electromagnetic wearable 3-DoF resonance human body motion energy harvester using ferrofluid as a lubricant. *Appl. Energy., 197*, 364–374.
31. Wang, Y., Zhang, Q., Zhao, L., & Kim, E. S., (2017). Non-resonant electromagnetic broad-band vibration-energy harvester based on self-assembled ferrofluid liquid bearing. *J. Microelectromechanical Systems, 26*(4), 809–819.
32. Zahn, M., (2001). Magnetic fluid and nanoparticle applications to nanotechnology. *J. Nanoparticle Research, 3*(1), 73–78.
33. Flament, C., Houillot, L., Bacri, J. C., & Browaeys, J., (2000). Voltage generator using a magnetic fluid. *European J. Phys., 21*(2), 145.
34. Zhang, L., Wu, H. B., Madhavi, S., Hng, H. H., & Lou, X. W., (2012). Formation of Fe2O3 microboxes with hierarchical shell structures from metal–organic frameworks and their lithium storage properties. *J. Am. Chem. Soc., 134*(42), 17388–17391.
35. Du, N., Chen, Y., Zhai, C., Zhang, H., & Yang, D., (2013). Layer-by-layer synthesis of γ-Fe$_2$O$_3$ @ SnO$_2$ @ C porous core–shell nanorods with high reversible capacity in lithium-ion batteries. *Nanoscale, 5*(11), 4744–4750.
36. Yan, Y., Du, F., Shen, X., Ji, Z., Sheng, X., Zhou, H., & Zhu, G., (2014). Large-scale facile synthesis of Fe-doped SnO$_2$ porous hierarchical nanostructures and their enhanced lithium storage properties. *J. Mater. Chem. A., 2*(38), 15875–15882.
37. Han, S., Hu, L., Liang, Z., Wageh, S., Al-Ghamdi, A. A., Chen, Y., & Fang, X., (2014). One-step hydrothermal synthesis of 2D hexagonal nanoplates of α-Fe$_2$O$_3$/graphene composites with enhanced photocatalytic activity. *Advanced Functional Mater, 24*(36), 5719–5727.
38. McDowell, M. T., Lee, S. W., Harris, J. T., Korgel, B. A., Wang, C., Nix, W. D., & Cui, Y., (2013). In situ TEM of two-phase lithiation of amorphous silicon nanospheres. *Nano Letters, 13*(2), 758–764.
39. Kim, J. K., Jung, J. Y., & Kang, Y. T., (2007). Absorption performance enhancement by nanoparticles and chemical surfactants in binary nanofluids. *International J. Refrigeration, 30*(1), 50–57.
40. Demirbas, M. F., (2006). Thermal energy storage and phase change materials: An overview. *Energy Sources, Part B: Economics, Planning, and Policy, 1*(1), 85–95.

41. Wu, S., Zhu, D., Zhang, X., & Huang, J., (2010). Preparation and melting/freezing characteristics of Cu/paraffin nanofluid as phase-change material (PCM). *Energy and Fuels*, *24*(3), 1894–1898.
42. Philip, J., Shima, P. D., & Raj, B., (2007). Enhancement of thermal conductivity in magnetite based nanofluid due to chainlike structures. *Appl. Phys. Lett.*, *91*(20), 203108.
43. Philip, J., Shima, P. D., & Raj, B., (2008). Nanofluid with tunable thermal properties. *Appl. Phys. Lett.*, *92*(4), 043108.
44. Philip, J., Shima, P. D., & Raj, B., (2008). Evidence for enhanced thermal conduction through percolating structures in nanofluids. *Nanotechnology*, *19*(30), 305706.
45. Shima, P. D., Philip, J., & Raj, B., (2009). Role of microconvection induced by Brownian motion of nanoparticles in the enhanced thermal conductivity of stable nanofluids. *Appl. Phys. Lett.*, *94*(22), 223101.
46. Shima, P. D., & Philip, J., (2011). Tuning of thermal conductivity and rheology of nanofluids using an external stimulus. *J. Phys. Chem. C.*, *115*(41), 20097–20104.
47. Parekh, K., & Lee, H. S., (2010). Magnetic field induced enhancement in thermal conductivity of magnetite nanofluid. *J. Appl. Phys.*, *107*(9), 09A310.
48. Parekh, K., & Lee, H. S., (2012). Experimental investigation of thermal conductivity of magnetic nanofluids. *AIP Conference Proceedings*, *1447*(1), 385–386.
49. Parekh, K., (2014). Thermo-magnetic properties of ternary polydispersed $Mn_{0.5}Zn_{0.5}Fe_2O_4$ ferrite magnetic fluid. *Solid State Communications*, *187*, 33–37.
50. Fang, X., Xuan, Y., & Li, Q., (2009). Anisotropic thermal conductivity of magnetic fluids. *Progress in Natural Sci.*, *19*(2), 205–211.
51. Krichler, M., & Odenbach, S., (2013). Thermal conductivity measurements on ferrofluids with special reference to measuring arrangement. *J. Magn. Magn. Mater.*, *326*, 85–90.
52. Katiyar, A., Dhar, P., Nandi, T., & Das, S. K., (2016). Magnetic field induced augmented thermal conduction phenomenon in magneto-nanocolloids. *J. Magn. Magn. Mater.*, *419*, 588–599.
53. Altan, C. L., & Bucak, S., (2011). The effect of Fe3O4 nanoparticles on the thermal conductivities of various base fluids. *Nanotechnology*, *22*(28), 285713.
54. Seshadri, I., Gardner, A., Mehta, R. J., Swartwout, R., Keblinski, P. T., Borca, T., & Ramanath, G., (2011). *J. Appl. Phys.*, *110*, 093917.
55. Patel, J., Parekh, K., & Upadhyay, R. V., (2015). Maneuvering thermal conductivity of magnetic nanofluids by tunable magnetic fields. *J. Appl. Phys.*, *117*(24), 243906.
56. Liu, Y. D., Zhou, Y. G., Tong, M. W., & Zhou, X. S., (2009). Experimental study of thermal conductivity and phase change performance of nanofluids PCMs. *Microfluidics and Nanofluidics*, *7*(4), 579.
57. Jaisankar, S., Ananth, J., Thulasi, S., Jayasuthakar, S., & Sheeba, K. N., (2011). A comprehensive review on solar water heaters. *Renewable and Sustainable Energy Reviews*, *15*(6), 3045–3050.
58. Mahian, O., Kianifar, A., Kalogirou, S. A., Pop, I., & Wongwises, S., (2013). A review of the applications of nanofluids in solar energy. *International J. Heat Mass Trans.*, *57*(2), 582–594.
59. Tyagi, H., Phelan, P., & Prasher, R., (2009). Predicted efficiency of a low-temperature nanofluid-based direct absorption solar collector. *J. Solar Energy Engineering*, *131*(4), 041004.

60. Pode, V., & Minea, R. S., (1900). *"Politehnica"*. University Timişoara, Faculty of Industrial Chemistry and Environmental Engineering, Victoriei Sq. 2, Timişoara, România.
61. Luminosu, I., Minea, R., & Pode, V., (1987). *Chem. Bull. Univ. Politehnica Timişoara*, 32, 77–81.
62. Avram, M., Pop, M., & Lucaci, A., (1979). *Fizica Moleculei, Lit. Univ. Timişoara*. Timişoara, Romania.
63. Luminosu, I., C., De Sabata, A., Sabata, D. E., & Jurca, T., (2009). *Buletinul AGIR Nr. 2–3* April–September.
64. Smit, J., & Wijn, H. P. J., (1959). *"Ferrites" Philips Technical Library*. Netherland.
65. Nkurikiyimfura, I., Wang, Y., & Pan, Z., (2013). Heat transfer enhancement by magnetic nanofluids—a review. *Renewable and Sustainable Energy Reviews*, 21, 548–561.
66. Resler, E. L., & Rosensweig, R. E., (1964). Magnetocaloric power. *AIAA J.*, 2(8), 1418–1422.
67. Lange, A., (2002). Kelvin force in a layer of magnetic fluid. *J. Magn. Magn. Mater.*, 241(2/3), 327–329.
68. Bahiraei, M., & Hangi, M., (2015). Flow and heat transfer characteristics of magnetic nanofluids: a review. *J. Magn. Magn. Mater.*, 374, 125–138.
69. Lian, W., Xuan, Y., & Li, Q., (2009). Characterization of miniature automatic energy transport devices based on the thermomagnetic effect. *Energy Conversion Management*, 50(1), 35–42.
70. Lian, W., Xuan, Y., & Li, Q., (2009). Design method of automatic energy transport devices based on the thermomagnetic effect of magnetic fluids. *International J. Heat Mass Trans.*, 52(23/24), 5451–5458.
71. Xuan, Y., & Lian, W., (2011). Electronic cooling using an automatic energy transport device based on thermomagnetic effect. *Appl. Thermal Engineering*, 31(8/9), 1487–1494.
72. Li, D., (2003). *China Science Press*, 165–166.
73. Donea, J. F., Lanza, E., & Van Der Voort, (1968). *Rep. EUR 4039e. Euratom, Ispra Establishment*. Varese, Italy.
74. Rosensweig, R. E., Nestor, J. W., & Timmins, R. S., (1965). *Mater. Assoc. Direct Energy Convers., Proc. Symp. AI Ch.E-I. Chem. E. Ser.*, 5, 104.
75. Popplewell, J., Charles, S. W., & Chantrell, R., (1977). *Energy Conuers.*, 16, 133.
76. Roth, J. R., Rayk, W. D., & Reiman, J. J., (1970). *NASA Tech. Memo*, 2106.
77. Danowitz, A., Kelley, K., Mao, J., Stevenson, J. P., & Horowitz, M., (2012). CPU DB: Recording microprocessor history. *Queen-Processors*, 10, 1–18.
78. Faghri, A., (2012). Review and advances in heat pipe science and technology. *J. Heat Transfer*, 134(12), 123001.
79. Mitomi, M., & Nagano, H., (2014). Long-distance loop heat pipe for effective utilization of energy. *International J. Heat Mass Trans.*, 77, 777–784.
80. Shuchi, S., Mori, T., & Yamaguchi, H., (2002). Flow boiling heat transfer of binary mixed magnetic fluid. *IEEE trans. Magn.*, 38(5), 3234–3236.
81. Shuchi, S., Sakatani, K., & Yamaguchi, H., (2004). Boiling heat transfer characteristics of binary magnetic fluid flow in a vertical circular pipe with a partly heated region. *Proc. Institution of Mechanical Engineers, Part C: J. Mechanical Engineering Sci.*, 218(2), 223–232.

82. Iwamoto, Y., Yamaguchi, H., & Niu, X. D., (2011). Magnetically-driven heat transport device using a binary temperature-sensitive magnetic fluid. *J. Magn. Magn. Mater.*, *323*(10), 1378–1383.
83. Yamaguchi, H., & Iwamoto, Y., (2017). Energy transport in cooling device by magnetic fluid. *J. Magn. Magn. Mater.*, 431, 229–236.
84. Yamaguchi, H., & Iwamoto, Y., (2013). heat transport with temperature-sensitive magnetic fluid for application to micro-cooling device. *Magnetohydrodynamics*, *0024–998X*, 49.
85. Resler, E. L., & Rosensweig, R. E., (1964). Magnetocaloric power. *AIAA J.*, *2*(8), 1418–1422.
86. Resler, E. L., & Rosensweig, R. E., (1967). Regenerative thermomagnetic power. *J. Engineering for Power*, *89*(3), 399–405.
87. Yamaguchi, H., Sumiji, A., Shuchi, S., & Yonemura, T., (2004). Characteristics of thermo-magnetic driven motor using magnetic fluid. *J. Magn. Magn. Mater.*, *272*, 2362–2364.
88. Fumoto, K., Yamagishi, H., & Ikegawa, M., (2007). A mini heat transport device based on thermo-sensitive magnetic fluid. *Nanoscale and Microscale Thermophysical Engineering*, *11*(1/2), 201–210.
89. Fumoto, K., Ikegawa, M., & Kawanami, T., (2009). Heat transfer characteristics of a thermo-sensitive magnetic fluid in micro-channel. *J. Thermal Sci. Tech.*, *4*(3), 332–339.
90. Xuan, Y., & Lian, W., (2011). Electronic cooling using an automatic energy transport device based on thermomagnetic effect. *Appl. Thermal Engineering*, *31*(8/9), 1487–1494.
91. Lian, W., Xuan, Y., & Li, Q., (2009). Characterization of miniature automatic energy transport devices based on the thermomagnetic effect. *Energy Conversion Management*, *50*(1), 35–42.
92. Lian, W., Xuan, Y., & Li, Q., (2009). Design method of automatic energy transport devices based on the thermomagnetic effect of magnetic fluids. *International J. Heat Mass Trans.*, *52*(23/24), 5451–5458.
93. Mitamura, Y., & Okamoto, E., (2015). Numerical analysis of the effects of a high gradient magnetic field on flowing erythrocytes in a membrane oxygenator. *J. Magn. Magn. Mater.*, *380*, 54–60.
94. Afifah, A. N., Syahrullail, S., & Sidik, N. A., (2016). Magneto viscous effect and thermomagnetic convection of magnetic fluid: A review. *Renewable and Sustainable Energy Reviews*, *55*, 1030–1040.
95. Ishimoto, J., Okubo, M., Kamiyama, S., & Higashitani, M., (1955). *Jpn. Soc. Mech. Eng. Int. J. Ser. B38*, 382–387.
96. Kamiyama, S., & Ishimoto, J., (1995). Boiling two-phase flows of magnetic fluid in a non-uniform magnetic field. *J. Magn. Magn. Mater.*, *149*(1/2), 125–131.
97. Aursand, E., Gjennestad, M. A., Lervåg, K. Y., & Lund, H., (2016). A multi-phase ferrofluid flow model with equation of state for thermomagnetic pumping and heat transfer. *J. Magn. Magn. Mater.*, *402*, 8–19.
98. Rosensweig, R. E., (2006). Refrigeration aspects of magnetic particle suspensions. *International J. Refrigeration*, *29*(8), 1250–1258.
99. Lian, W., Xuan, Y., & Li, Q., (2009). Characterization of miniature automatic energy transport devices based on the thermomagnetic effect. *Energy Conversion Management*, *50*(1), 35–42.

100. Lian, W., Xuan, Y., & Li, Q., (2009). Design method of automatic energy transport devices based on the thermomagnetic effect of magnetic fluids. *International J. Heat Mass Transfer*, *52*(23/24), 5451–5458.
101. Xuan, Y., & Lian, W., (2011). Electronic cooling using an automatic energy transport device based on thermomagnetic effect. *Appl. Thermal Engineering*, *31*(8/9), 1487–1494.
102. Chaudhary, V., Wang, Z., Ray, A., Sridhar, I., & Ramanujan, R. V., (2016). Self pumping magnetic cooling. *J. Phys. D: Appl. Phys.*, *50*(3), 03LT03.
103. Yamaguchi, H., & Iwamoto, Y., (2017). Energy transport in cooling device by magnetic fluid. *J. Magn. Magn. Mater.*, *431*, 229–236.
104. Facilities Instructions, Standards, and Techniques, (1991). *Permissible Loading of Oil-Immersed Transformers and Regulators* (Vol. 1–5, pp. 1–28).
105. Raj, K., & Moskowitz, R., (1995). Ferrofluid-cooled electromagnetic device and improved cooling method. *US patent US5*, *462*, 685.
106. Segal, V., Nattrass, D., Raj, K., & Leonard, D., (1999). Accelerated thermal aging of petroleum-based ferrofluids. *J. Magn. Magn. Mater.*, *201*(1–3), 70–72
107. Segal, V., Rabinovich, A., Nattrass, D., Raj, K., & Nunes, A., (2000). Experimental study of magnetic colloidal fluids behavior in power transformers. *J. Magn. Magn. Mater.*, *215*, 513–515.
108. Segal, V., & Raj, K., (1998). An investigation of power transformer cooling with magnetic fluids. *Indian J. Engg. & Mater. Sci.*, *5*, 416–422.
109. Segal, V., Hjortsberg, A., Rabinovich, A., Nattrass, D., & Raj, K., (1998). *Conference Record of the IEEE International Symposium on Electrical Insulation* (pp. 619–622). Arlington, Virginia, USA.
110. Jeong, G. Y., Jang, S. P., Lee, H. Y., Lee, J. C., Choi, S., & Lee, S. H., (2013). Magnetic-thermal-fluidic analysis for cooling performance of magnetic nanofluids comparing with transformer oil and air by using fully coupled finite element method. *IEEE Trans. Magn.*, *49*(5), 1865–1868.
111. Parekh, K., & Upadhyay, R. V., (2004). Characterization of transformer oil based magnetic fluid. *Indian J. Engg. Mater Sci.*, *11*, 262–266.
112. Stoian, F. D., Holotescu, S., Taculescu, A., Marinica, O., Resiga, D., Timko, M., Kopcansky, P., & Rajnak, M., (2013). *The 8th International Symposium on Advanced Topics in Electrical Engineering*, 978–1–4673–5980–1.
113. Segal, V., (1999). *Colloidal Insulating and Cooling Fluid*. US Patent 5,863,455.
114. Staoin, F. D., (2014). *J. Phys. Conference Series*, *547*, 012044.
115. Cader, T., Bernstein, S., & Crowe, P. C., (1999). *Magnetic Fluid Cooler Transformer*. US Patent 5,898,353.
116. Morega, A. M., Morega, M., Mihoc, G., Iacob, C., Pîslaru-Dănescu, L., Stoica, V., Nouraş, F., & Stoian, F. D., (2010). *12th International Conference on Optimization of Electrical and Electronic Equipment*, 140–146.
117. Pîslaru–Danescu, L., Morega, A. M., Morega, M., Stoica, V., Marinica, O. M., Noura, F., Paduraru, N., & Borbáth, I. T., (2013). Borbáth *IEEE Transactions on Industry Appl.*, *49*(3), 1289–1298.
118. Rosensweig, R. E., (1987). Magnetic fluids. *Annual Review of Fluid Mechanics*, *19*(1), 437–461.

119. Curtis, R. A., (1971). Flows and wave propagation in ferrofluids. *Phys. Fluids.*, *14*(10), 2096–2102.
120. Love, L. J., Jansen, J. F., McKnight, T. E., Roh, Y., & Phelps, T. J., (2004). A magnetocaloric pump for microfluidic applications. *IEEE Trans. Nanobiosci.*, *3*(2), 101–110.
121. Patel, J., Parekh, K., & Upadhyay, R. V., (2016). Prevention of hot spot temperature in a distribution transformer using magnetic fluid as a coolant. *International J. Thermal Sci.*, *103*, 35–40.
122. Strek, T., & Jopek, H., (2007). Computer simulation of heat transfer through a ferrofluid. *Physica Status Solidi(b).*, *244*(3), 1027–1037.
123. Patel, J., Parekh, K., & Upadhyay, R. V., (2017). Performance of Mn-Zn ferrite magnetic fluid in a prototype distribution transformer under varying loading conditions. *International J. Thermal Sci.*, *114*, 64–71.

CHAPTER 4

Nanostructured Materials and Composites for Renewable Energy

IDOWU DAVID IBRAHIM,[1] CHEWE KAMBOLE,[2] AZUNNA AGWO EZE,[1]
ADEYEMI OLUWASEUN ADEBOJE,[2] EMMANUEL ROTIMI SADIKU,[3]
WILLIAMS KEHINDE KUPOLATI,[2] TAMBA JAMIRU,[1]
BABATUNJI WUNMI OLAGBENRO,[4] and OLUDAISI ADEKOMAYA[1]

[1]*Department of Mechanical Engineering, Mechatronics, and Industrial Design, Tshwane University of Technology, Pretoria, South Africa, E-mails: ibrahimid@tut.ac.za, ibrahimidowu47@gmail.com*

[2]*Department of Civil Engineering, Tshwane University of Technology, Pretoria, South Africa*

[3]*Department of Chemical, Metallurgical and Materials Engineering, Polymer Section, Tshwane University of Technology, Pretoria, South Africa*

[4]*Department of Industrial and Production Engineering, University of Ibadan, Oyo State, Nigeria*

ABSTRACT

Alternative renewable and sustainable energy has gained worldwide interest due to the high-energy demand and environmental factors. The increasing urbanization, industrialization, population growth, depletion of ozone, and the hazardous effects of the current methods of energy generation from fossil fuels (coal, oil and natural gas), have contributed significantly to researches into renewable energy. Research has also shown the importance of nanostructure and composite materials in resolving issues surrounding renewable energy. This chapter focuses on nanostructured and nanocomposite materials that are used for renewable energy. The chapter presents nanomaterials (NMs) and their applications for renewable energy; which was drawn from the various literatures that are relevant to nanostructure and composites for renewable energy.

4.1 INTRODUCTION

In the past two decades, over 80% of used energy was from fossil fuels, which comprise of coal, natural gas, and oil. It has been predicted that a time will come when some of these resources will be depleted completely, since they are not renewable. The depletion time is assumed to be different, depending on the model that is used to estimate the period [1–3]. The increasing growth in population and industrialization has necessitated research into alternative energy solutions [2, 4]. Nuclear energy has been one of the alternative energy solutions. However, the health concerns surrounding the use of uranium (nuclear power) for power generation has led to a gradual shift from the use of this resource. Therefore, nuclear power may not be a feasible long-term option for power generation [5, 6]. Buller and Strunk [7] reported that the combination of hydroelectric, geothermal, and wind energy is not sufficient enough to meet the global energy demand. They further reported that diversifying totally into the renewable energy from the sun can meet the energy demand worldwide; adding that if the total solar radiation from the sun within 1 hour can be converted and stored, it would be enough to meet the energy demand for one full year. A similar explanation was made by Ganesh [2] and Lewis and Nocera [1]. Energy generation, conversion, and storage go side-by-side, and they are almost inseparable. Figure 4.1 shows the conversion and storage process of solar energy, as explained by Buller and Strunk [7].

In order to meet up the energy demands, the material aspect of the energy mix has to be critically considered. Research has shown the importance of nanostructured and composite materials in resolving issues surrounding renewable energy. Different materials studied, include ceramics, polymers, metals, and composites [8–10]. The dynamism in the energy sector has led to the use of different materials, as singly applied materials or as an upgrade to hybrid composites and nanocomposites. The suffix "nano" referred to the dimension of the materials. A material is assumed to be in nano-scale if at least one or more of its dimensions are ≤100 nm. Nanomaterials (NMs) with improved chemical and physical properties have found useful applications in the energy and environmental sectors [11]. Therefore, this chapter discusses the relevant nanostructured and composite materials that are used in renewable energy applications like solar cells (SCs), fuel cells and supercapacitor and hydrogen storage devices, and so on.

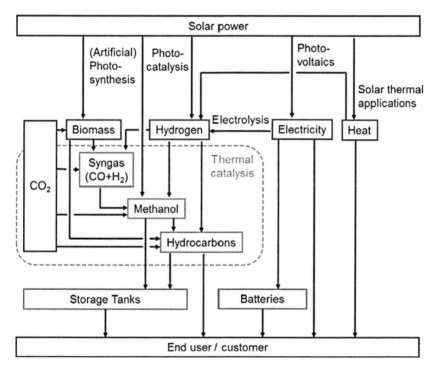

FIGURE 4.1 Energy conversion and storage of different energy sources influenced by solar power.

Source: Reprinted with permission from Ref. [7]. © 2016 Elsevier.

4.2 POTENTIAL MATERIALS USED FOR RENEWABLE ENERGY

At least 80% of the global sources of energy are derived from carbon-based materials (coal, oil, and natural gas) [13]. This has a huge negative impact on the environment. As the demand for energy increases due to a global increase in population growth, globalization, and modernization, there is an increasing concern on the environmental impact of this demand. Energy generation has been reported, by various researchers, to have a great impact on the environment. Different materials from micro- to nanoparticles have been studied and are considered suitable for solving the environmental challenges posed by energy generation. Continuing research and advances have further led to the discovery of renewable energy and techniques which have some remedial effects on the environment. These include biofuels, fuel cells, SCs, adsorption or separation techniques, and photocatalytic oxidation [11]. Environmental awareness and legislations are also being put in place to further promote and preserve non-renewable resources and the environment.

Global development is a product of sustainable and affordable energy. There is a huge global dependence on petroleum and electricity. The world population increased by 400%, and the global energy demand increased 16 times in the 20th century [14]. The outrageously huge energy demand has led to the excessive utilization of energy from fossil fuel that is detrimental to the environment. The projected energy demand by the year 2050 ranges between 1.2 and 1.3 terawatts (TW), which requires a further production of at least 10 TW of clean energy for the sustenance of life [14].

The global energy consumption was approximately 14 TW in the year 2006 alone, while an estimation of 20 TW per year of energy consumption by the year 2030 has been projected [13]. Excessive exploitation of the fossil fuel has contributed to greenhouse gases, thereby, leading to global warming. The effect of the global warming on the environment is the change in climatic conditions, which is currently been experienced globally. The surface of the Atlantic Ocean now has a higher temperature than it has ever been recorded in the last 1000 years [14]. In order to address the future demand of energy, alternative energy sources must be sought. The challenge of producing the required 10 TW of clean energy required may be tackled by renewable energy, nuclear power, and carbon-neutral energy (use of carbon sequestration and fossil fuel) [14].

The knowledge of the characteristics and properties of materials in their molecular form has helped greatly in predicting the performance of materials with adequate knowledge of their grain size, shape, configuration, crystallization and reactions [11]. The development of a conceptual production of nanostructured materials in an environmentally sustainable and cost-efficient manner is paramount in the generation of nanostructured materials or composites for energy generation [11].

The knowledge of space science and material chemistry may be required to solve the challenges of applying NMs where bulk materials are used, as there exists an effect of spatial confinement on the surfaces of materials and their properties [15]. As the demand for sustainable energy increases, alternative energy sources are being sought. NMs have emerged as the new platform or source for light-harvesting to enhance the delivery of renewable energy [14]. The continued reliance on carbon derived energy sources are inimical to the environment and human health. Hence, the interest and need in an environmentally-friendly and less toxic alternative means of deriving and storing of energy [13]. It was evaluated that approximately 1500 TW per year of energy is derivable from solar source alone. Photovoltaic (PV) devices can be used to directly convert energy from the sun to electrical energy. Nano-crystalline titania (TiO_2) film sensitizes monolayer dye molecules at photoanode to produce what is called dye-sensitized solar cells (DSSCs). This is an economical alternative to the normal solid PVs [13].

Currently, the efficiency of converting solar energy is ~11.3% for electrolytic DSSCs and ~5.1% for DSSCs in solid state [13]. Attention has shifted to the possibility of purifying the environment by using nanostructured semiconductor photocatalysts, such as Fe_2O_3, TiO_2, ZnO, CdS, and WO_3. The TiO_2 semiconductor stands out from the list of photocatalytic semiconductors because it is chemically inert with useful and well-defined band structure, readily and commercially available and it is also photostable. The arrangements of the nanostructure or nanoarchitecture adaptations of TiO_2 include nanoparticles (NPs), nanostructured films or coatings, nanorods, nanotubes, nanowires, nanoporous or mesoporous structures and various TiO_2-based composites [11]. Advances in TiO_2 also include application in PVs and from photocatalysis to photochromic or electrochromic sensors. Application of TiO_2 photo-catalyst will enhance abundant, clean, and safe solar energy for energy generation and aid environmental sustainability [11].

Hydrogen is an exceptional source of clean unpolluted energy; hence, it is admirable as a future sustainable fuel source. Hydrogen can be neatly obtained from water through the photocatalytic process. This is an artificial photosynthesis, similar to the chemical photosynthetic production of green plants where splitting water molecules are reduced through electrons which exist in the conduction band to form hydrogen gas (H_2). The molecules of water are oxidized by holes in the valence band that produces oxygen gas (O_2). Extra catalysts may speed up the two half-reactions to produce an ideal nanostructured semiconductor photocatalyst for solar photocatalytic water splitting by considering the following properties [11]:

i. The engineering of absorption of solar energy by suitable bandgap;
ii. High crystallinity to the facilitation of migrating electron or hole pairs at the surface by high crystallinity;
iii. Large and active surface area for activation of redox reactions at the catalyst surface.

Sol-gel is a chemical method for synthesizing or fabricating greatly porous and amorphous graphitic carbon. Due to the bulkiness and porosity of the carbon produced, it can be passed through a selection of catalysts, precursors, and solvents. Porous carbon structure and surface chemistry are manipulated by impregnating in chemical or solvent exchange, controlled removal of the solvent, activation, and pyrolysis. Controlled chemistry and the porous structure, enhance the properties of highly porous carbon for energy applications. Carbon cryogels, consisting of tuned porous structure with good surface chemistry, possess better cyclic stability, higher specific capacity, and the potential for high working voltage when used as electrodes for supercapacitors. Confinement of hydrides inside the pores of highly porous carbon results in coherent nanocomposites, the reaction pathway and dehydrogenation temperatures are significantly changed. Hence, there are the reductions in the activation energy and dehydrogenation temperature, when the pore size is reduced, and so is the carbon surface chemistry [16].

4.3 APPLICATION OF NANOSTRUCTURE AND COMPOSITES MATERIAL FOR RENEWABLE ENERGY

The most challenging issues of the 21st century are the energy demand and environmental degradation. The dependence on oil and electricity has made energy a vital component of our day-to-day needs, as exemplified by the fact that 13 TW of energy is required to maintain the life of 6.5 billion people, worldwide [14]. However, renewable energy such as solar radiation is the best alternative to meet this required energy need. Solar radiation requires a novel plan to converge the incident photons with better effectiveness [14, 17, 18]. Presently, significant advances have been made in the energy conversion and storage devices, such as fuel cells and SCs, batteries, and supercapacitors, respectively [19]. Nanostructured materials emerge as the most promising materials in enhancing the conversion and storage of renewable energy. Nanostructured materials and composites have shown great promises for the production of clean and renewable energy. Figure 4.2

presents different areas where nanostructure and composites materials have been largely applied in the field of renewable and clean energy.

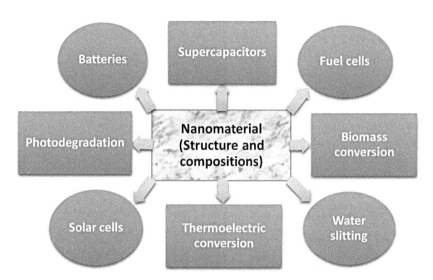

FIGURE 4.2 Applications of nanostructure and composites for renewable energy.

4.3.1 NANOSTRUCTURES FOR RENEWABLE ENERGY APPLICATION

Researches on renewable energy, such as solar and wind energy, depend considerably on the natural environment, such as the daytime, nighttime, wind, etc. and this is not stable and/or uniformly distributed. Hence, in order to harness the excess energy generation for optimal stabilization of the power grid connections, there are needs for the high-performance energy storage system to store the energy generated. In this regard, carbon NMs, titanium oxide, metallic sulfides, and many other nanostructured materials have been widely studied for applications in renewable energy conversions, storage devices and in improving the storage capacity of Li-ion batteries [20]. The carbon NMs that are found in nanostructures for renewable energy applications include the following: fullerenes, carbon nanotubes (CNTs), graphene, and their nitrogen-doped derivatives. The development of polymeric PV cells is made possible as a result of the photo-induced charge transfer fullerenes, which is used to store light energy as electron relays for the production of electricity [20].

Fullerenes (a carbon nanomaterial), has photon-induced charge transport capacities that can be vital for the improvement of polymeric PV cells. A nitrogen-doped derivative of carbon NMs is the product of incorporating nitrogen into carbon NMs. These can be achieved by way of introducing nitrogen-containing precursors, e.g., ammonia (NH_3) into carbon NMs [21–25]. The incorporation of nitrogen into carbon NMs can create electron change that offers desirable electronic structures for several possible applications of practical significance [26]. CNTs are among the vital carbon NMs that have a lot of possible applications in renewable energy devices. CNTs can show semiconducting or metallic performance, depending on their diameter and nature of the order of arrangement of the carbon atoms on the walls [27] and an electrical conductivity up to 5000 S/cm [28]. The improved electrical conductivity characteristics of the CNTs guarantee the high charge transportability. Owing to the existence of tubes, CNTs have porosity in the mesopore (the equivalent of 2 to 50 nm) [29, 30]. The high surface area and exceptional mesoporosity nature of CNTs make them highly electrochemically available to the electrolyte. In addition to the improved electrical conductivity and porosity, CNTs also possess a high thermal conductivity, up to 6000 $Wm^{-1}k^{-1}$, high thermal stability which can be up to 2800°C in space and improved mechanical properties (tensile strength of 45×10^{12} Pascal or 45 Tera Pascal) [31]. The aforementioned properties make CNTs very attractive for possible applications in sensors [32], hydrogen storage system [33], piezoelectric and thermoelectric energy harvesting devices [34], organic PV cells [35], fuel cells [36, 37], batteries [38], and supercapacitors [39]. CNTs also exhibit a higher tensile strength and Young's modulus when compared to carbon fibers. This makes it suitable for application in flywheel energy storage (FES) [40]. The combination of different NMs can equally have a positive effect on the performance of renewable energy devices. Studies have shown that a combination of CNTs and TiO_2 films enhance the carrier transport in photoelectrochemical SCs [41, 42]. When electrode materials are coated with CNTs, the charge transport can be made possible by the high conductive CNTs and the nanoscale thickness. Coating on the CNTs will permit rapid rate abilities; hence, CNTs are broadly used as the current collecting materials [43–46]. Two main types of CNTs stand-out as having high structural precision; they are single-walled nanotubes (SWNTs) and multiple walled nanotubes (MWNTs). The SWNTs are made up of single graphite sheet that is perfectly wrapped into a cylindrical tube, while the

MWNTs is made up of a collection of nanotubes that are cyclic-nested, similar to rings of a tree stem [30]. The MWNTs and SWNTs have the same electrical properties. This is due to the fact that in MWNTs, the coupling between the cylinders is weak and electronic transport in their metallic form moves without scattering over their lengths, enabling them to carry high currents with essentially no heating [30, 47, 48]. Phonons (charge carrier) also propagate easily along the nanotubes. The measured room temperature (RT) thermal conductivity for individual MWNTs are greater than that of natural diamond and the basal plane of graphite [30, 49]. CNTs have also been employed to fabricate composite electrodes with activated carbons, conjugated polymers, or metal oxide, in order to improve the performance (mainly energy density) of CNTs supercapacitors [20]; owing to the large surface area, high carrier transport mobility and excellent thermal and mechanical stability. However, CNTs-based capacitors do not reveal satisfactory capacitance for the expected device performance. This is because of the high contact resistance between the CNTs-based electrode and the current collector, the inefficient interaction between CNTs-based capacitors and electrolyte and the instability of double layers. Meanwhile, graphene has recently been studied as an alternative to the carbon-based electrode in supercapacitors [20, 50]. CNTs have been used as field emission electron source [51, 52] and it has provided stable emission, extended working lifetimes, and low emission threshold potentials [51, 53]. Nanotube materials can be used as supercapacitance, which have enormous capacitances in comparison with those of common dielectric-based capacitors; also as electromechanical actuators that may perhaps be used in electric motors and robots. On the other hand, there are some divisive reports on the use of nanotubes for hydrogen storage devices [54–56]. Among the nanostructured materials that are applied in the field of renewable energy are aerogels, nanostructured metallic sulfide, TiO_2, etc. Aerogels have been investigated as thermal insulators [57], translucent glazing [58] or in flat plate solar collectors [59]. It has been reported that by using nanostructured materials, the Li-ion diffusion length is decreased and the inner stress can be made less effective and thus improved rate ability, and recyclability can be expected [40]. Nanostructured metallic sulfides also play an important role in the area of renewable energy application, such as sensors, SCs, light-emitting diodes (LEDs), Lithium-ion batteries (LIBs), fuel cells, non-volatile memory devices and thermoelectric devices [19, 60–62].

Another nanostructured material that has been used to improve the property of the metallic compound is a silver nanoparticle. Andrade et al., [63], reported the loading of silver nanoparticle onto the surface of ZnO by photochemical reaction, which improved the photocatalytic activity of ZnO. Thiocyanate ion presence on the surface of the semiconductor prevented the uncontrollable growth of the silver nanoparticle, which could have resulted into various morphologies (nanorod, nanoplates, and nanosphere) and large degrees of polydispersity. Furthermore, TEM images of hybrid ZnO/Ag nanostructures of different concentrations of silver nitrate have influenced the average particle distribution and size. As the silver content increased, the samples average diameters increased, as shown in Figure 4.3.

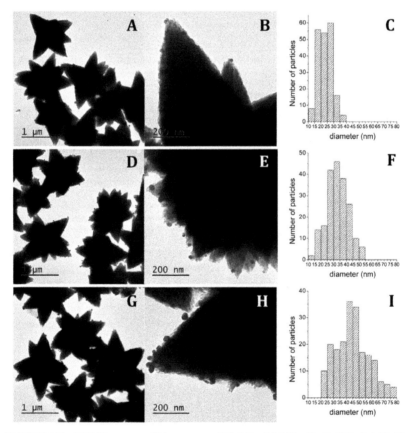

FIGURE 4.3 TEM images and histograms of ZnO/Ag0.05 (A–C), ZnO/Ag0.1 (D–F) and ZnO/Ag0.2 (G–I).

Source: Reprinted with permission from Ref. [63]. © 2017 Elsevier.

4.4 DEVELOPMENT OF NANOSTRUCTURED AND COMPOSITE MATERIALS

The field of energy conversion can be subdivided into three categories, namely: thermal-chemical energy conversion, electrochemical energy conversion, and solar energy conversion. There exists a noticeable connection between these three forms of energy conversions. Buller and Strunk [7] explained a scenario that linked the three methods together. They cited the conversion of CO_2 to methanol for a hypothetical photo-electrochemical cell under solar radiation and working at high temperatures. Such a device clearly cuts across all the three categories.

The energy sector generates most of the carbon dioxide (CO_2) emissions in the world, as a result of human activities of exploiting and using fossil and mineral fuels, nuclear and hydroelectric energy [64]. Since the 1960s, the global energy consumption, along with CO_2 emission, has been increasing drastically. Consequently, the global community has become more concerned about "greenhouse gases" and their damaging effects on the environment. This has necessitated the need to find ways of developing clean and renewable energy systems, such as solar and fuel cells on one hand and advanced energy storage devices, such as supercapacitors and batteries, on the other hand.

Furthermore, nanotechnology has brought about a cutting-edge technology that enables the creation of new materials and systems for energy generation and storage. Following the discovery of buckminsterfullerene (C_{60}) in 1985, numerous innovative carbon NMs and technologies have been developed. These technologies are considered key enablers in transforming the future of energy [20]. Several types of researches have also been developed around ZnO nanostructures [65–69].

4.4.1 NANOTECHNOLOGY IN SOLAR ENERGY

Nanotechnology has been a major developmental goal of many countries, and it has attracted huge investment. From literature, the nanotechnology market was reported to be worth US$2.6 trillion in 2014 as against US$7.9 billion in 2012 [70, 71]. Nanotechnology has been largely explored for solar energy generation, conversion, and storage. In addition, nanostructured materials are considered to have created significant improvement when it comes to renewable sources of energy. It has been reported that by 2050, the world's reserves of fossil fuels would have significantly diminished [72, 73] and by 2040, 50% of the global primary energy is expected to be generated from renewable sources [74].

4.4.1.1 NANOMATERIALS (NMS) IN PHOTOVOLTAIC (PV) SOLAR CELLS (SCS)

The PV SCs have undergone rigorous changes over time. The first generation (1G) PV SCs were crystalline silicon-based. They exhibited a high efficiency but worked with expensive thick wafers of several hundreds of microns for effective photon absorption. The costly vacuum processes for the manufacture of defect-free crystalline films are equally of great concern. The second-generation (2G) SCs incorporated thin-film inorganic semiconductors, integrating materials, such as amorphous silicon (a-Si) and micromorph silicon (a-Si/μc-Si), copper indium gallium selenide (CIGS) and cadmium-telluride (CdTe). The 2G cells were developed to surmount the high cost of 1G solar cell through the utilization of thin-film technology. However, this development did not lead to a reduction in the cost of the electricity produced. This challenge, therefore, led to the exploration of the third-generation (3G), which aims at generating electricity at a lower cost than the 2G SCs. The 3G or emerging SCs, include dye-sensitized, quantum dot (QD), organic or polymer and inorganic-organic perovskite solar cells (PSCs). Recent advances in nanotechnology and the development of molecular semiconductors, such as QDs, semiconductor NPs, nano-thin films, and nanostructures, are advantageous for revolutionizing the emerging SCs technology [75, 76]. Nanostructured thin film SCs offer the following advantages:

a. Their capability for multiple reflections, effective and considerably larger optical absorption path when compared to a cell with conventional film thickness.
b. They greatly reduce the travel path of light-generated electrons and holes. This results in a considerable reduction of recombination losses. This means that nanostructured SCs absorber layers can be as thin as 150 nm when compared to the several micrometers thick absorber layers in conventional thin film SCs.

c. The size of the NPs in thin SCs can be varied to enable the desired design of the energy bandgap. This enables some design flexibility in the absorber layers of SCs [77].

4.4.1.2 DYE-SENSITIZED SOLAR CELLS (DSSC)

DSSC is a promising alternative method of harvesting and converting solar energy. It is a representative of PV cells with basic components, such as electrolyte, photo-anode, and platinized counter electrode [78, 79]. Hussein [74] defines DSSC as a type of SCs that enables optical absorption and charge separation or injection, which is achieved by associating a dye-sensitizer (a light-absorbing material) with a wide bandgap semiconductor as the photo-anode (having nano-crystalline morphology). The level of interest that has been experienced with DSSCs is due to the low production cost, relatively high efficiency in terms of energy conversion, and easy to fabricate [80–84]. Approximately 13% power conversion efficiency (PCE) has been recorded with the DSSCs, and more researches are still ongoing to further improve the efficiency [85]. These improvements can be by increasing the absorption coefficient of the dye sensitization in order to have a wider solar spectrum range, light-harvesting efficiency improvement and constraining the method in which the carrier recombines [86–88]. In all this achievement, TiO_2 nanostructured material has been widely utilized based on its useful properties, offered by such material when it comes to energy conversion and storage [89, 90]. Some other metals that have been used include: but not limited to, silver (Ag), gold (Au) and platinum (Pt) for obvious reasons as explained by [78, 89, 91]. Other reports have succinctly explained the use of Au and Ag for DSSCs [92–98]. According to a report by Hagfeldt et al., [93], ZnO as a photo-anode in DSSCs was the very first oxide semiconductor that was used before TiO_2, which has better conversion efficiency. The reason for ZnO was due to the fact that it has better electron mobility when compared to TiO_2 and its highly crystalline structures that can easily be synthesized [99]. In a recent research work by Luo et al., [100], it was reported that porous ZnO nano-sheet framework on titanium (ZnO NSFs/Ti) was better than the commercial ZnO nanoparticle, resulting in a 3.23% PCE. This shows a good performance in comparison to the Pt-based photo-anode PV cell.

In a different research by Zhang et al., [89], it was reported that doped copper nanowires@TiO$_2$ (Cu NWs@TiO$_2$) helped to promote the charge production-separation and restrain recombination of the charge that is stimulated by tri-iodide ion and excited dyes and hence enhances the charge injection efficiency. It was further reported that the electron transfer tunnel that was formed, accelerated the charge transport rate and thus, significantly improved the collection efficiently as shown in Figure 4.4. The absence of Cu NWs@TiO$_2$ in the photo-anode caused the traps and scattering electrons to be randomly dispersed as observed in Figure 4.4a.

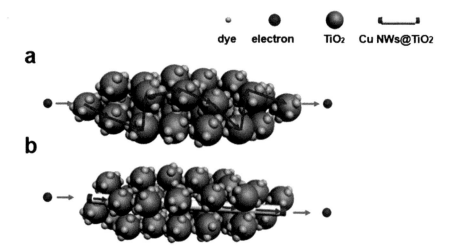

FIGURE 4.4 (See color insert.) Photoelectrons transporting process in a) dye-coated un-doped TiO$_2$ and b) doped Cu NWs@TiO$_2$ dye-coated TiO$_2$ photo-anode.

Source: Reprinted with permission from Ref. [89]. © 2017 Elsevier.

4.4.1.3 NANOPARTICLES (NPS) IN PHOTO-CATALYSIS

Catalysis that modifies the rate of photo-reaction is known as photo-catalysis. It is a chemical reaction which entails the absorption of light by adding catalysts that contribute to the chemical reaction and are not consumed during the process. Photo-catalysis in an ideal situation should be non-toxic, stable, low cost, and highly photoactive [101]. The conversion of solar energy into chemical energy by using the photocatalytic properties of some materials, either reduce or oxidize the materials for the attainment of certain materials. Photo-catalysis has been used for different products and applications like hydrocarbon, hydrogen evolution, bacteria and pollutant removal, air and water purification, sterilization, and photo-electrochemical conversion [102–108]. Similar to DSSC, TiO_2 is the most widely used in nanostructured materials in many photo-catalysis because of properties, such as chemical stability, good oxidizing potentials, transparent to visible light, low cost, non-toxic, superhydrophilicity and durability [108]. From literature, heterogeneous photo-catalysts semiconductors, such as ZnO and TiO_2, have been widely used due to their environmental protection abilities [109–112]. Figure 4.5 shows the various areas of application of TiO_2 photo-catalysis, especially in the energy and environmental sectors.

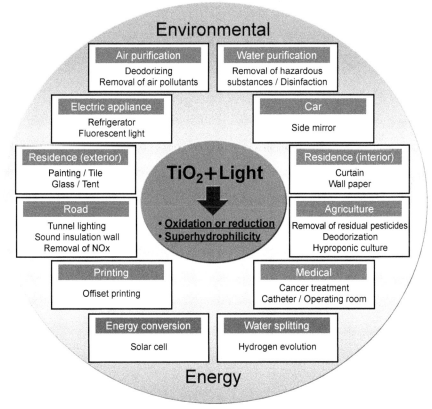

FIGURE 4.5 Areas of application of TiO_2 based photocatalysis.
Source: Reprinted with permission from Ref. [108]. © 2012 Elsevier.

Nanoparticle has a way of improving the properties of composites or hybrid materials due to the high contact surface area. Several nanostructured and composite materials have been produced either by doping or hybridization of two or more NPs for the purpose of photocatalysis [63, 113, 114]. In other studies, various nanostructured and composite materials, different from the most commonly used, i.e., ZnO and TiO$_2$ have been developed to enhance the performance of the photocatalysis processes [63, 115–119].

Andrade et al., [63] explained why zinc oxide (ZnO) had attracted such attention. It is due to its incredible properties emanating from the 3.37 eV wide direct bandgap and 60 meV binding energy. Andrade et al., further added that ZnO exhibits various synthetic conditions; its growth is easily controllable in solution and improved properties in catalysis applications. The drawback that has been experienced with ZnO is the rapid recombination rate of electrons and absorption within the ultraviolet (UV) range. The solutions that have been reported include surface modification by using noble metal NPs and through doping, either anionic or cationic [63, 120–123]. Figure 4.6 shows an example of the surface modification of ZnO with Ag NPs.

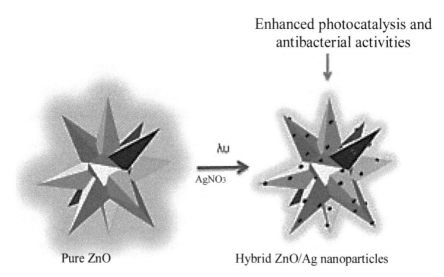

FIGURE 4.6 Hybrid star-shaped ZnO/Ag nanostructure [63].
Source: Reprinted with permission from Ref. [63]. © 2017 Elsevier.

4.4.2 NANOSTRUCTURED MATERIALS FOR HYDROGEN ENERGY PRODUCTION

Hydrogen gas is another area that has attracted considerable attention because it is a promising source of a clean and renewable form of energy due to the production of only water when burning. Hydrogen is known to possess high combustion enthalpy and extremely low density, which influences the high energy density that is usually experienced. Another important reason for the interest in hydrogen gas, as clean energy, is because it can be easily stored and used later [124, 125]. One of the most used photocatalysts for hydrogen generation from water is TiO_2, which can only function under energy-intensive UV light; while metal oxides (e.g., ZnO) and metal sulfides (e.g., CdS) can function under both the visible and UV lights for hydrogen production and water photolysis [126, 127]. Wang et al., [128], prepared composites of CdS-ZnS NPs for the production of photocatalytic hydrogen and organic degradation. It was reported that the deposition of ZnS on the CdS was to suppress recombination of electron/hole pairs produced on CdS; hence, hydrogen production was considerably fast, and the stability of CdS-ZnS was significantly improved when compared to the pure CdS. Figure 4.7 shows the mechanism of photocatalytic water splitting. A detailed explanation of the process involved is available in a report by Chiarello and Selli [129].

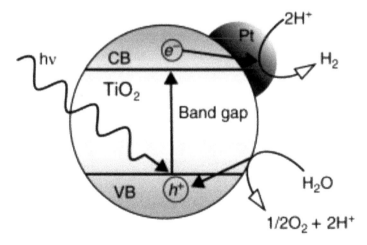

FIGURE 4.7 Mechanism of an illuminated Pt-modified TiO_2 semiconductor particle photocatalytic water splitting.

Source: Reprinted with permission from Ref. [129]. © 2014 Elsevier.

Majeed et al., [130], studied the role of nanoparticle size. NPs of between 2–5 nm sizes, was deposited on hexagonal CdS particle. The method used was the reduction of Au^{3+} with iodide ions. It was observed that the 4 nm particle size performed better than other sizes that were considered, leading to the highest production rate of hydrogen from the constituents: (92%) water-(8%) ethanol in Na_2S-Na_2SO_3 electrolyte medium. Other metal sulfides and metal oxides catalysts have been used to obtain similar result [102, 129, 131–133]. The drawback with the wide application of hydrogen gas as a form of energy source is as a result of the high fabrication cost [134]. In other studies, justifications have been made for the use of hydrogen-based energy, and they include long-term cost savings [134, 135], use of low-cost materials [136] and environmental issues surrounding carbon emission, known as greenhouse gases [135].

4.4.3 BIOMASS/BIOENERGY

With the global challenges of pollution from the energy industry, the need to replace fossil fuel energy resources with environmentally safer sources, is receiving a lot of attention. Biomass source is one of the best alternatives because biomass energy is renewable and environmentally safe [137].

Biomass is a term that refers to all organic materials from plants on land and all water-based vegetation, as well as organic wastes. Biomass energy resource may be considered as an organic matter, in which the energy is stored in chemical bonds. Digestion and combustion break the bonds between adjacent carbons, hydrogen and oxygen molecules of this matter, releasing their stored chemical energy. Bioenergy may be defined as the energy stored in materials that are produced from living matter. The types of bioenergy sources include biofuels, biogas, vegetable oil, bio-oil, and biodiesel [74]. Biofuels from agricultural crops, forests, and organic wastes are the basis for biomass technologies. These biofuels can be further used in fuel cells to generate electricity. Biomass fuel cells have, however, low energy density and relatively low conversion efficiency when compared to solar PV cells. Nonetheless, an advantage for the biomass cells is their ability to store solar energy for use on demand [64, 77].

Besides combustion, biomass thermochemical conversion processes, namely pyrolysis, gasification, and liquefaction can be used to produce bioenergy. Thermochemical biomass gasification converts biomass to a combustible gas mixture through partial oxidation at relatively high temperatures. Carbon monoxide and hydrogen (syngas) are the main products from this process. In order to enhance the process performance, nanocatalysts, namely: NiO, CeO_2, ZnO, and SnO_2, have been used. These NMs reduce the formation of tar, which is an unwanted process product. Furthermore, the use of nanoalloys, such as $(CeZr)xO_2$ and $Ni_3Cu(SiO_2)_6$, provide increased performances at relatively low gasification temperatures.

In biomass, liquefaction nanocatalysts have been successfully used to increase the liquids yields and also enhance product value-addition. The high temperatures associated with the liquefaction process, leads to an increase in gaseous products. Nanocatalysts lower the reaction temperatures, thereby causing increased liquids products. This improves the liquefaction process. In pyrolysis of spent tea, Co NPs lower the process temperature and increase the production yield of liquids by approximately 60% [137]. In the anaerobic digestion for biogas, the addition of micro/nano fly ash (MNFA) and micro/nano bottom ash (MNBA) increased the biogas production [138].

4.4.4 NANOFLUIDS IN GEOTHERMAL ENERGY

Geothermal energy is the thermal energy extracted from the earth crust. The extraction depths can range between 5 and 10 km. At these huge depths, the temperatures can be up to 1000°C. Nanofluids can, therefore, be used as cooling fluids for the drilling machinery that is subjected to high friction and high-temperature environment and other ancillary equipment, e.g., sensors and electronic components, which are subject to high temperatures. On the other hand, nanofluids can also be used directly as fluid in superconductors, to extract energy from the earth and convey it to a power plant system, for the production of work energy [139].

4.5 CONCLUSION

The environmental concern has attracted the attention of several individuals, organizations, and governments, which have equally influenced government decisions on the policies regarding the environment. The possible depletion of fossil fuel-based energy and increasing energy demand due to population growth, urbanization and industrialization have attracted many and different types of research on renewable and sustainable energy. The negative impact of the widest means of energy generation, i.e., fossil fuels on the environment, has been under the microscope for decades. The solution is to look into other more sustainable, renewable, and environmentally-friendly energies. The materials used for production, the method of production and storage of these energies are the key factors to consider in this regard. Nanostructured materials are very important materials, and they are used in production and storage of renewable energy, owing to their unique properties, which include: large surface area ratio to volume, high strength-to-weight, etc. Nonetheless, the current nanostructured materials need to be improved upon, while other possible nanostructures and composite materials demand critical and urgent research.

The use of CNTs as a hydrogen storage device is still not well understood. This is due to different contrary research reports. CNTs are, nevertheless, known to have the potential to absorb huge quantities of hydrogen. More research work is, therefore, needed in this regard. In summary, this chapter has discussed the different types of NMs used for renewable energy, the development of nanostructured and composite materials and the areas of applications of these materials in the generation of renewable energy. Nanostructured and composite materials are promising technologies, necessary for the improvement of renewable and sustainable energy generation and storage.

ACKNOWLEDGMENT

The financial support from CSIR-BIS South Africa is greatly appreciated by ID Ibrahim, AA Eze, and OS Adekomaya. C Kambole gratefully acknowledges the financial support from Copperbelt University, Kitwe, Zambia.

KEYWORDS

- **dye-sensitized**
- **nanostructure**
- **photo-catalysis**
- **photovoltaic cells**
- **renewable energy**
- **solar cells**

REFERENCES

1. Lewis, N. S., & Nocera, D. G., (2006). Powering the planet: Chemical challenges in solar energy utilization. *Proceedings of the National Academy of Sciences, 103*, 15729–15735.
2. Ganesh, I., (2014). Conversion of carbon dioxide into methanol–a potential liquid fuel: Fundamental challenges and opportunities (a review). *Renewable and Sustainable Energy Reviews, 31*, 221–257.
3. Shafiee, S., & Topal, E., (2009). When will fossil fuel reserves be diminished? *Energy Policy, 37*, 181–189.
4. Höök, M., & Tang, X., (2013). Depletion of fossil fuels and anthropogenic climate change-A review. *Energy Policy, 52*, 797–809.
5. Thomas, S., (2012). What will the Fukushima disaster change? *Energy Policy, 45*, 12–17.
6. Abbott, D., (2010). Keeping the energy debate clean: How do we supply the world's energy needs? *Proceedings of the IEEE, 98*, 42–66.
7. Buller, S., & Strunk, J., (2016). Nanostructure in energy conversion. *Journal of Energy Chemistry, 25*, 171–190.
8. Kim, J. Y., Kim, S. H., Lee, H. H., Lee, K., Ma, W., Gong, X., & Heeger, A. J., (2006). New architecture for high-efficiency polymer photovoltaic cells using solution-based titanium oxide as an optical spacer. *Advanced Materials, 18*, 572–576.
9. Li, G., Shrotriya, V., Huang, J., Yao, Y., Moriarty, T., Emery, K., & Yang, Y., (2005). High-efficiency solution processable polymer photovoltaic cells by self-organization of polymer blends. *Nature Materials, 4*, 864–868.
10. Han, G., Zhang, S., Li, X., & Chung, T. S., (2013). High performance thin film composite pressure retarded osmosis (PRO) membranes for renewable salinity-gradient energy generation. *Journal of Membrane Science, 440*, 108–121.
11. Hu, X., Li, G., & Yu, J. C., (2009). Design, fabrication, and modification of nanostructured semiconductor materials for environmental and energy applications. *Langmuir, 26*, 3031–3039.
12. Simões, F. R., & Takeda, H. H., (2017). 1 - Basic concepts and principles A2 - Róz, Alessandra Da, L. In: Ferreira, M., Leite, & F. D. L. O. N., (eds.), *Oliveira Nanostructures* (pp. 1–32). William Andrew Publishing.
13. Orilall, M. C., & Wiesner, U., (2011). Block copolymer based composition and morphology control in nanostructured hybrid materials for energy conversion and storage: Solar cells, batteries, and fuel cells, *Chemical Society Reviews, 40*, 520–535.
14. Kamat, P. V., (2007). Meeting the clean energy demand: nanostructure architectures for solar energy conversion. *The Journal of Physical Chemistry C, 111*, 2834–2860.
15. Arico, A. S., Bruce, P., Scrosati, B., Tarascon, J. M., & Van Schalkwijk, W., (2005). Nanostructured materials for advanced energy conversion and storage devices. *Nature Materials, 4*, 366–377.
16. Candelaria, S. L., Chen, R., Jeong, Y. H., & Cao, G., (2012). Highly porous chemically modified carbon cryogels and their coherent nanocomposites for energy applications. *Energy & Environmental Science, 5*, 5619–5637.
17. Green, M. A., (2004). Third generation photovoltaics: Advanced solar energy conversion. *Physics Today, 57*, 71, 72.

18. Barnham, K., Mazzer, M., & Clive, B., (2006). Resolving the energy crisis: Nuclear or Photovoltaics? *Nature Materials, 5*, 161–164.
19. Lai, C. H., Lu, M. Y., & Chen, L. J., (2012). Metal sulfide nanostructures: Synthesis, properties and applications in energy conversion and storage. *Journal of Materials Chemistry, 22*, 19–30.
20. Dai, L., Chang, D. W., Baek, J. B., & Lu, W., (2012). Carbon nanomaterials for advanced energy conversion and storage. *Small, 8*, 1130–1166.
21. Wang, E. G., (1999). A new development in covalently bonded carbon nitride and related materials. *Advanced Materials, 11*, 1129–1133.
22. Jiang, L., & Gao, L., (2003). Modified carbon nanotubes: An effective way to selective attachment of gold nanoparticles. *Carbon, 41*, 2923–2929.
23. Roy, S., Harding, A., Russell, A., & Thomas, K., (1997). Spectroelectrochemical study of the role played by carbon functionality in fuel cell electrodes. *Journal of the Electrochemical Society, 144*, 2323–2328.
24. Sidik, R. A., Anderson, A. B., Subramanian, N. P., Kumaraguru, S. P., & Popov, B. N., (2006). O2 reduction on graphite and nitrogen-doped graphite: Experiment and theory. *The Journal of Physical Chemistry B, 110*, 1787–1793.
25. Titirici, M. M., Thomas, A., & Antonietti, M., (2007). Aminated hydrophilic ordered mesoporous carbons. *Journal of Materials Chemistry, 17*, 3412–3418.
26. Su, D. S., Zhang, J., Frank, B., Thomas, A., Wang, X., Paraknowitsch, J., & Schlögl, R., (2010). Metal-free heterogeneous catalysis for sustainable chemistry. *Chem. Sus. Chem., 3*, 169–180.
27. Dai, L., (2006). *Carbon Nanotechnology: Recent Developments in Chemistry, Physics, Materials Science and Device Applications* (1st edn., pp. 1–12). Elsevier, Amsterdam.
28. Dresselhaus, M. S., Dresselhaus, G., & Eklund, P. C., (1996). *Science of Fullerenes and Carbon Nanotubes: Their Properties and Applications*. Academic Press.
29. Niu, C., Sichel, E. K., Hoch, R., Moy, D., & Tennent, H., (1997). High power electrochemical capacitors based on carbon nanotube electrodes. *Applied Physics Letters, 70*, 1480–1482.
30. Baughman, R. H., Zakhidov, A. A., & De Heer, W. A., (2002). Carbon nanotubes-the route toward applications. *Science, 297*, 787–792.
31. Collins, P. G., & Avouris, P., (2000). Nanotubes for electronics. *Scientific American, 283*, 62–69.
32. Dai, L., Soundarrajan, P., & Kim, T., (2002). Sensors and sensor arrays based on conjugated polymers and carbon nanotubes. *Pure and Applied Chemistry, 74*, 1753–1772.
33. Dillon, A., Jones, K., Bekkedahl, T., & Kiang, C., (1997). Storage of hydrogen in single-walled carbon nanotubes. *Nature, 386*, 377–379.
34. Kumar, A., Choudhury, S., Saha, S., Pahari, S., De, D., Bhattacharya, S., & Ghatak, K., (2010). Thermoelectric power in ultrathin films, quantum wires and carbon nanotubes under classically large magnetic field: Simplified theory and relative comparison. *Physica B: Condensed Matter, 405*, 472–498.
35. Berson, S., De Bettignies, R., Bailly, S., Guillerez, S., & Jousselme, B., (2007). Elaboration of P3HT/CNT/PCBM composites for organic photovoltaic cells. *Advanced Functional Materials, 17*, 3363–3370.

36. Liu, Z., Lin, X., Lee, J. Y., Zhang, W., Han, M., & Gan, L. M., (2002). Preparation and characterization of platinum-based electrocatalysts on multiwalled carbon nanotubes for proton exchange membrane fuel cells. *Langmuir, 18*, 4054–4060.
37. Yu, D., Nagelli, E., Du, F., & Dai, L., (2010). Metal-free carbon nanomaterials become more active than metal catalysts and last longer. *The Journal of Physical Chemistry Letters, 1*, 2165–2173.
38. Welna, D. T., Qu, L., Taylor, B. E., Dai, L., & Durstock, M. F., (2011). Vertically aligned carbon nanotube electrodes for lithium-ion batteries. *Journal of Power Sources, 196*, 1455–1460.
39. Lu, W., Qu, L., Henry, K., & Dai, L., (2009). High performance electrochemical capacitors from aligned carbon nanotube electrodes and ionic liquid electrolytes. *Journal of Power Sources, 189*, 1270–1277.
40. Liu, C., Li, F., Ma, L. P., & Cheng, H. M., (2010). Advanced materials for energy storage. *Advanced Materials, 22*, E28–E62.
41. Kongkanand, A., & Kamat, P. V., (2007). Electron storage in single wall carbon nanotubes. *Fermi Level Equilibration in Semiconductor–SWCNT Suspensions, ACS Nano, 1*, 13–21.
42. Kongkanand, A. R., Martínez, D., & Kamat, P. V., (2007). Single wall carbon nanotube scaffolds for photoelectrochemical solar cells. Capture and transport of photogenerated electrons, *Nano Letters, 7*, 676–680.
43. Liu, R., Duay, J., & Lee, S. B., (2011). Heterogeneous nanostructured electrode materials for electrochemical energy storage, *Chemical Communications, 47*, 1384–1404.
44. Zhang, H. X., Feng, C., Zhai, Y. C., Jiang, K. L., Li, Q. Q., & Fan, S. S., (2009). Cross-stacked carbon nanotube sheets uniformly loaded with SnO_2 nanoparticles: A novel binder-free and high-capacity anode material for lithium-ion batteries. *Advanced Materials, 21*, 2299–2304.
45. Hou, Y., Cheng, Y., Hobson, T., & Liu, J., (2010). Design and synthesis of hierarchical MnO2 nanospheres/carbon nanotubes/conducting polymer ternary composite for high performance electrochemical electrodes. *Nano Letters, 10*, 2727–2733.
46. Cui, L. F., Yang, Y., Hsu, C. M., & Cui, Y., (2009). Carbon-silicon core-shell nanowires as high capacity electrode for lithium ion batteries. *Nano Letters, 9*, 3370–3374.
47. Liang, W., Bockrath, M., Bozovic, D., Hafner, J. H., Tinkham, M., & Park, H., (2001). Fabry-Perot interference in a nanotube electron waveguide. *Nature, 411*, 665–669.
48. Frank, S., Poncharal, P., Wang, Z., & De Heer, W. A., (1998). Carbon nanotube quantum resistors. *Science, 280*, 1744–1746.
49. Kim, P., Shi, L., Majumdar, A., & McEuen, P., (2001). Thermal transport measurements of individual multiwalled nanotubes. *Physical Review Letters, 87*, 215502.
50. Si, Y., & Samulski, E. T., (2008). Exfoliated graphene separated by platinum nanoparticles. *Chemistry of Materials, 20*, 6792–6797.
51. De Heer, W. A., Chatelain, A., & Ugarte, D., (1995). A carbon nanotube field-emission electron source. *Science, 270*, 1179–1180.
52. Rinzler, A., Hafner, J., Nikolaev, P., & Lou, L., (1995). Unraveling nanotubes: Field emission from an atomic wire. *Science, 269*, 1550.
53. Saito, Y., & Uemura, S., (2000). Field emission from carbon nanotubes and its application to electron sources. *Carbon, 38*, 169–182.

54. Hirscher, M., Becher, M., Haluska, M., Quintel, A., Skakalova, V., Choi, Y. M., Dettlaff-Weglikowska, U., Roth, S., Stepanek, I., & Bernier, P., (2002). Hydrogen storage in carbon nanostructures. *Journal of Alloys and Compounds, 330*, 654–658.
55. Tibbetts, G. G., Meisner, G. P., & Olk, C. H., (2001). Hydrogen storage capacity of carbon nanotubes, filaments, and vapor-grown fibers. *Carbon, 39*, 2291–2301.
56. Zandonella, C., (2001). Is it all just a pipe dream? *Nature, 410*, 734–735.
57. Oelhafen, P., & Schüler, A., (2005). Nanostructured materials for solar energy conversion. *Solar Energy, 79*, 110–121.
58. Reim, M., Beck, A., Körner, W., Petricevic, R., Glora, M., Weth, M., Schliermann, T., Fricke, J., Schmidt, C., & Pötter, F., (2002). Highly insulating aerogel glazing for solar energy usage. *Solar Energy, 72*, 21–29.
59. Benz, N., Beikircher, T., & Aghazadeh, B., (1996). Aerogel and krypton insulated evacuated flat-plate collector for process heat production. *Solar Energy, 58*, 45–48.
60. Wu, Y., Wadia, C., Ma, W., Sadtler, B., & Alivisatos, A. P., (2008). Synthesis and photovoltaic application of copper (I) sulfide nanocrystals, *Nano Letters, 8*, 2551–2555.
61. Li, T. L., Lee, Y. L., & Teng, H., (2011). CuInS$_2$ quantum dots coated with CdS as high-performance sensitizers for TiO2 electrodes in photoelectrochemical cells. *Journal of Materials Chemistry, 21*, 5089–5098.
62. Bierman, M. J., & Jin, S., (2009). Potential applications of hierarchical branching nanowires in solar energy conversion. *Energy & Environmental Science, 2*, 1050–1059.
63. Andrade, G. R. S., Nascimento, C. C., Lima, Z. M., Teixeira-Neto, E., Costa, L. P., & Gimenez, I. F., (2017). Star-shaped ZnO/Ag hybrid nanostructures for enhanced photocatalysis and antibacterial activity. *Applied Surface Science, 399*, 573–582.
64. Serrano, E., Rus, G., & Garcia-Martinez, J., (2009). Nanotechnology for sustainable energy. *Renewable and Sustainable Energy Reviews, 13*, 2373–2384.
65. Perillo, P., Atia, M., & Rodríguez, D., (2017). Effect of the reaction conditions on the formation of the ZnO nanostructures. *Physica E: Low-Dimensional Systems and Nanostructures, 85*, 185–192.
66. Man, M. T., Kim, J. H., Jeong, M. S., Do, A. T. T., & Lee, H. S., (2017). Oriented ZnO nanostructures and their application in photocatalysis. *Journal of Luminescence, 185*, 17–22.
67. Veluswamy, P., Sathiyamoorthy, S., Chowdary, K. H., Muthusamy, O., Krishnamoorthy, K., Takeuchi, T., & Ikeda, H., (2017). Morphology dependent thermal conductivity of ZnO nanostructures prepared via a green approach. *Journal of Alloys and Compounds, 695*, 888–894.
68. Abbasi, H. Y., Habib, A., & Tanveer, M., (2017). Synthesis and characterization of nanostructures of ZnO and ZnO/Graphene composites for the application in hybrid solar cells. *Journal of Alloys and Compounds, 690*, 21–26.
69. Srinivasan, N., Revathi, M., & Pachamuthu, P., (2017). Surface and optical properties of undoped and Cu doped ZnO nanostructures. *Optik-International Journal for Light and Electron Optics, 130*, 422–426.
70. Liu, X., Jiang, S., Chen, H., Larson, C. A., & Roco, M. C., (2014). Nanotechnology knowledge diffusion: Measuring the impact of the research networking and a strategy for improvement. *Journal of Nanoparticle Research, 16*, 2613.

71. Karaca, F., & Öner, M. A., (2015). Scenarios of nanotechnology development and usage in Turkey. *Technological Forecasting and Social Change, 91*, 327–340.
72. Satyanarayana, K., Mariano, A., & Vargas, J., (2011). A review on microalgae, a versatile source for sustainable energy and materials. *International Journal of Energy Research, 35*, 291–311.
73. Demirbas, A., (2009). Global renewable energy projections. *Energy Sources, Part B, 4*, 212–224.
74. Hussein, A. K., (2015). Applications of nanotechnology in renewable energies-A comprehensive overview and understanding. *Renewable and Sustainable Energy Reviews, 42*, 460–476.
75. Abdin, Z., Alim, M., Saidur, R., Islam, M., Rashmi, W., Mekhilef, S., & Wadi, A., (2013). Solar energy harvesting with the application of nanotechnology. *Renewable and Sustainable Energy Reviews, 26*, 837–852.
76. Wang, Q., Xie, Y., Soltani-Kordshuli, F., & Eslamian, M., (2016). Progress in emerging solution-processed thin film solar cells–Part I: Polymer solar cells. *Renewable and Sustainable Energy Reviews, 56*, 347–361.
77. Sagadevan, S., (2013). Recent trends on nanostructures based solar energy applications: A review. *Rev. Adv. Mater. Sci., 34*, 44–61.
78. Xiao, Y., Lin, J. Y., Wu, J., Tai, S. Y., Yue, G., & Lin, T. W., (2013). Dye-sensitized solar cells with high-performance polyaniline/multi-wall carbon nanotube counter electrodes electropolymerized by a pulse potentiostatic technique. *Journal of Power Sources, 233*, 320–325.
79. Dong, H., Wu, Z., El-Shafei, A., Xia, B., Xi, J., Ning, S., Jiao, B., & Hou, X., (2015). Ag-encapsulated Au plasmonic nanorods for enhanced dye-sensitized solar cell performance. *Journal of Materials Chemistry A, 3*, 4659–4668.
80. Grätzel, M., (2001). Photoelectrochemical cells. *Nature, 414*, 338–344.
81. Grätzel, M., (2003). Dye-sensitized solar cells. *Journal of Photochemistry and Photobiology C: Photochemistry Reviews, 4*, 145–153.
82. Adachi, M., Murata, Y., Takao, J., Jiu, J., Sakamoto, M., & Wang, F., (2004). Highly efficient dye-sensitized solar cells with a titania thin-film electrode composed of a network structure of single-crystal-like TiO_2 nanowires made by the "oriented attachment" mechanism. *Journal of the American Chemical Society, 126*, 14943–14949.
83. Gao, F., Wang, Y., Shi, D., Zhang, J., Wang, M., Jing, X., Humphry-Baker, R., Wang, P., Zakeeruddin, S. M., & Grätzel, M., (2008). Enhance the optical absorptivity of nanocrystalline TiO_2 film with high molar extinction coefficient ruthenium sensitizers for high performance dye-sensitized solar cells. *Journal of the American Chemical Society, 130*, 10720–10728.
84. Fan, S. Q., Kim, C., Fang, B., Liao, K. X., Yang, G. J., Li, C. J., Kim, J. J., & Ko, J., (2011). Improved efficiency of over 10% in dye-sensitized solar cells with a ruthenium complex and an organic dye heterogeneously positioning on a single TiO_2 electrode. *The Journal of Physical Chemistry C, 115*, 7747–7754.
85. Mathew, S., Yella, A., Gao, P., Humphry-Baker, R., Curchod, B. F., Ashari-Astani, N., Tavernelli, I., Rothlisberger, U., Nazeeruddin, M. K., & Grätzel, M., (2014). Dye-sensitized solar cells with 13% efficiency achieved through the molecular engineering of porphyrin sensitizers. *Nature Chemistry, 6*, 242–247.

86. Yum, J. H., Baranoff, E., Wenger, S., Nazeeruddin, M. K., & Grätzel, M., (2011). Panchromatic engineering for dye-sensitized solar cells. *Energy & Environmental Science, 4,* 842–857.
87. Yuan, C., Chen, G., Li, L., Damasco, J. A., Ning, Z., Xing, H., Zhang, T., Sun, L., Zeng, H., & Cartwright, A. N., (2014). Simultaneous multiple wavelength upconversion in a core-shell nanoparticle for enhanced near infrared light harvesting in a dye-sensitized solar cell. *ACS Applied Materials & Interfaces, 6,* 18018–18025.
88. Dong, H., Wu, Z., Lu, F., Gao, Y., El-Shafei, A., Jiao, B., Ning, S., & Hou, X., (2014). Optics–electrics highways: Plasmonic silver nanowires@ TiO_2 core–shell nanocomposites for enhanced dye-sensitized solar cells performance. *Nano Energy, 10,* 181–191.
89. Zhang, Y., Zhou, N., Zhang, K., & Yan, F., (2017). Plasmonic copper nanowire@ TiO_2 nanostructures for improving the performance of dye-sensitized solar cells. *Journal of Power Sources, 342,* 292–300.
90. Cai, H., Tang, Q., He, B., & Li, P., (2014). Pt-Ru nanofiber alloy counter electrodes for dye-sensitized solar cells. *Journal of Power Sources, 258,* 117–121.
91. Hu, H., Shen, J., Cao, X., Wang, H., Lv, H., Zhang, Y., Zhu, W., Zhao, J., & Cui, C., (2017). Photo-assisted deposition of Ag nanoparticles on branched TiO_2 nanorod arrays for dye-sensitized solar cells with enhanced efficiency. *Journal of Alloys and Compounds, 694,* 653–661.
92. Brown, M. D., Suteewong, T., Kumar, R. S. S., D'Innocenzo, V., Petrozza, A., Lee, M. M., Wiesner, U., & Snaith, H. J., (2010). Plasmonic dye-sensitized solar cells using core-shell metal- insulator nanoparticles. *Nano Letters, 11,* 438–445.
93. Hagfeldt, A., Boschloo, G., Sun, L., Kloo, L., & Pettersson, H., (2010). Dye-sensitized solar cells. *Chemical Reviews, 110,* 6595–6663.
94. Meng, F., Luo, Y., Zhou, Y., Zhang, J., Zheng, Y., Cao, G., & Tao, X., (2016). Integrated plasmonic and upconversion starlike Y_2O_3: Er/Au@ TiO_2 composite for enhanced photon harvesting in dye-sensitized solar cells. *Journal of Power Sources, 316,* 207–214.
95. Dang, X., Yi, H., Ham, M. H., Qi, J., Yun, D. S., Ladewski, R., Strano, M. S., Hammond, P. T., & Belcher, A. M., (2011). Virus-templated self-assembled single-walled carbon nanotubes for highly efficient electron collection in photovoltaic devices. *Nature Nanotechnology, 6,* 377–384.
96. Erwin, W. R., Zarick, H. F., Talbert, E. M., & Bardhan, R., (2016). Light trapping in mesoporous solar cells with plasmonic nanostructures. Energy & Environmental Science, 9, 1577–1601.
97. Guo, K., Li, M., Fang, X., Liu, X., Sebo, B., Zhu, Y., Hu, Z., & Zhao, X., (2013). Preparation and enhanced properties of dye-sensitized solar cells by surface plasmon resonance of Ag nanoparticles in nanocomposite photoanode. *Journal of Power Sources, 230,* 155–160.
98. Zhou, N., López-Puente, V., Wang, Q., Polavarapu, L., Pastoriza-Santos, I., & Xu, Q. H., (2015). Plasmon-enhanced light harvesting: Applications in enhanced photocatalysis, photodynamic therapy and photovoltaics. *RSC Advances, 5,* 29076–29097.
99. Xu, F., & Sun, L., (2011). Solution-derived ZnO nanostructures for photoanodes of dye-sensitized solar cells. *Energy and Environmental Science, 4,* 818–841.

100. Luo, Q. P., Wang, B., & Cao, Y., (2017). Single-crystalline porous ZnO nanosheet frameworks for efficient fully flexible dye-sensitized solar cells. *Journal of Alloys and Compounds, 695*, 3324–3330.
101. Beydoun, D., Amal, R., Low, G., & McEvoy, S., (1999). Role of nanoparticles in photocatalysis. *Journal of Nanoparticle Research, 1*, 439–458.
102. Kudo, A., & Miseki, Y., (2009). Heterogeneous photocatalyst materials for water splitting., *Chemical Society Reviews, 38*, 253–278.
103. Maeda, K., (2011). Photocatalytic water splitting using semiconductor particles: History and recent developments. *Journal of Photochemistry and Photobiology C: Photochemistry Reviews, 12*, 237–268.
104. Inoue, T., Fujishima, A., Konishi, S., & Honda, K., (1979). Photoelectrocatalytic reduction of carbon dioxide in aqueous suspensions of semiconductor powders. *Nature, 277*, 637–638.
105. Sunada, K., Watanabe, T., & Hashimoto, K., (2003). Bactericidal activity of copper-deposited TiO_2 thin film under weak UV light illumination. *Environmental Science & Technology, 37*, 4785–4789.
106. McCullagh, C., Robertson, J. M., Bahnemann, D. W., & Robertson, P. K., (2007). The application of TiO_2 photocatalysis for disinfection of water contaminated with pathogenic micro-organisms: A review. *Research on Chemical Intermediates, 33*, 359–375.
107. Peller, J. R., Whitman, R. L., Griffith, S., Harris, P., Peller, C., & Scalzitti, J., (2007). TiO_2 as a photocatalyst for control of the aquatic invasive alga, Cladophora, under natural and artificial light. *Journal of Photochemistry and Photobiology A: Chemistry, 186*, 212–217.
108. Nakata, K., & Fujishima, A., (2012). TiO_2 photocatalysis: Design and applications. *Journal of Photochemistry and Photobiology C: Photochemistry Reviews, 13*, 169–189.
109. Salameh, C., Nogier, J. P., Launay, F., & Boutros, M., (2015). Dispersion of colloidal TiO_2 nanoparticles on mesoporous materials targeting photocatalysis applications. *Catalysis Today, 257*(Part 1), 35–40.
110. Zhang, Q., Fan, W., & Gao, L., (2007). Anatase TiO_2 nanoparticles immobilized on ZnO tetrapods as a highly efficient and easily recyclable photocatalyst. *Applied Catalysis B: Environmental, 76*, 168–173.
111. Iglesias-Juez, A., Kubacka, A., Colón, G., & Fernández-García, M., (2013). Photocatalytic nanooxides: The case of TiO_2 and ZnO, catalysis by nanoparticles. In: Suib, S., (ed.), *New and of the Future Developments in Catalysis Series* (pp. 245–263). Elsevier, Amsterdam.
112. Aguado, J., R. Van Grieken, Lopez-Munoz, M., & Marugán, J., (2002). Removal of cyanides in wastewater by supported TiO_2-based photocatalysts. *Catalysis Today, 75*, 95–102.
113. Mohammadi, S., Sohrabi, M., Golikand, A. N., & Fakhri, A., (2016). Preparation and characterization of zinc and copper co-doped WO_3 nanoparticles: Application in photocatalysis and photobiology. *Journal of Photochemistry and Photobiology B: Biology, 161*, 217–221.
114. Tsai, T. Y., Wang, H. L., Chen, Y. C., Chang, W. C., Chang, J. W., Lu, S. Y., & Tsai, D. H., (2017). Noble metal-titania hybrid nanoparticle clusters and the interaction

to proteins for photo-catalysis in aqueous environments. *Journal of Colloid and Interface Science, 490*, 802–811.
115. Wan, Y., Liang, C., Xia, Y., Huang, W., & Li, Z., (2017). Fabrication of graphene oxide enwrapped Z-scheme Ag_2SO_3/AgBr nanoparticles with enhanced visible-light photocatalysis. *Applied Surface Science, 396*, 48–57.
116. Liu, M., Zhang, D. X., Chen, S., & Wen, T., (2016). Loading Ag nanoparticles on Cd(II) boron imidazolate framework for photocatalysis. *Journal of Solid State Chemistry, 237*, 32–35.
117. Liu, Y., Hu, K., Hu, E., Guo, J., Han, C., & Hu, X., (2017). Double hollow MoS_2 nano-spheres: Synthesis, tribological properties, and functional conversion from lubrication to photocatalysis. *Applied Surface Science, 392*, 1144–1152.
118. Asapu, R., Claes, N., Bals, S., Denys, S., Detavernier, C., Lenaerts, S., & Verbruggen, S. W., (2017). Silver-polymer core-shell nanoparticles for ultrastable plasmon-enhanced photocatalysis. *Applied Catalysis B: Environmental, 200*, 31–38.
119. Huang, J., He, Y., Wang, L., Huang, Y., & Jiang, B., (2017). Bifunctional Au@TiO_2 core–shell nanoparticle films for clean water generation by photocatalysis and solar evaporation. *Energy Conversion and Management, 132*, 452–459.
120. Deng, Q., Duan, X., Ng, D. H., Tang, H., Yang, Y., Kong, M., Wu, Z., Cai, W., & Wang, G., (2012). Ag nanoparticle decorated nanoporous ZnO microrods and their enhanced photocatalytic activities. *ACS Applied Materials and Interfaces, 4*, 6030–6037.
121. Ibrahim, A. A., Dar, G., Zaidi, S. A., Umar, A., Abaker, M., Bouzid, H., & Baskoutas, S., (2012). Growth and properties of Ag-doped ZnO nanoflowers for highly sensitive phenyl hydrazine chemical sensor application. *Talanta, 93*, 257–263.
122. Cho, J., Lin, Q., Yang, S., Simmons, J. G., Cheng, Y., Lin, E., Yang, J., Foreman, J. V., Everitt, H. O., & Yang, W., (2012). Sulfur-doped zinc oxide (ZnO) nanostars: Synthesis and simulation of growth mechanism. *Nano Research, 5*, 20–26.
123. Bensouici, F., Bououdina, M., Dakhel, A. A., Tala-Ighil, R., Tounane, M., Iratni, A., Souier, T., Liu, S., & Cai, W., (2017). Optical, structural and photocatalysis properties of Cu-doped TiO_2 thin films. *Applied Surface Science, 395*, 110–116.
124. Xiong, J., Han, C., Li, Z., & Dou, S., (2015). Effects of nanostructure on clean energy: Big solutions gained from small features. *Science Bulletin, 60*, 2083–2090.
125. Ahmad, H., Kamarudin, S. K., Minggu, L. J., & Kassim, M., (2015). Hydrogen from photo-catalytic water splitting process: A review. *Renewable and Sustainable Energy Reviews, 43*, 599–610.
126. Bao, N., Shen, L., Takata, T., & Domen, K., (2007). Self-templated synthesis of nanoporous CdS nanostructures for highly efficient photocatalytic hydrogen production under visible light. *Chemistry of Materials, 20*, 110–117.
127. Wang, X., Shih, K., & Li, X., (2010). Photocatalytic hydrogen generation from water under visible light using core/shell nano-catalysts. *Water Science and Technology, 61*, 2303–2308.
128. Wang, X., Peng, W. C., & Li, X. Y., (2014). Photocatalytic hydrogen generation with simultaneous organic degradation by composite CdS–ZnS nanoparticles under visible light. *International Journal of Hydrogen Energy, 39*, 13454–13461.
129. Chiarello, G. L., & Selli, E., (2014). 8 - Photocatalytic production of hydrogen. In: *Advances in Hydrogen Production, Storage and Distribution* (pp. 216–247). Woodhead Publishing.

130. Majeed, I., Nadeem, M. A., Al-Oufi, M., Nadeem, M. A., Waterhouse, G., Badshah, A., Metson, J., & Idriss, H., (2016). On the role of metal particle size and surface coverage for photo-catalytic hydrogen production: A case study of the Au/CdS system. *Applied Catalysis B: Environmental, 182,* 266–276.
131. Kudo, A., Kato, H., & Tsuji, I., (2004). Strategies for the development of visible-light-driven photocatalysts for water splitting. *Chemistry Letters, 33,* 1534–1539.
132. Silija, P., Yaakob, Z., Suraja, V., Binitha, N. N., & Akmal, Z. S., (2011). An enthusiastic glance in to the visible responsive photocatalysts for energy production and pollutant removal, with special emphasis on titania. *International Journal of Photoenergy,* 2012.
133. Yao, W., Song, X., Huang, C., Xu, Q., & Wu, Q., (2013). Enhancing solar hydrogen production via modified photochemical treatment of Pt/CdS photocatalyst. *Catalysis Today, 199,* 42–47.
134. Yeh, G. T., Kao, Y. L., Yang, S. Y., Rei, M. H., Yan, Y. Y., & Lee, P. C., (2014). Low cost compact onsite hydrogen generation. *International Journal of Hydrogen Energy, 39,* 20614–20624.
135. Cantuarias-Villessuzanne, C., Weinberger, B., Roses, L., Vignes, A., & Brignon, J. M., (2016). Social cost-benefit analysis of hydrogen mobility in Europe. *International Journal of Hydrogen Energy, 41,* 19304–19311.
136. Dalapati, G. K., Chua, C. S., Kushwaha, A., Liew, S. L., Suresh, V., & Chi, D., (2016). All earth abundant materials for low cost solar-driven hydrogen production. *Materials Letters, 183,* 183–186.
137. Akia, M., Yazdani, F., Motaee, E., Han, D., & Arandiyan, H., (2014). A review on conversion of biomass to biofuel by nanocatalysts. *Biofuel Research Journal, 1,* 16–25.
138. Ganzoury, M. A., & Allam, N. K., (2015). Impact of nanotechnology on biogas production: A mini-review. *Renewable and Sustainable Energy Reviews, 50,* 1392–1404.
139. Wong, K. V., & De Leon, O., (2010). Applications of nanofluids: Current and future. *Advances in Mechanical Engineering.* http://journals.sagepub.com/doi/pdf/10.1155/2010/519659 (Accessed on 25 July 2019).

CHAPTER 5

Simulation and Modeling of Nanotechnology Aircraft Energy System Using MATLAB

INDRADEEP KUMAR

Chairman of Bibhuti Education and Research, Bhagalpur, Bihar, India; Research Scholar, Department of Mechanical Engineering, Vels Institute of Science, Technology and Advanced Studies (VISTAS), Chennai – 600117, India

ABSTRACT

The design methods based on aircraft models have been widely used in aircraft energy system conceptual design for decades and demonstrated very effective when it restricted to simple problems with very approximate analyses. These monolithic, large, design and analysis codes are genuinely multidisciplinary, but as analyses become more complex, such codes have grown so large as to be incomprehensible and hence difficult to maintain. This chapter explains the computational behavior of nanoparticles (NPs). Nanomaterials (NMs) constitute a prominent sub-discipline in the materials science. Conventional materials like glass, ceramics, metals, polymers, or semiconductors can be formed in nanoscale proportions. They have various microstructural distinctive attributes such as nano-discs, nanotubes, nanocoatings, quantum dots (QDs), nanocomposites, and nanowires. The unique properties of nanoparticle derived materials and devices depend directly on size and structure-dependent properties.

Nanoparticle size should be properly controlled to take its all advantage of the effects of quantum size in technical applications. It can be done if the flocks are controlled, that requires, the formation rate of the new particle is quantitatively resolute.

5.1 INTRODUCTION

In real-time measurements, distributions of the size of the particle and structure of particle are, thus, validate the techniques of the evolution of nanotechnology. More accurate and reliable models for simulating transport, coagulation of particle, deposition, and dispersion of the nanoparticles (NPs) and their cluster are needed in the formation of design tools for applications in technological, and in nanoparticle instrumentation, sensing, dilution, sampling, and focusing on their behavior in the chemically reactive environment. The computational method allows us to validate and explore hypotheses about the experimental observation that may not be approachable through conventional experimental techniques. Additionally, simulations with computer allow us for the theory of areas of interest in which experimental method can be applied.

Computational science provides many opportunities for nanotechnology. Soft computational techniques like cellular automata, genetic algorithm, and swarm intelligence, can impart various new properties like self-repair, growth, and complex-networks to the framework. So many books have successfully implemented the above techniques in the real-world problems, including complex control systems for manufacturing and control in the Aircraft energy system. Dealing in nanoscale-systems comprises to understand and development of nanotechnology computing method as well as the application of these methods in real-world, often to the problems of other areas. Techniques in nanotechnology systems comprise algorithms in machine learning, artificial intelligence (AI), reasoning, and natural computing. With some improvement in nanotechnology properties, these techniques can be implemented to control a flock of trillion nano-assemblers.

Fabrication techniques have allowed engineers and scientists to form magnetic materials in nanometer scale. These materials show very interesting magnetic properties that differ from the bulk materials.

This chapter deals with MATLAB applications in behavioral analysis of systems consisting of carbon nanotubes (CNTs) through molecular dynamics simulation. This chapter explains the applications of MATLAB for calculations of nano-systems, which contain CNTs. Also, some specialized applications of MATLAB for studying the behavior of nano-oscillators are introduced. In this, the authors describe the behavior of single-walled CNTs and functionalized carbon nanotubes (FCNTs) with four functional groups in water using the molecular dynamics simulation method. In these days, nanotechnology and nonmaterial's are hot topics in Aerospace framework.

This is defined as the nanotechnology is the study of matter on an atomic and molecular scale to develop material at a scale of 1 to 100 nanometers.

In this chapter we will see the "nanointerface" evaluated by the use of modeling and simulation with MATLAB, in combination with experimental statistics, to understand the interfacial behavior of the metal-oxide polymer interfaces for nanoelectronic devices. This chapter introduces new technology for the design of complex systems and describes the tools that may aid in the implementation of these schemes.

5.2 WHAT IS OPTIMIZATION AND SIMULATION?

Optimization and predictability in effectively every branch of science, there is a requirement for calculations for estimating the parameters of the system, to evaluate the extremum of an objective function. Since mathematical models are rarely used to guide experimentation and to adapt or forecast the behaviors, there is an equally robust need to finding the errors in these models and to be capable to make mathematically sound and precise statements regarding the accuracy of the predictions The degree of detail and dynamic complexity including wide-ranging, non-separable scales in both space as well as time make the formulation and solution of problems in this area fundamentally more challenging than most and arguably all problems successfully tackled to date. The payoff to developing the new set of computational models and tools would be better understanding, prediction, design, optimization, and control of the complex nanosystems with confidence in the reliability of the results. Successes in this field would almost certainly have an impact on the other fields that share the complexity of a large range of complex behavior on non-separable temporal and spatial scales.

5.2.1 WHY OPTIMIZATION IS REQUIRED?

Many problems mentioned above rely on formulating and solving optimization problems. For example, in order to improve the quantitative understanding of nanosystems at a fundamental level, it is necessary to have the ability to calculate the ground state of the system. This can be formulated as a minimum energy problem where the solution provides the configuration of the particles of the system at the lowest energy gives energy functional. Complicated nanosystems of interest may have millions to billions of particles, resulting in huge optimization problems characterized by an equally large number of local minima with energy levels near to the ground state.

5.2.2 WHY SIMULATION IS REQUIRED?

In real-time measurements, distributions of particle size and particle structure are; thus, validate techniques for the evolution of nanotechnology. Precise and genuine models for simulating transportation, coagulation, deposition, and dispersion of NPs and their cluster are required for the development of tools for applications in technological, including energy storage and transportation, nanoparticle instrumentation, fuselage, wings control surfaces and other parts focusing the nanoparticle behavior in the chemically and environmentally reactive framework. Computational techniques allow us to validate and explore hypotheses about the experimental observation that may otherwise not be approachable through conventional experimental techniques. Additionally, simulations with computer allow for the theory to propose areas of interest in which experimental techniques can be applied.

5.3 HISTORY

Computational science offers many opportunities for nanotechnology. Soft computing techniques such as cellular automata, genetic algorithms, and swarm intelligence, can impart required emergent properties like self-repair, growth, and complex networks to the framework. Many Books have successfully applied such techniques to real-world problems, including complex control systems in manufacturing and control in the Aircraft energy system. This chapter in nanoscale systems involves the understanding and development of nanotechnology computational techniques as well as the application of these techniques in the aerospace world especially in energy conservation, often to the problems of other areas. The techniques in nanotechnology systems comprise algorithms in machine learning, AI, knowledge representation reasoning, and natural computing. With some improvement in nanotechnology characteristics, these techniques can be applied to control a flock of a trillion nano-assemblers.

5.4 HOW THIS SERVES?

Advancement in computational techniques has significantly enabled the simulation of heat transfer, acoustics complex flows, and fluid-structure interaction phenomena. Furthermore, advances in high-performance computing also allowed more complex simulations to be performed in shorter turn-around times. The design and condition monitoring challenge requires an understanding of complex phenomena and conditions. The uncertainty can be reduced in engineering simulations by increasing the accuracy, while making computing more efficient, it's still a challenge. In this chapter, we will see results from the development and application of high-order CFD and multi-scale or multi-physics methods for nanotechnology in aerospace applications.

An optimization technique for nano-sized magnetic materials using MATLAB and fabrication techniques developed over the past few decades have given engineers and scientists to form magnetic materials in the nanometer range. Such materials exhibit very surprising magnetic properties that are very different from bulk materials. At such scales, the behaviors of the magnetic materials are inadequately explained by Maxwell's equations. The theory of micromagnetics of the material deals with the characteristics of these nano-sized magnetic materials, including the quantum mechanical effects that are very significant at nanoscale. Micromagnetics theory has been successfully predicting the shaping of domain walls in magnetic materials and also the shaping of surprising magnetic states such as the vortex states and leaf states. It's application in many aerospace engineering aspects such as the digitally data storage technique, development of aircraft parts, and navigation systems.

In aircraft energy storage system, it is very difficult to understand and compute performance, safety, nature, and other aspects of the systems before and after they are ready for testing. Simulated models are examined at all development stages to gain knowledge in order to make decisions.

Modeling and simulation in Aircraft energy system development, like, fuel storage system, hydraulic and electrical power systems, are a very important part in these days for the design process. Through modeling and simulation using MATLAB problem in a function or system are found early on in the process. A developed part of the aircraft energy storage system validation depends on results obtained from simulated models rather than costlier testing in flight tests. Hence, the requirement for integrated models of complex energy storage systems, and their validation is increasing day by day. It's not only one model is required, but many similar models with well-known and high accuracies and validity ranges are needed. The development of modeling and simulation and computer performance, these tools have allows large-scale simulation (Figure 5.1).

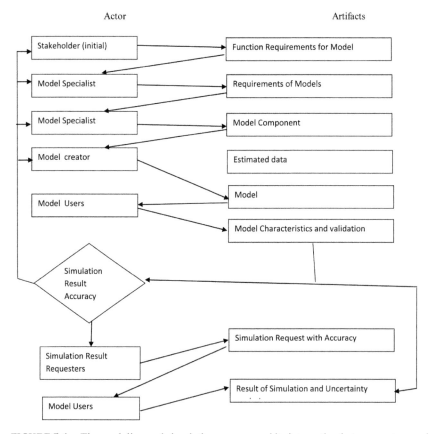

FIGURE 5.1 The modeling and simulation process and its interaction between actors and artifacts for an unreleased and a released.

This chapter includes two parts related to the topics. The first part explains the modeling technique; simulation performed a how to simulate the complete energy storage system with models from different-different tools, e.g., controlling software from one tool and the equipment model from another tool. The second part describes the use of modeling and simulation in the development of an aircraft energy storage system. The use of modeling and simulation in early system development phases has significantly increased. This has led to the recent need of uncertainty management, and the complete validation technique that complements the traditional validation based on measurement statics, knowledge of a model's maturity, and simulation result accuracy. A solution to this essential is dominant to advance in system development where the early design conclusion has to be made supported by a simulation technique.

Sensitivity analyses made to point out model parameters like inputs that have a very strong influence on the result of simulation or compare the assumed uncertainty statistics influence with similar model accuracy experience. Sensitivity analyses can, therefore, help the process to increase the knowledge of the models' validity.

5.5 AIM, LIMITATION, AND CONTEXT

The aim of the chapter is to introduce industrially useful methods for model validation when measurement data is not up to the mark or lacking. First, simulation methods are required, that enable simulation of the system with models of different domains. Second, the model constantly grows through the development process, and these validation methods must fit into the whole development procedure. Third, the industrial applicability of the technique and metrics by using uncertainty data must also be verified. Limitation that reduces the number of methods is that the selected model type and category, like mathematically based Aircraft energy storage system models, is computationally challenging, and also the model equations which are not explicitly approachable, and hence the number of model parameters get affected by uncertainty can be relatively high. The context in which the methods for supporting model validation must be developed to be useful is in area of a top-level view of domains which are always present when using the sought methods which support model validation and also exhibit the consequences if few domains are lacking. The areas should be considered in order to get a balanced context for the problem-solving technique. To acquire the efficient development surrounding with useful techniques, consideration must be taken to the development procedure, modeling and simulation methods, and the applied system. For example, the process is not present during the system development, process, and its repeatability will may be underdeveloped, which will exhibit disorder.

5.6 FRAME OF REFERENCE

This section provides an overview of types of model, simulation techniques, and their relation to engineering and aircraft energy system classification. The focus will be on mathematical models for Aircraft energy storage systems.

5.6.1 MODELS

Everything that represents the something can be considered as a model. In other words, a model is a simplification of something.

The largest possible system of all is the universe. Whenever we decide to cut out a piece of the universe such that we can clearly say what is inside that piece (belongs to that piece), and what is outside (does not belong to that piece), we define a new system. A system is characterized by the fact that we can say what belongs to it and what does not, and by the fact that we can specify how it interacts with its environment. System definitions can furthermore be hierarchical. We can take the piece from before, cut out a yet smaller part of it, and we have a new system.

The model simplifies and helps us to better understand the real object. Models can be of at least four types [49, 68]:

- Physical;
- Verbal;
- Schematic; and
- Mathematical.

Physical models are those who recall the system in terms of physically appearance. Typical physical models are scale models used in, e.g., wind tunnels, test rigs, or prototypes, etc.

Verbal models are simpler than the remaining three model types and provide textual or oral interpretation. Verbal models are often used earlier in projects when very little is known. The drawback of verbal models is their ambiguity, which may cause difficulties in interpretation, of programming and modeling languages. The outputs from the mathematical models in the aircraft energy storage system are pressure, temperature, and flow of fuel in the system. Mathematical models can be an attempt to recall the system in terms of behavior and are, therefore, also named as behavior models. Which type of model is required, however, is usually read from the context. Another, technical, way to categorize the models is as specifying or descriptive and analytical models [3]. The motive of specifying or descriptive the models is to provide a manufacturing schedule, like drawing of the detail, verbal specification, or modeled software specification. The purpose of analytical model's however, is to contribute in the system analysis for taking the necessary design decisions which are input to the specifying models, such as the necessary analytical models, like CFD models, and hence design iterations are very important. If much more of the design occurs in analytical models, like the selection and evaluation of the system concepts, the analytical model corresponding to the selected system can then be considered as a specifying or descriptive model in terms of the system layout. The specifying model may have different-different resolutions, and still, it's useful from the first day of its construction, while the analytical model must be complete before use and the weakest link determines its quality. The benefits of modeling are several, such as:

- Models require organizing and quantifying information and, in the process, often indicate areas where more information is required.
- It provides a systematic approach to problem-solving.
- It increases the understanding of the problem.
- Models entail us to be very distinct about objectives.
- It serves as a consistent tool for analysis.
- Models give a standardized format for analyzing the problem.

Verbal modeling is hence, the process of establishing and reproducing the interrelationships between two or more entities of a system in a model.

5.7 MODELING AND SIMULATION OF AIRCRAFT ENERGY SYSTEMS

This section presents an overview of the modeling and simulation work for the Aircraft energy systems. Described methods and tools are also applicable to passenger aircraft, automotive industry, and other complex products. The Aircraft energy storage systems contain fuel, hydraulic, and auxiliary power systems and also ECS. Aircraft energy systems have many modeling and simulation provocation such as less compressible air or compressible fluids which give rigid differential equations, saturation and nonlinear cavitations, gravitational force effects; It is also a complex system of which requires a model with integrated system software. The dynamic models based on the physical differential equations generally have been used.

5.7.1 SIMULATION OF MATHEMATICAL MODELS

Model is the simplified representation of the real-world system, and simulation is an execution of a model using input data to extract the desired information from the model. Simulation method is an experiment performed on a model. Simulation can be for a real or hypothetical situation. By the analogy, an experiment in the real world on the model that needs atmospheric conditions and delivers required data, the model also needs desired inputs and delivers simulated data. The simulation enables the prediction of the behavior of the model from a set of various parameters and given initial conditions.

There are many reasons to simulate instead of setting up an experiment on a system in the real world [15]. Some main reasons are:

- It is so expensive to perform experiments on real systems.
- It is so dangerous. The pilot may practice a dangerous maneuver before doing it on the plane.
- The system may not yet exist.

Other important features that simulation gives are:

- Variables cannot be accessible in the real system, but it can be observed in a simulation.
- It is easy to very use and modify models and to change the parameters and perform new simulations. With system design optimization, different-different variants can be evaluated.

- The time scale of the system may be increased or decreased. For example, a flight of several hours can be simulated in a few minutes.

The most important data flows in a model while simulations are:

- Parameters remain constant during a simulation, but it can be changed in-between.
- Constants are not accessible in the simulation.
- Inputs are the variables which affect the model.
- Outputs are the variables which are observed.

In contrast to the static model, the output from a dynamic model can be a function of the history of the models' inputs and not only a function of the current inputs. For example, the pressure in the fuel tank of an aircraft is a function of (depends on) the previous flight conditions. The pressure will be different when the aircraft is climbing, diving, or is in steady level flight just before the observed time. Dynamic models are generally represented with differential equations. Setting of the time derivative to zero in a dynamic model will become a static model. The sampling of an analog signal corresponds to a conversion of "continuous time" to "discrete time."

5.7.2 MODELING AND SIMULATION TOOLS

This section describes a few aspects of modeling and simulation tools. Computational technique of design is a growing field whose development is combined with the continuous improvement of computers performance and their computational ability. Figure 5.2 shows the simulation of nanostructured interfaces is controlled at the nanoscale. There is no clear precision of the term in computational design technique, and it is incorporated differently in different domains because of its broad implications. However, computational design technique is characterized by operating on computer-based models in different-different ways in order to get the various information [17] (Figures 5.3 and 5.4).

Two examples of tools/languages for aircraft energy system development are:

- MATLAB is an object-oriented language for physical systems modeling of a complex model. It accepts multi-domain modeling like, in aviation sector modeling of mechatronics systems, which is the combination of electrical, mechanical, and control systems as well as hydraulic subsystems. MATLAB uses the equation-based modeling; the physical behavior of a model is described by the differential, algebraic, and discrete equations; and
- MATLAB models are causal and uses the power port technique. A modeling and simulation environment is needed to solve actual problems. The environment gives a customizable set of block libraries that allow users to design and simulate. There are both commercial and free modeling and simulation environments for MATLAB. Dymola, OpenMATLAB, and SimulationX, are well known commercial MATLAB tools.

FIGURE 5.2 **(See color insert.)** Simulation of nanostructured interfaces are controlled at the nanoscale.

Simulation and Modeling Using MATLAB   133

5.7.3 MODEL-BASED SYSTEMS ENGINEERING (SE)

The International Council on Systems Engineering (INCOSE) defines Systems Engineering (SE) as "It is an interdisciplinary approach and means to enable the realization of successful systems." SE with model-based technique is called Model Based System Engineering (MBSE). MBSE uses "modeling" of the system in, for an example, the design of the complex systems and of its desiredness to facilitate a more efficient design process.

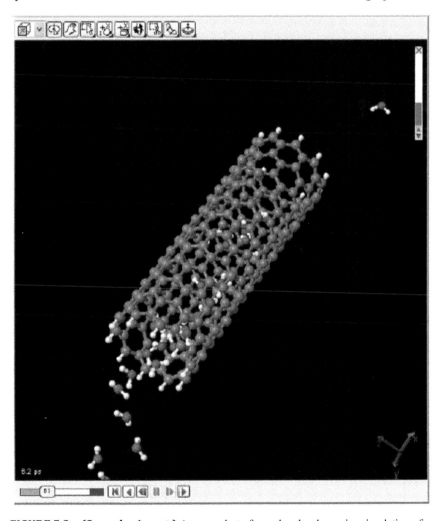

FIGURE 5.3 (See color insert.) A screenshot of a molecular dynamics simulation of a carbon nanotube using the 3D molecular dynamics simulator.

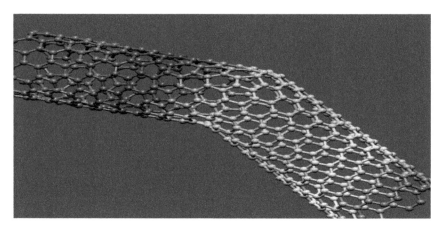

FIGURE 5.4 Study of bending properties of CNTs by molecular dynamics simulation for aerospace applications.

There are various definitions of how to models are used in system development, depending upon the design process it is used and in which technical domain, for example:

- Model-based systems engineering;
- Model-based design;
- Model-based development;
- Model-based definition;
- Model-based engineering;
- Model-driven architecture;
- Model-driven design;
- Model-driven engineering.

As modeling and simulation grow, the earlier definitions are exchanged between the domains and are used in widely or slightly different senses. Some of the definitions are the principles, and some are the means to perform those principles, for example, the principle of MBSE is used to perform MBD. Some of the benefits of a model-centric approach are:

- **Documentation:** The model is the only information source. Documentation and code are by-products and are generated according to the model.

- **Design by Simulation:** Executable models increases the understanding of the behavior of the model and hence lead to better design quality.
- **Test and Verification:** Continuous tests and verifications can be done on the model.
- Auto-generated code decreases the number of factors involved and maintains consistency between the specifying model and implementable code and its documents.

Some more benefits of MBSE are:

- Early identification of the requirements;
- Reuses of models.

MBSE is incorporated by placing the model at the center, which gives the following fundamentals:

- The interaction between developers and the various tasks is completed using the model.
- The model may have more than one purpose and can be needed by more than one person also.

5.7.4 THE MODEL-BASED DESIGN PROCESS

Engineering design is a technique for solving problems, where a set of doubtful objectives need to be balanced without violating the set of constraints. By employing modern modeling, simulation, and optimization techniques, vast improvements can be achieved in all parts of the design process. A great deal in this chapter has been done in the field of engineering design and has approaches to different design processes and methods. They all describe a phase-type process of different granularity with phases such as: concept design, preliminary design, specification, prototype development, redesign, detail design, and production, speaks of the design paradox, where very little is known about the design problem at the beginning, but we have full design freedom. As time in the design process lapse, knowledge and data about the problem are gained, but the design freedom is lost due to the design decisions made during the process.

To further stress the importance of the early phases of the design process, it is here that most of the cost is committed (Figure 5.5).

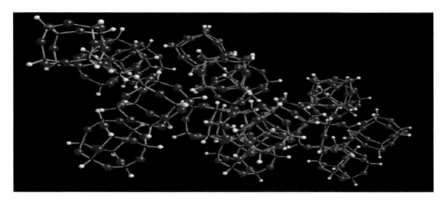

FIGURE 5.5 Screen shot of Atomistic simulation of a nanostructure.

Computer models are a vital tool when evaluating the performance of the concept proposals. Most often, the modeling activities begin with simple stationary models of the systems, perhaps in MS Excel or equivalent, [7]. However, most designs are actually modifications of already existing systems, if so, all the required models are already present in the early stages of the design process and relatively easy to modify for reuse. This also applies to other test facilities, such as test rigs, simulators, and test a/c, with the exception that these are not always so easy to modify.

5.7.5 AIRCRAFT ENERGY SYSTEM SIMULATION AND SIMULATORS

Some models can be executed together in a so-called simulator. If the simulated system is an aircraft, then it is called a flight simulator. Simulation is an important activity for system level analysis or verification with the main objective being to reduce risk and cost. For a whole Aircraft energy system, this activity is characterized as large-scale-simulation with specific prerequisites. Some definitions related to simulation are provided below [3]:

1. **Mid-Scale Simulation:** The activity performed when some simulation models of aircraft subsystems, developed with different modeling techniques, are integrated into a bigger model, complex enough not to be simulatable in a desktop M&S tool.

2. **Large-Scale Simulation:** The activity performed when several simulation models of the aircraft subsystems are integrated, and specific arrangements for performance or interoperability exist. Examples of such arrangements are real-time execution, "pilot-in-the-loop simulation" or "hardware-in-the-loop simulation" (HILS) configurations (Figure 5.6).

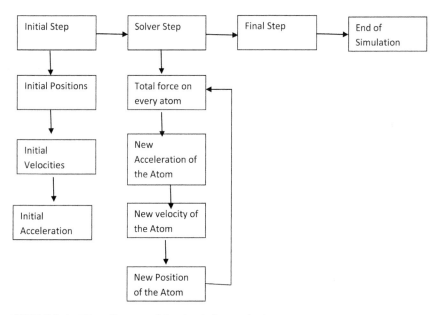

FIGURE 5.6 Flow diagram of the simulation method.

5.7.6 VERIFICATION AND VALIDATION

Verification and validation (V&V) are two terms that are often mixed up. The general definitions of V&V are:

- **Verification:** Did I build the thing right?
- **Validation:** Did I build the right thing?

Verification tasks are often independent of context and can often be objectively answered with a yes or a no, e.g., formal verification. With M&S tools, such as Dymola and Simulink, the number of computer programming and implementation errors have been reduced. Model validation, on the other hand, cannot be performed in early development phases like the concept phase due to the requirement of system experiment data. However, sensitivity analyses can still be made that point out model component parameters that have a strong influence on the simulation result or compare the assumed uncertainty data influence with similar model accuracy experience. Later in the development process, when more knowledge and measurement data is available, model validation with measurement data can begin. Model validation is context-dependent, e.g., for a specific simulation task, a model can be validated in parts of the flight envelope for some model outputs. For another simulation task with its context, a complete other model validation status can exist. In Ref. [38], a broader aspect and on a higher level has been taken concerning M&S result credibility. Eight factors have been defined as a five-level assessment of credibility for each factor, Verification Validation, Input Pedigree Results, Uncertainty, Results, Robustness, Use History M&S, Management, People, and Qualifications. The approach clearly demonstrates a large number of factors that affect a model's credibility, which inevitably means that in the case of large models, it is an extremely time-consuming task to make them credible.

- The equipment model, with pumps, valves, and pipes for performance evaluation and dimensioning. The models are usually based on physical differential equations. Black-box models can be used for some equipment of minor interest such as sensors. Tables can be used for highly nonlinear equipment such as compressors and turbines.
- The embedded software model, for control and monitoring of the equipment. The system software model can be hosted as a subroutine and be generated automatically from the software model for facilitating the integrating of new code in the case of the redesign. The effects of the control units' hardware are rarely modeled.
- The environmental models with input data to the two categories above. The pilot, flight case, and the atmosphere will be found here.

5.8 APPLICATIONS

MATLAB applications in energy storage, behavior analysis of the systems consisting of CNT by the molecular dynamics simulation. Figure 5.3

shows a screenshot of a molecular dynamics simulation of a CNT using the 3D Molecular Dynamics Simulator of MW. Applications of MATLAB in calculations for nano-systems, which mostly contain CNTs. The molecular dynamic simulation and analytical studies of CNTs (Figure 5.7).

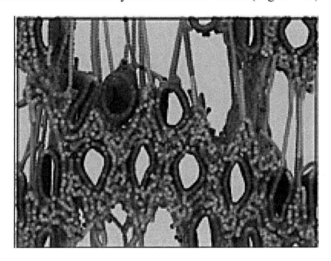

FIGURE 5.7 Example of computational modeling of CNT.

5.9 CONCLUSION

This chapter shows the modern design requirements with the help of simulation to assure the properties of aircraft like faster, miniature, highly maneuverable, self-healing, a lightweight warrant for materials, intelligence guided, smart, eco-friendly, with extraordinary mechanical and multifunctional properties in the aviation sector.

MATLAB allows reuse of models, which means, methods, and measurements for model development and validation, for example, should be used in many life cycle phases.

Using MATLAB, results in unavoidable model handovers, because of the long life of models. People from many disciplines, with and without modeling and simulations experience can use and make design decisions based on modeling and simulations outcomes. This means that methods and model quality measurements should be comprehensible and transparent, in order to be accepted by any organization. Measurements should also be directly used as feedback in the model's improvement process.

The data used for validation, like uncertainty data and its context, must be strongly connected to it. This increases the speeds of the process of the user gaining knowledge and an insight into the model validity. By showing the uncertainties visible to the user, the required and continuous model improvement process is supported.

ACKNOWLEDGMENT

I would like to give thanks to my Parents Mr. Ramesh Chandra Lal Das and Mrs. Mira Devi and My brothers Mr. Dilip Das and Mr. Pradeep Kumar with my nephew D. Amay.

KEYWORDS

- carbon nanotubes (CNTs)
- composite material
- modeling
- nano-structure
- nano-materials
- nano-science
- nano-technology
- optimization
- simulation
- simulation and modeling (S&M)

REFERENCES

1. Akaike, H., (1974). A new look at the statistical model identification. *IEEE Transactions on Automatic Control, 19*, 716–723.
2. Anderberg, O., (2011). SAAB AB. *Private Communications, Linköping Studies in Science and Technology, Thesis No. 1497* (pp. 1–64).
3. Andersson, H., (2009). Aircraft energy systems modeling: Model based systems engineering in avionics design and aircraft simulation. *Linköping Studies in Science and Technology, Thesis No. 1394* (pp. 1–118). Linköping.
4. ARINC 653, (2006), *Avionics Application Software Standard Interface, ARINC 653, Specification* (Part 1–3). Aeronautical Radio Incorporated, Annapolis, Maryland, USA.

5. Backlund, G., (2000). *The Effects of Modeling Requirements in Early Phases of Buyer-Supplier Relation, Linköping Studies in Science and Technology, Thesis No. 812* (pp. 1–72). Linköping.
6. Becker, M. C., Salvatore, P., & Zirpoli, F., (2005). The impact of virtual simulation tools on problem solving and new product development organization. *Research Policy*, 34(9), pp. 1305–1321.
7. Biltgen, P. T., (2008). Uncertainty quantification for capability-based systems-of-systems design. *26th International Congress of the Aeronautical Sciences* (Vol. 1, pp. 1–10). Anchorage USA, ICAS2008–1.3.3.
8. Carloni, L. P., Passerone, R., Pinto, A., & Sangiovanni, V. A. L., (2006). Languages and tools for hybrid systems design. *Foundations and Trends in Electronic Design Automation*, 1(1/2), pp. 1–193.
9. Cassandras, C. G., & Lafortune, S., (1999). *Introduction to Discrete Event Systems* (2nd edn., pp. 1–772). Springer, New York, USA, ISBN 0-7923-8609-4.
10. Chen, W. J., & Sudjianto, A., (1991). Analytical variance-based global sensitivity analysis in simulation-based design under uncertainty. *ASME Journal of Mechanical Design, xiv, 127*, 1–384, 875–886.
11. Council for Regulatory Environmental Modeling, (2009). *Guidance Document on the Development, Evaluation and Application of Environmental Models*. US Environmental Protection Agency.
12. RESCENDO, (2009). Collaborative and robust engineering using simulation capability enabling next design optimizations: a research project funded by the European Union Seventh Framework Programme (FP7/2007–2013). Deliverable "D5.4.1 BDA Quality Laboratory State-of-the-Art." Release 1.0. Courtesy of EADS.
13. Cullen, A. C., & Frey, H. C., (1999). *Probabilistic Techniques in Exposure Assessment*. Plenum Press: New York.
14. Fabrycky, W. J., & Blanchard, B. S., (1991). *Life-Cycle Cost, and Economic Analysis* (pp. 1–81). Prentice Hall, Englewood Cliffs, NJ.
15. Fritzson, P., (2004). *Principles of Object-Oriented Modeling and Simulation with MATLAB 2.1*. Wiley-IEEE Press, New York, USA.
16. Gains, B., (1979). General system research, Qua Vadis. *General Systems Yearbook*, 24, 1–9.
17. Gavel, H., (2007). On aircraft fuel systems: Conceptual design and modeling. *Linköping Studies in Science and Technology, Thesis No. 1067* (Vol. 1, pp. 1–90). Linköping.
18. Gavel, H., & Andersson, J., (2003). Using optimization as a tool in fuel system conceptual design. *SAE World Aviation Congress and Display 2003, Paper No 2003–01–3052* (Vol. 7, Issue, 2, pp. 127–130). Montreal, Canada.
19. Gavel, H., Krus, P., Andersson, J., & Johansson, B., (2005). Probabilistic design in the conceptual phase of an aircraft fuel system. *7th AIAA Non-Deterministic Design Forum, Paper No AIAA- 2005–2219*. Austin, USA.
20. Graf, S., (2008). OMEGA: Correct development of real time and embedded systems. *Software and Systems Modeling*, 7(2), 1–46.
21. Giese, T., Oehler, B., & Sielemann, M., (2010). A systematic approach to optimize conventional environmental control architectures. *Deutscher Luft- und Raumfahrtkongress*, Hamburg.

22. Gunnarson, R., (2011). *Vetenskapsteori. Dept of Prim Health Care Göteborg University-Research Methodology Web Site.* http://infovoice.se/fou.–03–08 (Accessed on 25 July 2019).
23. Hayden, R., (2010). *Specialists' Meeting on System Level Thermal Management for Enhanced Platform Efficiency, NATO AVT-178.* Bucharest Romania.
24. http://functional-mockup-interface.org/ (Accessed on 25 July 2019). FMI.
25. http://www.dynasim.com/ (Accessed on 25 July 2019). Dynasim AB.
26. http://www.incose.org (Accessed on 25 July 2019).
27. http://www.incose.org/practice/whatissystemseng.aspx (Accessed on 25 July 2019).
28. http://www.itisim.com/simulationx_505.html (Accessed on 25 July 2019).
29. http://www.mathworks.com/products/rtw/ (Accessed on 25 July 2019).
30. http://www.mathworks.com/products/rtwembedded/ (Accessed on 25 July 2019).
31. http://www.mathworks.com/products/sfcoder/ (Accessed on 25 July 2019).
32. http://www.mathworks.com/products/simscape/ (Accessed on 25 July 2019).
33. http://www.mathworks.com/products/simulink/ (Accessed on 25 July 2019). The MathWorks, Inc.
34. http://www.mathworks.com/products/stateflow/ (Accessed on 25 July 2019).
35. http://www.MATLAB.org/ (Accessed on 25 July 2019). MATLAB Association.
36. http://www.MATLAB.org/documents (Accessed on 25 July 2019). MATLAB Association.
37. http://www.openMATLAB.org/ (Accessed on 25 July 2019).
38. http://standards.nasa.gov/documents/detail/3315599 (Accessed on 25 July 2019). *NASA-STD-7009 Standard for Models and Simulations*
39. Moir, I., & Seabridge, A., (2004). *Design and Development of Aircraft Energy Systems: An Introduction* (Vol. 1, pp. 1–334). ISBN 156347722X.
40. Karlsson, L., (2010). *Simulator Testing Basic Course* (Vol. 7, No. 2, pp. 127–130). SAAB AB, Linköping.
41. Krus, P., (2008). *Engineering Design Analysis and Synthesis* (Vol. 1, pp. 1–357). Liu-IEI—07/0010—SE, Linköping–11–04.
42. Krus, P., (2000). Post optimal system analysis using aggregated design impact matrix 2000. *ASME International Design Engineering Technical Conferences & Computers and Information in Engineering Conference* (pp. 74–81). Baltimore, USA.
43. Krus, P., (2008). Computational tools for Aircraft energy system analysis and optimization. *26th International Congress of the Aeronautical Sciences* (pp. 31–40). Anchorage, USA, ICAS 2008–1.8.2.
44. Kuhn, R. A., (2008). Multi level approach for aircraft electrical systems design. *6th International MATLAB Conference, Bielefeld, Germany* (Vol. 1, pp. 95–101).
45. Larsson, E., (2007). Modeling of a fuel system in MATLAB – applied to an unmanned aircraft. *Master Thesis, ISRN LITH-ISY-EX--07/4128--SE, Department of Electrical Engineering*. Linköping University.
46. Larsson, J., & Krus, P., (2003). Stability analysis of coupled simulation. *ASME International Mechanical Engineering Congress* (Vol. 1, pp. 861–868). Washington D.C.
47. Lind, I., & Andersson, H., (2011). Model based systems engineering for aircraft energy systems – How does MATLAB based tools fit? *8th International MATLAB Conference, Dresden* (pp. 52–59). Germany.

48. Liscouet-Hanke, S., (2008). *A Model-Based Methodology for Integrated Preliminary Sizing and Analysis of Aircraft Power System Architectures, PhD Thesis* (pp. 1–174). Institute National des Sciences Appliquées de Toulouse.
49. Ljung, L., & Glad, T., (1991). *Model by Geochsimulering, Student Litterateur* (Vol. 1, pp. 1–350). ISBN 9144024436.
50. Malak, R. J. Jr., (2005). *A Framework for Validating Reusable Behavioral Models in Engineering Design, Master's Thesis, G.W* (pp. 1–51). Woodruff School of Mechanical Engineering, Georgia Institute of Technology, Atlanta, GA.
51. Malak, R. J., & Predis, C. J. J., (2007). Validating behavioral models for reuse. *Res. Eng. Design, 18,* pp. 111–128.
52. Mavris, D. N., & DeLaurentis, D. A, (2000). A probabilistic approach for examining aircraft concept feasibility and viability. *Aircraft Design, 3,* pp. 79–101.
53. Muessig, P. R., Laack, D. R., & Wrobleski, J. J., (2000). An integrated approach to evaluating simulation credibility. In: *Proceedings of the 2000 Summer Computer Simulation Conference* (pp. 98–109). Vancouver, British Columbia.
54. NASA-STD-7009, (2008). *Standard for Models and Simulations.* National Aeronautics and Space Administration Washington, DC 20546–0001, November 7th, available online at http://standards.nasa.gov (Accessed on 25 July 2019).
55. National Science Foundation, (1996). *"Research Opportunities in Engineering Design."* NSF Strategic Planning Workshop Final Report (NSF Grant DMI-9521590), USA.
56. Neelamkavil, F., (1987). *Computer Simulation, and Modeling* (Vol. 19, Issue: 3, pp. 113–168). John Wiley & Sons Inc.
57. Oberkampf, W. L., Sharon, M., De Land, et al., (1961). Error and uncertainty in modeling and simulation. *Reliability Engineering & System Safety, 75*(3), pp. 333–357.
58. Paynter, H. M., (1961). *Analysis and Design of Engineering Systems* (pp. 1–303). Class notes for M.I.T., course 2,751, MIT Press.
59. Potts, C., (1993). Software engineering research revisited. *IEEE Software, 10*(5), 19–28.
60. Saltelli, A., Chan, K., & Scott, E M., (2008). *Sensitivity Analysis, Probability and Statistics Series* (Vol. 1, pp. 1–305). John Wiley & Sons: New York, NY.
61. Sargent, R. G., (2001). Some approaches and paradigms for verifying and validating simulation models. *Winter Simulation Conference, WSC'01, 2,* pp. 106–114.
62. Sargent, R. G., (2007). Verification and validation of simulation models 2007. *Winter Simulation Conference,* pp. 124–137.
63. Schlesinger, S., et al., (1979). Terminology for model credibility. *Simulation, 32,* pp. 103–104.
64. Scholtz, D., (2002). *Aircraft Energy Systems – Reliability, Mass Power and Costs.* European Workshop on Aircraft Design Education.
65. Sinha, R., Liang, V., Paredis, C., & Khosla, P., (2001). A survey of modeling and simulation methods for design of engineering systems. *Journal of Computing and Information Science in Engineering, 1,* pp. 84–91.
66. Smith, P., Prabhu, S., & Friedman, J., (2007). *Best Practices for Establishing a Model-Based Design Culture, 2007–01–0777.* SAE World Congress & Exhibition, Detroit, MI, USA.
67. Sobol, I. M., (1993). Sensitivity analysis for nonlinear mathematical models. *Mathematical Modeling & Computational Experiment, 1,* 407–414.

68. Stevenson, W. J., (2002). *Operations Management* (7th edn.). Boston: McGraw-Hill/Irwin.Chapter1.
69. Ullman, D., (1992). *The Mechanical Design Process*. McGraw-Hill Inc. New York.
70. www.dynasim.com/ (Accessed on 25 July 2019). Dynasim AB.
71. www.incose.org/practice/whatissyste (Accessed on 25 July 2019).

CHAPTER 6

Promising Nanomaterials and Their Applications in Energy Technology

TATHAGAT WAGHMARE

Faculty of Food Bio-Engineering Technology Department of Microbiology and Biotechnology, Szent Istvan University, Budapest – 1118, Hungary, Europe

ABSTRACT

As of late, different types of nanomaterials (NMs) and semiconductor photocatalytic materials has pulled in overall consideration in light of tremendous applications of these nanomaterials in tackling worldwide vitality and ecological issues by powerful usage of solar energy.

Nonetheless, the photocatalytic action of these materials is still a long way from that required by commercialized applications, and in this way, an additional way of thinking towards a breakthrough research is very much essential. Nanostructured photocatalytic materials [1] with novel properties (e.g., upgraded light retention, quantum imprisonment, high particular surface region and Reasonable surface-to-volume proportion, progressive permeable structure, and so forth) expands their applications in different fields such as water and air sanitization, photocatalytic purifications, photocatalytic hydrogen generation, and colored-sensitized solar cells (SCs) [24].

6.1 INTRODUCTION

Nanomaterials (NMs), in different forms, are reported to be used in various energy-related applications. Nanoparticles (NPs), nanorod, nanofibers, thin-film, nanocluster, nanocomposites, nanocrystals, and hybrids all are applicable in various energy-related applications like energy storage and energy conversion. Solid-state lighting, in its status, is based on light-emitting diodes (LEDs), which are eco-friendly and less power consuming. Compared to the macrostate, LEDs made of NMs are more power-efficient, and hence they have huge commercial value.

The energy conversion devices are based on the conversion of any other types of energy to electrical power by electrochemical changes through electrolyte, cathode, and anode. The transformation of biomass to powers and synthetic concoctions offers incredible potential to lessen energy dependence on oil and decrease ozone harming substance emanations. Batteries include the generation and capacity of electrical charge, the exchange of cations and electrical present, each in view of electrochemical responses and chemical reactants. Battery execution depends on the mind-boggling procedures and variables that influence the vehicle of charge in the reactants, and over the interface between the chemical phases.

The general energy is incessantly creating and, as shown by the guesses of the Worldwide Essentialness Association, it is depended upon to climb by approx. 50% until 2030. At present, more than 80% of the essential energy demand is secured by petroleum derivatives. Even though their stores will keep going for the following decades, they won't have the capacity to cover the overall energy utilization over the long haul. In perspective of conceivable climatic changes because of the expansion in the air CO_2-content and in addition the possible shortage of petroleum products, it turns out to be certain that future energy supply must be ensured through expanded utilization of sustainable power sources.

Although their reserves will last for the next decades, they will not be able to cover the worldwide energy consumption in the long run. In view of possible climatic changes due to the increase in the atmospheric CO_2-content as well as the conceivable scarcity of fossil fuels, it becomes clear that future energy supply can only be guaranteed through increased use of renewable energy sources.

With energy recuperation through sustainable sources like sun, wind, water, tides, geothermal or biomass the worldwide energy demand could be met many circumstances over; as of now anyway, it is yet wasteful and excessively costly much of the time, making it impossible to assume control noteworthy parts of the energy supply.

Because of the standard change reactions on the business segments, it is unsurprising that expenses for non-sustainable power sources will rise, while through and through diminished expenses are typical for practical power sources. As of now today, wind, water and sun are monetarily aggressive in a few locales. Nonetheless, to take care of energy and atmosphere issues, it isn't just important to monetarily use inexhaustible other options to petroleum products, yet to upgrade the entire esteem included chain of energy, i.e., from improvement and change, transport and capacity up to the purchasers' usage.

Headway and augmentations in capability in conjunction with a general reducing of essentialness energy usage are sincerely required in all fields to accomplish the high focuses inside the given time since the worldwide population is expanding and taking a stab at greater flourishing.

Nanotechnologies as key and cross-sectional advancements demonstrate the standout potential for convincing mechanical accomplishments in the essentialness part, thusly impacting noteworthy duties to financial imperativeness to supply. The scope of conceivable nano-applications in the energy segment involves slow short and medium-term enhancements for a more effective utilization of regular and sustainable power sources, and additionally, new whole deal approaches for imperativeness recovery and utilize. Advancement of nanotechnology and the paralleled inventions of new materials have created a broad spectrum of energy applications. All spheres of industries will be financially benefited from the low cost, eco-friendly, and sustainable. Nanotechnology headways could influence each bit of the regard-included chain in the imperativeness division.

6.2 ENERGY CHANGE

The change of fundamental energy sources into power, warm and dynamic energy requires most extraordinary viability. Due to the discovery of many new materials and tremendous research orientation on nanotechnology, both energy conversion efficiency and energy storage capacity have been remarkably increased. So, the gases coming from fossil fuels and carbon dioxide released from steam control plants can be reduced satisfactorily.

Higher power plant efficiencies, in any case, require higher working temperatures and thusly warm safe turbine materials. Upgradation is possible by exploiting NMs because they can resist failure at high temperature. Sharp edges of turbines in the charge plants or engines can be protected and allowed to expose at high temperature by applying nanocoatings. Lightweight and strong NMs like titanium aluminides can be utilized for the purpose.

Nano-improved layers can widen the degree of possible results for separation and climate fair accumulating of carbon dioxide for control age in coal-let go control plants, to render this basic technique for control.

The energy yield from the change of synthetic energy through power modules can be wandered up by Nano-sorted out cathodes, driving forces and layers, which realizes financial application possible results in automobiles, structures and the activity of versatile gadgets.

In all industrial applications, energy conversion as thermoelectric change is largely inspiring. The old semiconductors may be replaced with nanostructured components layer-by-layer to build a structure that can adequately give an opportunity to use further the waste heat.

6.3 ENERGY DISPERSION

With respect to decrease of energy misfortunes in current transmission, trust exists that the unprecedented electric conductivity of NMs like carbon nanotubes (CNTs) can be used for application in electric links and electrical cables. Besides, there are nanotechnological approaches for the advancement of superconductive materials for lossless current conduction.

6.4 ENERGY STOCKPILING

The usage of nanotechnologies for the update of electrical energy stores like batteries and super-capacitors ends up being outright encouraging. Because of the high cell voltage and the exceptional energy and power thickness, the lithium-particle innovation is viewed as the most encouraging variation of electrical energy stockpiling.

Nanotechnologies can upgrade point of confinement and prosperity of lithium-molecule batteries decisively, as through new terminated, warm protected and still versatile separators and world-class anode materials. The organization Evonik pushes the commercialization of such frameworks for the application in the mixture and electric vehicles (EV) and in addition for stationary energy stockpiling.

Over a long time, even hydrogen is in every way a promising imperativeness store for naturally welcoming energy supply. Aside from fundamental nanostructure alterations, the effective stockpiling of hydrogen is viewed as one of the basic elements of progress while in transit to a conceivable hydrogen administration.

Current materials for engineered hydrogen storing don't meet the solicitations of the car business, which requires a hydrogen-accumulating point of confinement of up to 10 wt%.

Different NMs bury Alia in view of nonporous metal-natural mixes; give improvement possibilities, which appear to be monetarily feasible at any rate about the activity of energy components in convenient electronic gadgets.

Another imperative field is thermal energy stockpiling. Fascinating, from a monetary perspective, are likewise adsorption stores considering nonporous materials like zeolites, which could be connected as heat stores in area warming lattices or in industry. The adsorption of water in zeolite permits the reversible stockpiling and arrival of warmth.

Over the long haul, choices are given for remote energy transport, e.g., through laser, microwaves or electromagnetic reverberation. Future power scattering will require control structures giving unique load and frustration organization; ask for driven essentialness supply with versatile esteem instruments and the probability of feeding through numerous decentralized sustainable power sources.

Nanotechnologies could contribute definitively to the acknowledgment of this vision, bury Alia, through nano-tactile gadgets and power-electronical parts ready to adapt to a great degree complex control and checking of such matrices.

6.5 ENERGY UTILIZATIONS

To achieve sensible energy supply, and parallel to the streamlined headway of available energy sources, it is critical to upgrade the profitability of energy use and to avoid pointless vitality use. This applies to all branches of industry and private family units. Nanotechnologies give many ways to deal with vitality sparing.

Cases are the diminishing of fuel usage in vehicles through lightweight advancement materials in perspective of nanocomposites, the change in fuel start through wear-sheltered, lighter engine fragments and nanoparticular fuel included substances or even NPs for enhanced tires with low moving obstruction [5].

Extensive vitality reserve funds are feasible through tribological layers for mechanical parts in plants and machines. Building advancement moreover gives inconceivable conceivable outcomes to essentialness venture stores, which could be tapped.

6.6 GRAPHENE NANOTECHNOLOGY IN ENERGY

The authors note, notwithstanding, that before graphene-based NMs and gadgets find broad business utilize, two vital issues must be understood: one is the readiness of graphene-based NMs with all around characterized structures, and the other is the controllable fabrication of these materials into functional devices [2].

6.6.1 SOLAR CELLS (SCS)

Graphene can possibly be utilized for minimal effort, adaptable, and exceedingly productive photovoltaic (PV) gadgets because of its magnificent electron-transport properties and to a great degree high transporter portability. As of late, a few graphene-based sun-powered cells have been accounted for, in which graphene fills in as various parts of the cell. Graphene is an ideal 2D material which can be assembled into film electrodes with good transparency, high conductivity, and low roughness [3].

6.6.2 LITHIUM-ION BATTERIES (LIBS)

The energy densities and exhibitions of rechargeable lithium-particle batteries – which are utilized broadly in a compact hardware, for example, PDAs, smartphones, cameras, and so forth. Rely upon the physical and concoction properties of the cathode materials. In this manner, numerous exploration endeavors have been made to plan novel nanostructures and to investigate new terminal materials to accomplish higher limit and to build the battery's charge rate, progressively also employing graphene in the form of nanosheets, paper, and CNT or fullerene hybrids [6].

6.6.3 SUPERCAPACITORS

Rather than the ordinary high-surface-territory materials, the viable surface zone of graphene materials as capacitor anodes does not rely upon the appropriation of pores in a strong state, which is unique in relation to the current supercapacitors created with initiated carbons and CNTs of graphene materials ought to depend exceedingly on the layers.

6.6.4 CATALYSIS

Graphene has as of late gotten exceptional enthusiasm for the field of catalysis in view of its one of a kind two-dimensional structure with its high surface zone, extraordinary electronic and ballistic transport properties.

There are numerous graphene-based nano-materials, such as functionalized graphenes, doped graphene, and graphene/metal or metal oxide composites, have been developed [3] (Table 6.1).

Nanotechnology's incredible supportability guarantee is to realize the truly necessary power move in the sustainable power source: another age of exceedingly productive PVs, nanocomposites for more grounded and lighter breeze energy rotor; but also, a new class of materials called nanomembranes for carbon capture. Carbon capture and storage can be enabled by carbon nanomembrane at fossil fuel power plants [5].

Investment in the energy sector will be fruitful by choosing the best NMs so that not only the power storage and power conversion devices will be upgraded, but also suitable nanostructures will facilitate the conveyance by diminishing fuel consumption and carbon emission. Sustainable power sources can be realized by decentralized power sources with renewable energy where the availability problem can be overpowered by using integrated energy storage units using NMs. Sun control windows or electrochromic windows can effectively concentrate solar energy and amplify considerably to achieve in house ambient warmth. Parallel to these applications, nanocatalysis may be the process for improving fuel creation, and advanced NMs can redesign the concept of batteries and supercapacitor for energy storage.

Nanomembranes, on the other hand, are present in all living cell. These are useful for transportation of food or oxygen to different cells. Mimicking their structure, nanomembranes using nanocomposites can be prepared, which has lots of potential in water treatment and sensors.

TABLE 6.1 Instances of Potential Uses of Nanotechnology Along the Regard Included Chain in Energy Fragment

	Nanomaterials and Nanocomposites That Facilitate Energy Harvesting
	Renewable
Solar Energy	CNT and graphene composites (with and without doping), silicon nanowire, silver nanoparticles, quantum dots like CdSe, Nano heterostructures have the great possibility in photovoltaics.
Wind Energy	CNT or graphene composites for rotor blades (these are light and strong), different nanoparticle made anti-corrosion coating for bearings, wind turbine transmission, and blades.
Geo-Thermal	Drilling machine can be coated with frictionless nanocomposites. Some nanoparticles can store heat.
Bio-Fuel Driven Energy	Lightweight strong container can be made of nanocomposites. Yield should be released in a controlled way with nanosensor. Surface Plasmon resonance (SPR) in nanosensors can control the pest.
Hydro Power	Nanocoating reduces energy loss, decreases reactivity, and hence reduces friction and corrosion.
	Fossil Fuels
	Drilling machines can be made frictionless using nanocoating and less power wastage. Nanomaterials reduce the extraction cost and increase the mileage of machines. Also, nanotechnology gifts us some alternative fuels like ethanol.
	Nuclear Power
	Nanocomposites are radiation protective. They reduce the operating temperature, improves moderator material. Heat resistant materials can be prepared with nanocomposites.
	Energy Conversion
Fuel Cells	Nanocomposites fuel cell such as oxide fuel cell, organic polymer membrane, proton exchange membrane cells is a direct result of nanoscience in energy conversion. CNT and their composites are used as catalytic electrodes. They are used in electronics, mobile.
Thermoelectric Conversion	Silver, lead, bismuth nanocomposites having high thermal conductivity can be used to convert waste heat into electricity. For personal electronics, body heats can also be transformed. High efficiency of conversion is observed with nanocomposites.
Photocatalysis	Photocatalytic hydrogen generation is possible with graphene/metal compound composites. Nano-catalysts have high efficiency in generating hydrogen.

Promising Nanomaterials and Their Applications

TABLE 6.1 (Continued)

Turbines	Nanocoating in turbine blades resists them from corrosion resulting from vibration and rotation. CNT/polymer composites have high potential in preparing wind blades.
Combustion Engines	Carbon nanocomposites for the engine, corrosion-resistant nanocoating, composites of polymer with adhesive may be useful.
Motors for Kinetic Energy	Polymer epoxy-based nanocomposites, Co nanocomposites are cost-effective, energy-efficient materials for electro engines (Ex. In Ship engines.)

Energy Distribution

Magnetically Tunable Distribution System	Magnetic properties of nanocomposites are exploited for smart grids. Polymer-based magnetic nanoparticle composites are prepared with intrinsic magnetic properties. Decentralization is achieved by adjusting grid components by nanosensors.

Power Transmission

Super Conducting Wires and Cables	No transmission loss wires may be made with CNTs and polymer nanocomposites. Magnetic ceramic nanocomposites potentially can be used too for high-temperature use.
Power Lines	Superconducting cables based on carbon nanotubes (long term)
Wireless Transmission	High frequency, energy-efficient wireless communication is possible with CNT polymer composites. Different organic and inorganic materials are used for nanocomposites preparation.
High Voltage Transmission	Magnetic nanoparticles are used for transmission. Polymer nanocomposites serve as insulation material for long-term anti-corrosion effect.
Heat Transfer	Nanocomposite based energy exchangers can be made.

Energy Storage

Electrical Energy	Improved Li-biobatteries by high surface are highly porous nanostructured materials. Tin, titanium-based cathode materials and anodes from carbon, silicon, and different metals are prepared. Nano clay for separating foil, adopters. Nano graphene composites for gadgets, vehicle, and adaptable load administration in control network. Super Capacitors: Nano materials for electrodes (carbon aerogels, CNT, Metal (-oxide) and electrolytes for higher energy densities).

TABLE 6.1 (Continued)

Chemical Energy	Hydrogen Nano-permeable materials (Organometals, Metal hydrides) for application in smaller-scale energy units for portable gadgets or autos (long term).
Thermal Energy	Stage change materials: Typified PCM for aerating and cooling of structures. Zeolites for reversible warmth stockpiling in structures and warming nets."

Energy Usage

Highly Efficient Lighting	A highly efficient light-emitting system with tailored nanocomposites. Plasmonic cavities improve light efficiency in LEDs. Less heat wastage using special nanomaterials.
Thermal Insulation	Nano foams and gels are used to modify thermal properties like conductivity or water intake, etc. Structure building and aerospace need such type of insulation.
Impact on Industries	Almost all sectors of industry are benefitted; e.g., aviation, transportation, aerospace, food processing, packaging, renewable energy, etc. Concept of low-cost commodities is in progress everywhere

These are dimension less than 100 nm [5]. These are porous structures only a few molecules thick, can be thought of comprised of molecularly engraved materials and ideal for biosensing purpose in treating diseases. Natural aquaporins in them form water channel and have huge application in nanomedicine. NMs-based composites are also in focus on preparing absorbents, and particles like TiO_2 are examined continuously to eliminate contamination.

Nanosilver earthenware channels have officially discovered their direction onto advertise applications because of their antibacterial and antiviral activity.

KEYWORDS

- **carbon nanotubes**
- **light emitting diodes**
- **nanomaterials**
- **nanoparticles**
- **photovoltaic**
- **solar cells**

REFERENCES

1. Yu, J. G., Jaroniec, M., & Lu, G. X., (2012). "TiO_2 photocatalytic materials." *International Journal of Photoenergy* (p. 5). Article ID 206183.
2. Xiang, Q. J., Yu, J. G., & Jaroniec, M., (2012a). "Graphene-based semiconductor photocatalysts." *Chemical Society Reviews, 41*, pp. 782–796.
3. Xiang, Q. J., Yu, J. G., & Jaroniec, M., (2012b). "Synergetic effect of MoS2 and graphene as cocatalysts for enhanced photocatalytic H2 production activity of TiO2 nanoparticles." *Journal of the American Chemical Society, 134*, pp. 6575–6578.
4. Zhou, X. M., Liu, G., Yu, J. G., & Fan, W. H., (2012). "Surface plasmon resonance-mediated photocatalysis by noble metal based composites under visible light." *Journal of Materials Chemistry, 22*, pp. 21337–21354.
5. Zhang, J., Yu, J. G., Jaroniec, M., & Gong, J. R., (2012). "Noble metal free reduced graphene oxide-ZnxCd1−xS nanocomposite with enhanced solar photocatalytic H2—production performance." *Nano Letters, 12*, pp. 4584–4589.

6. Zhang, J., Yu, J. G., Zhang, Y. M., Li, Q., & Gong, J. R., (2011). "Visible light photocatalytic H2-production activity of CuS/ZnS porous nanosheets based on photoinduced interfacial charge transfer." *Nano Letters, 11*, pp. 4774–4779.
7. Li, Q., Guo, B., Yu, J., et al., (2011). "Highly efficient visible-light-driven photocatalytic hydrogen production of CdS-cluster-decorated graphene nanosheets." *Journal of the American Chemical Society, 133*(28), pp. 10878–10884.
8. Liu, S. W., Yu, J. G., & Jaroniec, M., (2011). "Anatase TiO2 with dominant high-energy 001 facets: synthesis, properties, and applications." *Chemistry of Materials, 23*, pp. 4085–4093.
9. Energy Information Administration (EIA), (2007). *International Energy Outlook*, p. 07. http://www.eia.doe.gov/oiaf/ieo/index.html (Accessed on 25 July 2019).
10. Suehiro, S., (2007). *Energy Intensity of GDP as an Index of Energy Conservation* (Vol. 11, pp. 4774–4779). Institute of Energy Economics Japan Report.
11. Berger, M., (2012). *Nanotechnology Applications Could Provide the Required Energy Breakthroughs*, 0605.
12. Joachim, C., (2005). To be nano or not to be nano? *Nature Materials, 4*(2), 107–109.
13. Yunus, I. S., Harwin, K. A., Adityawarman, D., & Indarto, A., (2012). Nanotechnologies in water and air pollution treatment. *Environmental Technology Reviews, iFirst*, 1–13.
14. Hartono, (2010). Nanowerk.com/nanotechnology-in-energy.php. *Mineral Energy, 8*(1), 10–16.
15. Luther, W., (2008). *Application of Nano-Technologies in the Energy Sector* (Vol. 41, pp. 782–796). Germany: Hessian Ministry of Economy, Transport, Urban and Regional Development.
16. Tatsumisago, M., (2012). *Solid-State Lithium Batteries Using Glass Electrolytes*, p. 10.
17. Zhao, X., Hayner, C. M., Kung, M. C., & Kung, H. H., (2011). In-plane vacancy-enabled high-power Si-graphene composite electrode for lithium-ion batteries. *Advanced Energy Materials, 1*(6), 1079–1084.
18. Bullis, K., (2010). *Higher-Capacity Lithium-Ion Batteries Technology Review*, p. 6–11.
19. Liang, S., Zhu, X., Lian, P., Yang, W., & Wang, H., (2011). Superior cycle performance of Sn@C/graphene nanocomposite as an anode material for lithium-ion batteries. *Journal of Solid State Chemistry, 184*(6), 1400–1404.
20. Schroder, K., (2012). *Understanding the Formation and Composition of the Solid-Electrolyte Interphase at Silicon Surfaces*, p.10.
21. Nazri, G. A., & Pistoia, G., (2003). *Lithium Batteries: Science and Technology* (pp. 621–637). New York: Springer.
22. Rice, B. M., & Jow, T. R., (2012). *Energy and Energetics* (p. 5). U.S. Army Research Laboratory.
23. Smithsonian Institution, (2011). *Fuel Cell Basics*, pp. 11–24.
24. Antolini, E., & Perez, J., (2011). The renaissance of unsupported nanostructured catalysts for low-temperature fuel cells: From the size to the shape of metal nanostructures. *Journal of Materials Science, 46*(13), 4435–4457.

25. Guo, S., & Sun, S., (2012). Fe Pt nanoparticles assembled on graphene as enhanced catalyst for oxygen reduction reaction. *Journal of the American Chemical Society, 134*(5), 2492–2495.
26. Dai, L., Chang, D. W., Baek, J. B., & Lu, W., (2012). Carbon nanomaterials for advanced energy conversion and storage. *Small, 8*(8), 1130–1166.
27. Chu, K. L., Gold, S., Subramanian, V., Lu, C., Shannon, M. A., & Masel, R. I., (2006). A nanoporous silicon membrane electrode assembly for on-chip micro fuel cell applications. *Journal of Microelectromechanical Systems, 15*(3), 671–677.
28. Bagotsky, V. S., (2009). *Fuel Cells in Electrochemistry Encyclopedia*. Yeager Center for Electrochemical Sciences (YCES) Report.

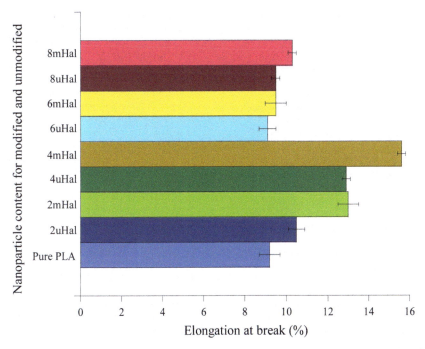

FIGURE 1.1 % Elongation-at-break of unmodified and modified Hal nanotube-reinforced PLA nanocomposites. Adapted from Ref. [32].

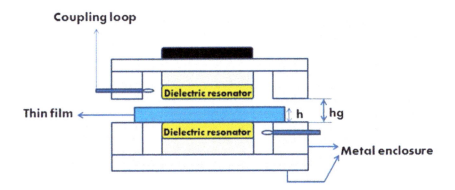

FIGURE 2.3 Typical cross-sectional view of SPDR fixture.

B   *Nanomaterials-Based Composites for Energy Applications*

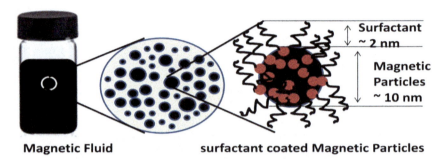

FIGURE 3.1 Components of magnetic fluid–a schematic representation.

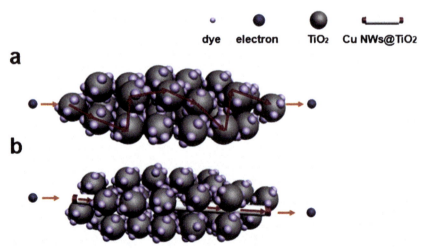

FIGURE 4.4 Photoelectrons transporting process in a) dye-coated un-doped TiO_2 and b) doped Cu NWs@TiO_2 dye-coated TiO_2 photo-anode. Adapted from Ref. [89].

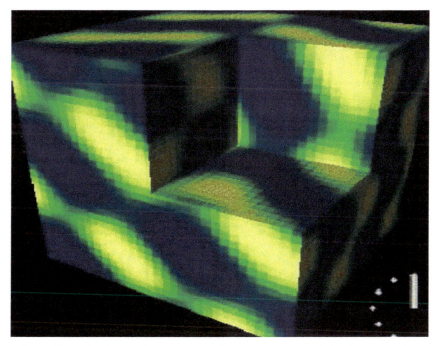

FIGURE 5.2 Simulation of nanostructured interfaces are controlled at the nanoscale.

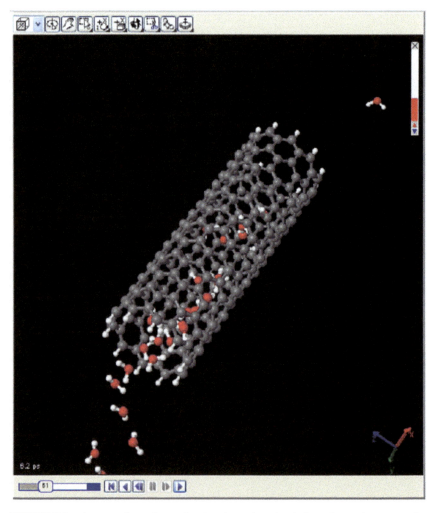

FIGURE 5.3 A screenshot of a molecular dynamics simulation of a carbon nanotube using the 3D molecular dynamics simulator.

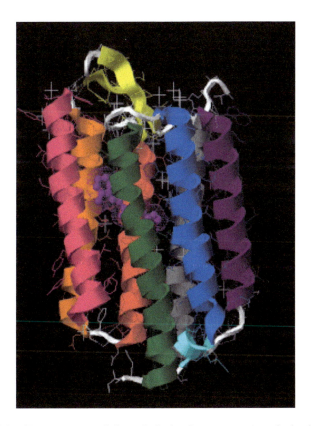

FIGURE 7.1 3D structure of bacteriorhodopsin monomer, retinal chromophore surrounded with seven transmembrane helices (Source: Wikipedia).

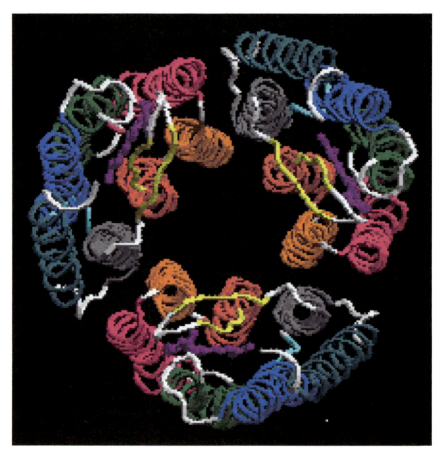

FIGURE 7.2 The crystal packing of bacteriorhodopsin trimer with one retinal molecule (purple) in each subunit viewed from the extracellular side (Source: Wikipedia).

Nanomaterials-Based Composites for Energy Applications

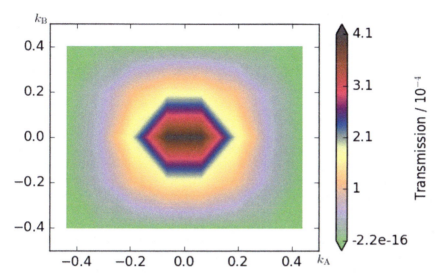

FIGURE 10.10 Contour plot of transmission coefficients along the reciprocal vectors k_A and k_B for AlNi nanohybrid.

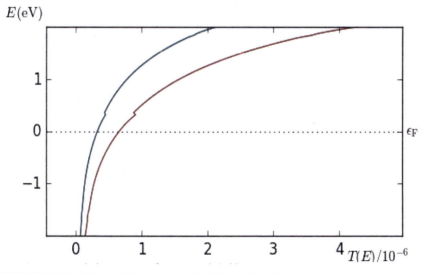

FIGURE 10.11 Transmission spectra of AuZn nanohybrid.

H   *Nanomaterials-Based Composites for Energy Applications*

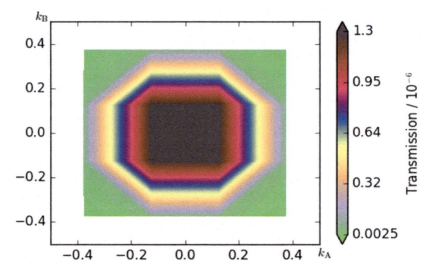

FIGURE 10.12 Contour plot of transmission coefficients along the reciprocal vectors k_A and k_B for AuZn nanohybrid.

PART II
Nanocomposites for Sustained Energy

CHAPTER 7

Nano-Bio Hybrid Platform to Meet the Energy Challenge

MIHIR GHOSH

Department of Organic Chemistry, Weizmann Institute of Science, 234 Herzl Street, Rehovot – 7610001, Israel,
E-mail: mihir.ghosh@weizmann.ac.il

ABSTRACT

Searching for advanced nano-bio hybrid photocatalyst with high activity and durability to harvest solar energy is extremely desirable but it still remains a significant challenge. Bacteriorhodopsin (BR) functions as a photoinduced proton pump by converting solar energy into proton gradient across the membrane. As well as BR has the capacity to survive under strenuous circumstances, including exposure to light and oxygen, temperatures, a broad range of pH values and high ionic circumstances. Due to unique optoelectronic properties, BR became a promising material for nano-bio hybrid photocatalyst. In this chapter, we summarized the photovoltaic background before the BR becomes a topic of interest and summarized the structural characteristics and photocycle of BR and we give an overview of the recent progress of utilizing the BR that has been designed for the environmentally friendly bio-inspired photo-energy conversion schemes. The interfacing of BR with various nanostructures including colloidal nanoparticles and nanoparticulate thin films has developed novel functional materials. BR-based solar energy conversion devices, such as photoelectrochemical cells and biomolecule-sensitized solar cells are cost-effective, green and quite efficient but they are still in the evolution stage. This hybrid bio-nanomaterials with outstanding chemical, physical and biological properties suggest the potential to develop new bio-photonic devices. Hence, in order to achieve more power conversion efficiency, further studies are required to work on this platform.

7.1 INTRODUCTION

The solution to all global challenges are around us; therefore, the aim of science is to mimic natural phenomena. For instance, plants give us the inspiration for energy storage, humpback whales for efficient wind-power, the eye of a camera, birds for airplanes, lotus leaf for hydrophobic surface, the brain to a supercomputer, termites to create sustainable buildings, morpho butterfly for the visual technology and many more. Nature, besides stirring scientists to mimic natural phenomena, also provides biomolecules having their inherent functionality which can be employed for advanced bio-inspired hybrid photocatalysts [1, 58]. Nanostructured materials are acquiring the significance for the energy creation and storage. With the advent of innovative nanostructured materials and nanotechnology remarkable hike forward to meet energy challenge and the artificial synthesis is anticipated. Based on the nature's fundamental solar energy transformation mechanism biological molecules and complexes can be broadly used in two main groups: the first one utilizes chlorophyll related antenna complexes and the second group is based on rhodopsin, a membrane-embedded protein with a chromophore retinal (vitamin A aldehyde). Besides the device applications, significant studies have been explored towards energy-related applications of various heterostructures in recent years over the metalloporphyrins [2–5], and free base porphyrins [5, 6] as chlorophyll plays a vital role for photosynthesis, which allows plants to absorb energy from light and it is structurally analogous and created through the same metabolic pathway as other porphyrin pigments such as heme. Structurally chlorophyll pigments represent magnesium ion-centered circulated tetrapyrrole complexes that provide the reaction center of the plants a charge separation capability and serve as a principal antenna for light harvesting.

The extent and quality of light are different in every part of the world, which results in the advancement of different organisms according to the conditions of the environment. These organisms contain properties that might form the origin of a novel kind of solar cell skilled of converting energy to a working form of chemical energy, as it contains the natural photoreceptor coupled with the proteins. It shows exceptional photoreaction performance with efficient and very fast charge transfer processes on light absorption and an excellent photoresponse at a very low light intensity, which is one the important necessity of solar light-harvesting systems. Purified samples can be isolated very easily of these organisms and can be modified according to our need by chemical, physical, and also by applying the genetic engineering technique. They can be synthesized easily in our laboratory without using any hazardous chemicals which leads to environmental pollution. The most sensitive photoreceptor found till date in our nature is rhodopsin protein originated in our nature. These rhodopsins are known to belong, two distinct protein families, archaeal and visual [7]. Archaea is one kind of microorganisms without having any cell nucleus or any membrane-bound organelles. Archaealrhodopsins, found in extreme halophiles, function as light-driven chloride ion pumps (halorhodopsins), proton pumps (bacteriorhodopsins or BR) or photosensory receptors (sensory rhodopsins). Whereas, the visual rhodopsins, are found in human eyes throughout the animal kingdom are photosensory pigments, although both protein families do not show any significant sequence similarity. Microorganisms that contents rhodopsin genes range from the freshwater, surface, and deep seawater, salt flats, soil, glacial sea habitats to plant and human tissues as fungal pathogens. It encompasses an extensive phylogenetic range of microbial life that contains halobacteria, cyanobacteria, green algae, proteobacteria, and fungi. Archaeans contain natives of some of the most extreme environments on the earth. Few of them alive near crack holes in the deep sea at temperatures well over 100°C. Some lives in hot springs or in acid waters or very high alkaline. Archaea have been found booming inside the digestive tracts of termites, cows, and marine life. Archaea may be the only organisms that can survive in extreme environments such as hypersaline water or thermal vents. It may be lavish in the atmospheres that are inimical to all other life forms.

Some archaeans can persist the dehydrating effects of extremely saline waters, and one of the salt-loving groups of archaea comprises Halobacterium. The light-sensitive pigment BR gives Halobacterium its color and delivers it with chemical energy. BR has an attractive purple color with the capability to pump protons to the outside of the cell membrane.

7.2 PHOTOVOLTAICS (PV) BACKGROUND

Before we discuss about the BR, we should know about some existing solar PVs and their potentiality in solar energy harvesting systems. Existing crystalline silicon solar cell technologies despite having efficiency 24.2% and many more benefits still not well accepted due to some gross deficiencies. Silicon solar cells (SCs) involve very high-cost materials and fragility. Moreover, over the years, despite the introduction of the thin-film technologies, neither cheaper manufacturing cost nor the greater efficiency silicon solar cell could be accomplished for widespread acceptance in global power generation. Professor Michael Graetzel of Swiss Federal Institute of Technology reported in 1991 a new type of solar cell [8], dye-sensitized solar cell (DSSC), based on dye-sensitized nanocrystalline oxide films which fruitfully mimic the light reaction takes place in green leaves and algae during natural photosynthesis; however these cells are out of any biologically derived materials. The transparency and low costing of Gratzell cells are the promising alternatives of existing silicon-based PVs [9]. DSSC comprises a non-crystalline porous semiconductor electrode absorbed the dye, counter electrode, and electrolyte. DSSC is constructed on a perception of charge separation at an interface of two materials of different conduction mechanism and the match of the energy gaps of the two materials. Effective electron transfer from the highest occupied molecular orbital of dye to the conduction band of mesoporous semiconductor TiO_2 surface with minimum back electron transfer is the basic factors to define DSSC PV performance [10]. According to Prof. Gratzel, a two-level tandem cell (a tandem cell consists of two or more sub-cells together that convert more of the sunlight spectrum into electricity) embodiment could reach up to 46%. However, the authenticity of this comment is still in under question, as till date, the efficiency of DSSC reached up to only 14.1%. DSSC improvements are buoyed by the potential developments in the areas of dyes, photoanodes redox, couples, electrolytes, and tandem cell configurations. ZnO-based DSSC technology alternate to TiO_2 is considered as one of the most promising materials for SCs. ZnO holds the energy band structure and physical properties analogous to those of TiO_2, but its electronic kinesis is higher by 2–3 orders of magnitude [11] Consequently, ZnO shows faster electron transport with reduced recombination loss. Nanostructured photoelectrodes (nanowire [12, 13], nanotubes [14], nanobelts [15], nanosheets [16, 17], nanotips [15], and nanoforest [18]) have been well studied and anticipated to increase the electron diffusion length in photoelectrode films by providing a direct conduction pathway for the prompt collection of photo-generated electrons. A new hike in the research of perovskite solar cells (PSCs) is observed in the PV cells. The performance of metal halide PSCs has made rapid increases in energy conversion efficiency from under 4% in 2010 with a record efficiency of more than 20% in 2017 [19]. Because of the high absorption coefficient, only about 500 mm thickness is needed to absorb solar energy. Although the hybrid material has high efficiency, still the organic component in it is volatile, and durability of the semiconductor presents a major technical hurdle to commercialization. After the research over the years over the PVs lack of appreciable solar energy harvesting efficiency, photostability, cheaper manufacturing cost, and implementation barriers in PVs engineering are still under challenge to meet the energy crisis.

The advance development of biological systems has created natural and sustainable nanoscale materials with skills beyond that of current technology, including a wider light absorbance spectrum capability. Making a composite of biological materials with inorganic nanomaterials (NMs) opens new possibilities for protein-sensitized solar cell (PSSC). The novelties that make the PSSC stand apart are: (a) their light-harvesting protein with broadband absorption, including the IR region, faster electron injection rates, thermal robustness, high quantum efficiency and nontoxicity; (b) well-organized one-dimensional wide gap semiconductors that collects and transports electron are vastly superior to nanoparticles (NPs). Therefore, easily it can be assumed that the fast electron injection rate of light-harvesting proteins coupled to inorganic NMs can provide substantial advantages in a PV cell. Notwithstanding, this excellent advantage we need more approaches to couple proteins to make NMs based composites for solar energy harvesting. This bio-nano hybrid complex is materials of active interest not only in PV engineering, but also its immense potential in the diverse application in biotechnology, cancer therapy, bio-mimic devices, sensors and electro-optics [20–23]. In the phenomenal development of protein science, there are potential candidates that can be used to fabricate protein-based nanostructured PV cell out of which BR proves to be the most promising. This light-driven proton pumps under constant wavelength illumination reveal a stationary photocurrent by the transformation of light energy to electrochemical energy in the form of a proton gradient across the membrane [24]. BR offers recompenses for its ease of generation. Purple membrane can be routinely isolated from halophilic bacteria via simple cell lysis by exposing cells to low salt medium followed by sucrose gradient centrifugation. One of the most important factors to increase the PV performance of a solar cell is the quantum efficiency. According to the literature value, the quantum efficiency of the BR photochemical cycle range from 0.25 to 0.79, and the sum of forward and back photoreaction is 1 [25]. Preliminary studies also indicate that energy bandgap of BR matches very good semiconductors, therefore, it is very likely to use BR as a composite-based SCs. As well as BR takes the most promising place in the field of PV devices due to their thermal robustness, which is the mandatory requirement for any photo-excitable molecule to be used in the solar cell. Therefore, nanoelectronic devices based on more thermally stabile BR mutant is a field of research interest in different laboratories. It is easy to handle as well as its durability of temperature up to 80°C (in water) and up to 140°C (dry), stability in a long pH range (1 to 12), high ionic strength (3(M) NaCl), maximum charge separation on photon absorption, broadband absorption and mostly durability to sunlight exposure in the presence of oxygen for years give its most promising place in the field of solar light-harvesting systems. The unique photochromic retinal protein BR has emerged as an outstanding material for biomolecular photonic applications due to its unique properties and advantages.

7.3 BACTERIORHODOPSIN (BR)

BR is a compacted molecular machine that pumps protons across the membrane are isolated from the cell membrane of extreme photo-organo-heterotrophic halobacteria *Halobacterium halobium* [26]. Halobacterium is a rare organism that flourishes in tough environments, and it is competent to sustain in every hardship. Halobacterium is a member of the Archaea, which are often found in harsh environments-many are extremophiles. Plants, algae, and some bacteria undergo photosynthesis for their growth; however, halobacterium is not photosynthetic. We know that mitochondria and bacteria during the respiration use protons as a form of positive electrical energy to drive the ATPase (Adenosinetriphosphatase), an enzyme which widths the membrane and spins like an electric motor when the protons run through it and arrest the rotational energy as chemical energy by synthesizing ATP. Thus, we have the transformation or transduction of energy from electrical, to rotary mechanical, to chemical, and these process energies the living cells. Nevertheless, halobacterium has a unique mechanism of making ATP by exploiting sunlight to create an electrical potential difference.

To synthesize ATP in normal aerobic conditions halobacterium undergoes normal respiration using oxygen from nature. However, in an extreme environment, oxygen starvation must be a common occurrence in halobacterium lives. Therefore, it synthesizes a new membrane component, the purple pigment BR, which is dumped in the dense patches of the cell membrane, developing purple membrane patches. Purple membrane patches occur in a form that gives its inelasticity and makes it prospective for incorporation in a physical device designed to arrest solar light. The light sensitivity of BR and its color result from the presence of retinal chromophore covalently bound to an opsin apoprotein. Vertebrate retinas

Nano-Bio Hybrid Platform to Meet the Energy Challenge 167

have rhodopsin as a visual light-sensitive pigment whereas in lower organisms BR plays a role for rhodopsin for photoenergy conversion as well as photoreceptors. Seven transmembrane α-helices protein segment belongs to the G-protein coupled receptor (GPCR) organize the basic structure of BR. The helices (named A-G) are organized in an arc-like structure and firmly surround a retinal molecule that is covalently attached via a Schiff base to a preserved lysine residue on helix G (Figure 7.1).

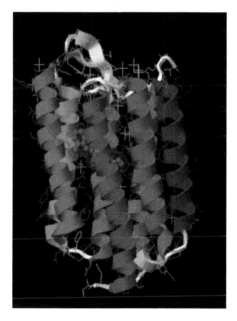

FIGURE 7.1 (See color insert.) 3D structure of bacteriorhodopsin monomer, retinal chromophore surrounded with seven transmembrane helices (Source: Wikipedia).

Transmembrane lipids of BR possess L-glycerol with inverted stereochemistry linked to branched isoprene chains that deliver the BR the more stability compared to other archaeans. Polypeptide chain of BR consists 248 amino acids out of which 67% are hydrophobic, aromatic amino acids and remaining 33% are hydrophilic residues of glutamic acids, aspartic acids, lysine and arginine [27]. These amino acid residues attribute to trimer formation with an average diameter of 0.5 μm and thickness of 5–6 nm by their spatial coordination and each of the trimer is bounded by six other trimers that result in a regular hexagonal lattice [28] (Figure 7.2).

168 Nanomaterials-Based Composites for Energy Applications

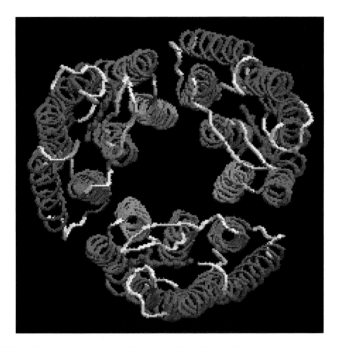

FIGURE 7.2 (See color insert.) The crystal packing of bacteriorhodopsin trimer with one retinal molecule (purple) in each subunit viewed from the extracellular side (Source: Wikipedia).

It possesses light-adapted (LA, λ_{max} –568 nm) and dark-adapted (DA) state (λ_{max} –558 nm), which are inter-convertible. LA state form of BR is a biologically active form and roles as light-driven proton pump consisting almost exclusively in the all-*trans* configuration of the retinal chromophore, whereas in the DA state retinal is found in a mixture of both 13-*cis* (60%) and all-*trans* (40%) configurations. Thermal structural changes due to thermal changes in the protein reveal to the photocycle, to be precise spectroscopically distinguishable intermediates (Figure 7.3).

Schiff base associates the retinal chromophore to the opsinapoproteinviaLysine-216 amino acid residue. Schiff base is a molecule having N atom bound to a C atom on one side by a C=N double bond and to a carbon on the other side by a C-N single bond. A proton can reversibly attach to the N atom, which acts as a proton stockpile. BR proton pumps with a molecular weight ~ 26.5 kDa are relatively small protein architectures. When the retinal chromophore absorbs a photon, it becomes energized and experiences a conformational change that leads to a change to an advanced energy state which results in the deposited proton to be unconfined. Therefore, in the presence of light, the purple membrane protons get released into the extracellular side of the cell membrane. These protons do not go away, but are directed back into the cell. Protons are positively charged, and intercellular side of the membrane is negatively charged with an electrochemical potential up to 250 millivolts. Therefore, BR is capable of carrying ions against an electrochemical potential with 10,000-fold difference in proton concentration on either side of the membrane [29]. This electrochemical potential is used in photophosphorylation to produce ATP. As the ATP synthesis is done in BR without having no chlorophyll molecule; therefore, the mechanism is known as "non-chlorophyll photosynthesis."

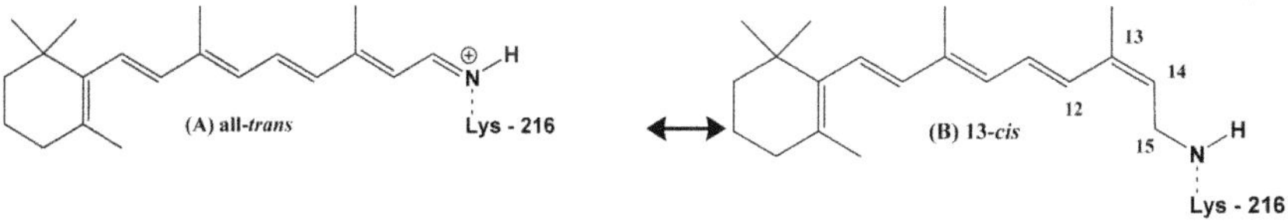

FIGURE 7.3 (A) Light adapted state (all-*trans* conformation) of retinal is covalently linked with lysine residue of helix G forming protonated Schiff base. (B) Dark state (predominant 13-*cis* conformation).

The BR photocycle is initiated by photon illumination, and absorption of the light by retinal that generates an electrochemical gradient followed by the translocation of a proton across the membrane form cytoplasmic to the extracellular side of the cell. The quantum efficiency of BR was determined to be 0.64 ± 0.04 [30]. All-*trans* to 13-*cis* photo-isomerization takes place within sub-picoseconds by the absorption of light with λ_{max} 570 nm. Due to the precise, amino acids and chromophore framework arrangement, the protonated Schiff base then discharges the proton to the Aspartic acid residue followed by changes in protonable groups within the apo-protein molecule to the extracellular side of the cell. In the photocycle of the BR, a cascade of photochemical reactions takes place lasting from 3 milliseconds to 1 picosecond with the development of J, K, L, M, N, and O impermanent intermediates, which is shown in Figure 7.4.

Here it is important to reemphasize L, M, and N are blue-shifted intermediates whereas K and O are red-shifted intermediates. With the formation and decay of M intermediates, de-protonation and re-protonation follow respectively, and which results in a concentration gradient in between the internal and external surface of the membrane. As a consequence that illuminated BR cell becomes eligible to synthesize ATP, i.e., conversion of light energy to chemical energy without having any chlorophyll molecule inside the cell.

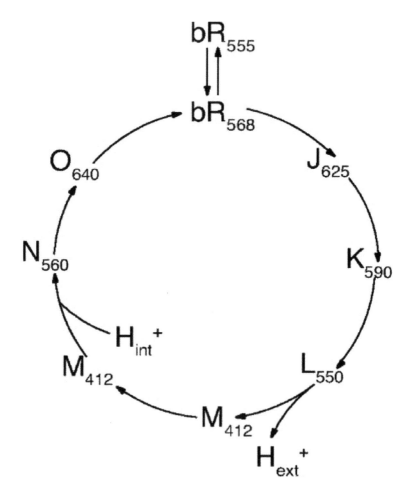

FIGURE 7.4 A photocycle scheme of BR. Here J, K, L, M, N, and O represent the spectral intermediates. The subscripts represent the corresponding absorption maximum of the photocycle intermediates.

Source: Reprinted with permission from Ref. [31]. © 2018 National Academy of Sciences, U.S.A.

7.4 RECENT STUDIES IN BR BASED HYBRID NANO-BIO SOLAR CELLS (SCS)

The journey begins in the year of 2001 by Lanyi et al., [32], with their announcement that BR can be exploited in macro- or nano-scale devices due to its capability of continuous, renewable conversion of light to chemical energy. In nanoscale, proteoliposomes containing proton pumps are the most promising, general-purpose bio-devices for efficient, continuous, and renewable generation of energy in biotechnology. These systems are substantial at technological scales. With the emerging understanding over the BR PV technology, it is possible to re-engineer them to improve their expediency in precise applications, they announced in their article. The study by Li and his coworkers [33] pushed one step forward towards the design of BR-based heterostructures for intelligent optoelectronic and nonlinear optical devices. Oriented BR film monolayers were deposited on hydrophobic or hydrophilic silicon (Si) substrates through the Langmuir–Blodgett technique. The extracellular surface ($-NH_2$ terminal) and cytoplasmic surface (-COOH terminal) of BR façade the silicon directly, giving oriented patterns of p-Si/extracellular – cytoplasmic/indium tin oxide (ITO) and Si/ cytoplasmic – extracellular/ITO heterostructure, respectively. PV features and interfacial charge separation of p-Si/BR/ITO were measured by surface photovoltage spectroscopy (SPS). The PV response value versus the external potential of the p-Si/cytoplasmic-extracellular/ITO heterostructure showed good results. SPS results reveal that the Schottky-like contact formed on the Si/BR/ ITO heterostructure, and BR orientation and its photoinduced field has a major contribution over the microscopic charge separation on the surface of the p-Si side. Energy band diagram for BR and p-Si are shown in Figure 7.5.

According to Jin et al., report, the highest photocurrent density values up to the year 2008 is around 0.2–40 pA cm^{-2} monolayer^{-1} in thin-film systems for BR based biomolecular electronics. However, Yen group [34] first fabricated a solution-based electrochemical cell without the fabrication of BR film and external bias. They observed more pronounced plasmonic field augmentation of the stationary photocurrent in their solution-based electrochemical cell. They proved that the photocurrent density is greatly improved when the electrochemical cell is open to blue light in the mien of the plasmonic Ag NPs in the BR solutions. The light-adapted BR exhibits a retinal absorption band approximately at 568 nm [35]. After illumination, the retinal chromophore isomerizes from its all-trans conformation to the 13-*cis* conformation. This is accompanied by the evolution of a series of spectrally discernible intermediates of K$_{590}$, L$_{550}$, M$_{412}$, N$_{560}$, and O$_{640}$ [36] (Figure 7.6).

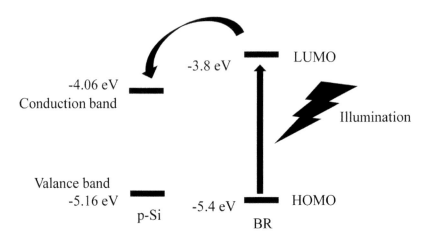

FIGURE 7.5 Energy level diagram and photoinduced charge carrier injection within BR p-Si hybrid. HOMO/LUMO of BR were determined to be –3.8 eV and –5.4 eV respectively. Conduction band of p-Si was determined to be –4.06 eV. The energy difference between the BR LUMO and the conduction band of p-Si makes the electron injection feasible.

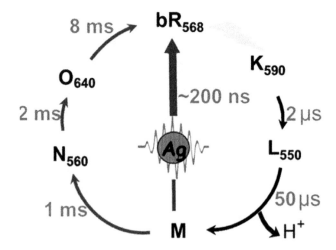

FIGURE 7.6 The plasmonic field augmentation of the blue light effect on the BR photocycle in the presence of Ag nanoparticle.

Source: Reprinted with permission from Ref. [39]. © 2010 American Chemical Society.

A proton gets released to the extracellular side of the protein, which takes place during the conversion of L to M states [37]. Consequently, a reprotonation takes place to the Schiff base which consequences in the generation of the N intermediate subsequently the generation of the O intermediate involving the 13-*cis* to all-*trans* isomerization of the retinal chromophore [38]. But when this photocycle takes place under blue light, the long-lived blue light-absorbing M_{412} intermediate is transformed to BR_{568} in a shortcut manner avoiding the photointermediate states of N_{560} and O_{640}. This shortcut significantly cuts the conventional photocycle time and impressively alters its normal or regular path of the kinetics of the photocycle. As it was already established that an intense plasmonic field is brought forth, upon the resonant excitation of the metallic NPs. Therefore, Chu et al., [39] exploiting the concept of plasmonic Ag NPs in BR studied the AgNPs plasmonic field effect on the spectroscopy and kinetics of the BR proton pumping photocycle. They used different sizes of Ag NPs in their study. They found that the 40 nm Ag NPs is about four hundred times more efficient than the 8 nm Ag NPs, as there is considerably bigger field strength and enhanced overlap between their extinction band and M intermediate absorption band.

The application of BR as a photosensitizer in BSSCs was first reported by Thavasi et al., [40]. They did their experiment for wild type BR and triple mutant of BR, [E9Q/E194Q/E204Q] BR, in mixture with widespread gap semiconductor TiO_2 for its aptness as competent light-harvesting agent in the solar cell. Wild type BR and triple mutant of BR reveals thermal stability and make it more suitable as a photosensitizer in SCs. Using molecular modeling, they have shown in their article that binding of BR to the exposed oxygen atoms of anatase TiO_2 surface, which is promising for electron transfer and concentrating by short and limited distance interactions. They successfully fabricated and measured the photocurrent density–photovoltage characteristics of the nanocrystalline TiO_2 film with wild type BR and triple mutant of BR. They demonstrated that the electrostatic potential map of the triple mutant showed a superior prospect to bind capably to TiO_2, which directed to the greater photoelectric response of the bio-solar cell composed of the triple mutant BR, than that fabricated by the wild type BR. Under illumination intensity 40 mWcm^{-2} triple mutants can produce the open-circuit photovoltage 0.35 Volt and short-circuit photocurrent density of 0.09 mAcm^{-2}.

Allam et al., [41] reported that the assembly and use of BR/TiO_2 nanotube array hybrid electrode system, fabricated using 3-mercaptopropionic as an anchoring agent onto the TiO_2 surface. A 7 mm long nanotube array of TiO_2 films were fabricated via the anodization of titanium next the sensitization of the electrodes with BR. They have announced in their article that the hybrid electrodes under AM 1.5 illumination accomplished a photocurrent density of 0.65 mAcm^{-2} with 50% increase compared to pure TiO_2 nanotubes (0.43 mA cm^{-2}) fabricated and tested under the same conditions. In the presence of redox electrolyte, the photocurrent increased to 0.87 mA cm^{-2}. They expected that the proton pumping activity of BR is the reason behind the enhancement of the photocurrent generation. This improvement in photocurrent is based on the band alignment of the HUMO-LUMO levels of BR compared to the band edge positions of TiO_2, which is shown in Figure 7.7.

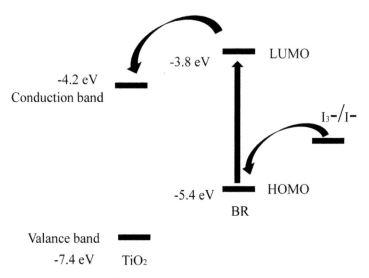

FIGURE 7.7 Energy level diagram and a possible mechanism of charge carrier injection in the BR/TiO_2 photoanode cell in the presence of redox electrolyte solution. The energy difference between the BR LUMO and the conduction band of TiO_2 makes the electron injection feasible.

Almost similar improvement in photocurrent densities like Allam et al., was reported by S. Balasubramanian et al., [42] for Pt/TiO_2 NPs amended with BR using a long time process based on the absorption of the samples in BR solution overnight. Their photoelectrochemical and transient absorption studies point toward the efficient charge transfer between BR protein molecules and TiO$_2$ NPs. However, N. Naseri group [43] declared a fast and simple one-step method for the synthesis of different TiO$_2$ nanostructures morphologies (nanoparticulate and nanotubular films) with BR species, which amplified the photocurrent density of the systems several times. I–V characteristic study of modified photoanodes reveals that the TiO$_2$ NPs with BR led to a seven times more significant photoresponsive enhancement as compared with that of the nanotubular films due to more surface areas for BR to get attached to the NPs.

Janfaza et al., [44] declare the performance of BR, which was evaluated as a sensitizer in DSSCs. BR was efficiently adsorbed on nanostructured TiO$_2$, and the achievability and proficiency of BR for use in bio-SCs were analyzed. The reason behind the successful adsorption over the TiO$_2$ is revealed from the AFM results. The morphology of the TiO$_2$ is very untidy with irregular pores that help to absorb the BR over the nanostructured surfaces. Using platinum as a counter electrode under AM1.5 irradiation, a fill factor of 0.62, a J$_{sc}$ of 0.28 mA cm^{-2}, a V$_{oc}$ of 0.51 Volt and an efficiency of 0.09% they achieved. To cut expenses, carbon counter electrode was used as a substitute of the platinum electrode. Short-circuit current of 0.21 mA cm^{-2} and open-circuit voltages of 0.52 V were obtained based on their carbon electrode (Figure 7.8).

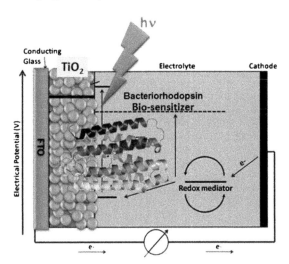

FIGURE 7.8 Schematic diagram of BR sensitized solar cell. The counter electrode (FTO/Pt)) and the working electrode (FTO/TiO$_2$/BR) are composed to form a bio-nanohybrid solar cell.

Source: Reprinted with permission from Ref. [48]. © 2013 Springer.

From the ultrafast spectroscopic studies, it was exposed that there is a very important blue-shifted I_{460} intermediate state in the photocycle of BR after the absorption of a photon. By the addition of gold nanoparticle to BR leads to control over the I_{460} intermediate state kinetics in the photocycle depending on the size and concentration of the gold NPs. Using a pump-probe, all optical switching at lower intensities and higher switching contrast compared to Cu-Pc doped thin films and graphene oxide thin films publicized by Roy et al., [45].

Guo et al., [46] first reported granum-like heterogeneous multilayers BR/AuNPs photo-voltaic stack system in the year of 2015. As granum consists of a stack of approximately 10 to 100 thylakoids having pigments which provide a large surface area to capture solar radiation capably. They present BR/AuNPs heterogeneous multilayers to reproduce this stack structure and serve as a PV stack system where solid BR layers acts as photon acceptors, and AuNPs layers increase the photocurrent density by their native surface plasmon field effect which accelerates the photocycle of BR (Figure 7.9).

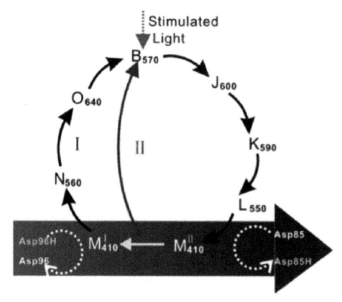

FIGURE 7.9 A bypass photocycle model of BR with several intermediate states in the whole photocycle process. After coated with AuNPs, the photocycle gets shortened from M state to B state directly owning to the plasmonic field effect of AuNPs.

Source: Reprinted with permission from Ref. [46]. © 2015 Elsevier.

Upon a cooperative coating of AuNPs and the stacking layers, the photocurrent can be effectively regulated to reach about 350 nAcm^{-2}. The proposed bypass photocycle model is strongly supported by tested accumulation of M_{412} and O_{640} intermediates during the BR photocycle process.

The achievability and proficiency of using two bacterial pigments BR and bacterioruberin found in *H. salinarum* membrane was used and investigated by Molaeirad et al., [47] for study over bio solar cell. These two pigments were adsorbed on nanoporous TiO_2 films effectively and employed as molecular sensitizers in DSSC with efficient photocurrent generation. The co-sensitization of these two bacterial pigments are balanced in their spectral responses, reveals very good PV performance compared with that of an individual DSSC. The PV performance of DSSCs based on BR and bacterioruberin sensitizers under AM1.5 irradiation showed a V_{oc} of 0.57 V, J_{sc} of 0.45 mA cm^{-2}, a fill factor of 0.62, and overall efficiency of 0.16%. They concluded that bio-SCs that use of BR and bacterioruberin are more proficient than those fabricated using only BR in DSSCs.

Experimental results of Mohammadpour et al., [48] exhibited that superficial treatment of TiO_2 nanoparticle and optimum BR loading conditions could enhance the charge transfer rate in the TiO_2/BR interface. Stuffing the nanostructured photoelectrode with more than one layer of BR can considerably lower the charge transfer rate in the TiO_2/BR interface. They showed that $TiCl_4$ treatment of TiO_2 nanoparticle surface and controlling the BR loading time could boost the charge transfer rate in a TiO_2/BR interface. Under AM1.5 irradiation, the solar-light-to-electricity conversion efficacy of the designed solar cell (Figure 7.10), showed a V_{oc} of 0.533V and J_{sc} of 1 mA cm^{-2}, and an efficiency of 0.35%.

Lu et al., [49] first reported transient spikes from the BR photocycle are prompted with NIR irradiation by incorporating BR with upconversion nanoparticles (UCNPs). They used lanthanide-doped nanocrystals as UCNPs, which emit high-energy photons under excitation by the near-infrared (NIR) light, to incorporate with BR to have NIR triggered photoelectrical response. NaYF4:Yb, Er nanocrystal is one of the supreme proficient NIR to visible up-conversion phosphors. In their study, the upconversion ability of NaYF4:Yb, Er nanocrystal was united with the BR-initiated photocycle to fabricate a system with NIR-activated photoelectric responses. A group of Indian scientist [50] also investigated the interaction between one-dimensional single-walled carbon nanotubes (SWNT) and BR as a stable electron donor-acceptor system. Due to unique optoelectronic properties, SWNTs are considered as a promising material for various photonic devices, electro-optical and biosensors. They reported optically persuaded rapid aggregation and the ensuing separation of selective SWNTs functionalized with BR, and the aggregation rate is dependent on the absorption of BR. Optically separated, bio-nano hybrid complexes show stable, favored binding with SWNTs of specific diameters. In the same year 2016, a group of Russian scientists [51] also announced that there is a large influence of Ag NPs over the photocycle of photosensitive protein BR. Spherical silver nanoparticles (AgNPs) have a surface possess plasmon resonance at 400 nm and gets red-shifted upon AgNP aggregation and/or change in their shape. Resonance energy transfer from AgNPs to BR is used to enhance the performance of light energy converters based on BR by accelerating its photocycle. They have shown that AgNPs alter the photocycle of the BR molecules located in surface-enhanced Raman scattering (SERS) active spots. AgNPs suppresses the photocycle by freezing the state of retinal in which BR has bound AgNPs.

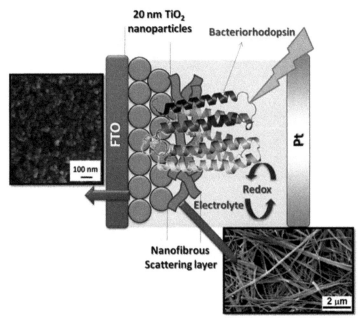

FIGURE 7.10 Schematic diagram of BR sensitized solar cell utilizing BR as biophotosensitizer and the photoelectrode containing nanoparticular TiO_2 as active medium and nanofibers as scatters.

Source: Reprinted with permission from Ref. [48]. © 2015 American Chemical Society.

As it was already established that the retinal, embedded inside the seven-transmembrane protein pocket, absorbs light at about 570 nm, which leads to the pumping of protons across the membrane from the internal cytoplasm to the external medium. The proton transfer within BR is accomplished by charge separation followed by de- and reprotonation of the retinal [52]. As the retinal chromophore is located at the middle of the BR and 2.5 nanometers from both sides of BR surfaces [53], therefore it can be considered that the distance is within Forster radius (5 nm) in energy transfer process [54]. This phenomenon is associated either with fluorescence enhancement, fluorescence quenching, or nonradiative energy transfer from nanoparticle to the retinal and is commonly related to the Förster resonance energy transfer (FRET) process [55]. By exploiting the basic phenomenon of FRET Roy et al., [56] announced that spin-dependent charge transfer between the substrate and the retinal reins the photoluminescence intensity from the NPs. They show in their experiment that the photoluminescence intensity of CdSe NPs gets quenched in wild type BR thin film on a ferromagnetic Ni-alloy substrate. They observed the momentous change in the fluorescence intensity from the CdSe NPs when spin-specific charge transfer occurs between the Ni alloy substrate and the retinal. This feature entirely dissolves when retinal protein covalent bond was hewed in wild-type BR and a BR bearing a locked retinal chromophore (Figure 7.11).

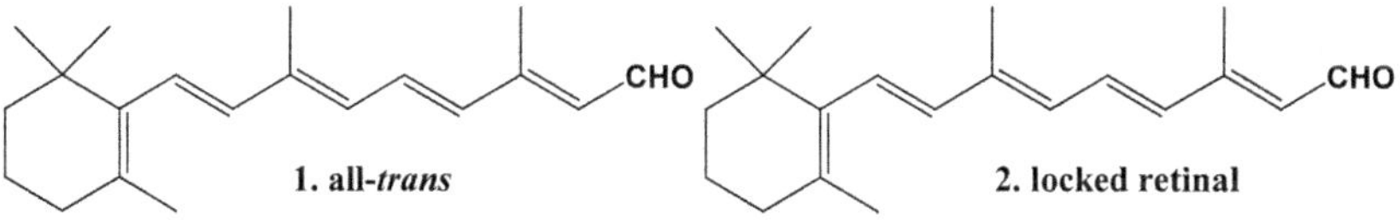

FIGURE 7.11 Structure of the retinal chromophore in 1) Wild type BR; 2) BR bearing locked retinal chromophore. In the latter, no isomerization can occur upon photoexcitation.

The article published by Krivenkov et al., [57] reported that the possibility of extending the spectral region of BR activity by means of the FRET from CdSe/ZnS quantum dots (QDs) to BR upon one- and two-photon excitation. As CdSe/ZnS QDs possess a broad spectrum of one-photon absorption and a large two-photon absorption cross-section. It is shown that QDs and BR-containing purple membranes can be used as a basis for creating bio-nanohybrid complexes bound by electrostatic interaction between a negatively charged surface of QDs and the positive charge on the surface of the purple membrane. In addition, the large two-photon absorption cross-section of QDs, which is two orders of magnitude greater than those of BR and organic dyes, opens up a means for selective two-photon excitation of synthesized bio-nano hybrid complexes. On the basis of the results, it is possible to extend the spectral region in which BR converts light energy into electrical energy can be extended from the UV to the IR region, which opens up new opportunities for the use of this promising material in PVs and optoelectronics.

7.5 CONCLUSION

BR and its membrane complex (purple membrane) are evolved to maintain and tolerate their photoactivity under strenuous circumstances, including exposure to light and oxygen, temperatures, a broad range of pH values and high ionic strength. We described the previous PV background before the nature's purple solar membrane becomes the topic of research interest and summarized the structural characteristics, photocycle and in the end we overviewed current successful examples of utilizing the BR that has been designed for the environmentally friendly bio-inspired photoenergy conversion schemes in the demand of renewable energy source, inexpensive, earth-abundant, robust and clean zero-emission fuel. As nature functions as an integrated manner; therefore, it is very important to mimic nature's integration phenomenon to meet global energy challenge and to ensure human welfare.

KEYWORDS

- dye-sensitized solar cell
- nanomaterials
- nanoparticles
- perovskite solar cells
- photovoltaics
- protein sensitized solar cell
- solar cells

REFERENCES

1. Tian, J., Sang, Y., Yu, G., Jiang, H., Mu, X., & Liu, H., (2013). A Bi_2WO_6-based hybrid photocatalyst with broad spectrum photocatalytic properties under UV, visible, and near-infrared irradiation. *Advanced Materials, 25*, 5075–5080.
2. Yella, A., Lee, H. W., Tsao, H. N., Yi, C., Chandiran, A. K., Nazeeruddin, M. K., Diau, E., Yeh, C. Y., Zakeeruddin, S. M., & Grätzel, M., (2011). Porphyrin-sensitized solar cells with cobalt (II/III)–based redox electrolyte exceed 12 percent efficiency. *Science, 334*, 629–634.
3. Ghosh, M., Mora, A. K., Nath, S., Hajra, A., & Sinha, S., (2014). Photoinduced electron transfer in metallo-octaethylporphyrin (donor)-2-nitrofluorene (acceptor) systems in polar acetonitrile liquid medium. *J. Photochem. Photobiol. A: Chem., 290*, 94–100.
4. Ghosh, M., Roy, B., Majhi, K., Mora, A. K., Nath, S., & Sinha, S., (2015). Fluorescence quenching of 9-cyanoanthracene by metallo-octaethylporphyrins in cyanobenzene. *J. Porphyr. Phthalocya., 19*, 1063–1071.
5. Roy, B., Ghosh, M., & Sinha, S., (2014). Solvent dependent photophysical properties of free base tetrapyridylporphyrin. *J. Mol. Liq., 200*, 323–328.
6. Ghosh, M., Mora, A. K., Nath, S., Chandra, A. K., Hajra, A., & Sinha, S., (2013). Photophysics of Soret-excited free base tetraphenylporphyrin and its zinc analog in solution, *Spectrochim. Act. A Mol. Biomol. Spectrosc., 116*, 466–472.
7. Spudich, J. L., Yang, C. S., Jung, K. H., & Spudich, E. N., (2000). Retinylideneproteins: Structures and functions from Archaea to Humans. *Annu. Rev. Cell Dev. Biol., 16*, 365–392.
8. O'Regan, B., & Gratzel, M., (1991). A low cost, high efficiency solar cell based on dye sensitized colloidal TiO2 films. *Nature, 5*, 77–740.
9. Hagfeldt, A., Boschloo, G., Sun, L., Kloo, L., & Pettersson, H., (2010). Dye-sensitized solar cells. *Chem. Rev., 11*, 6595–6663.
10. Gratzell, M., (2001). Photoelectrochemical cells. *Nature, 414*, 338–344.
11. Pan, H., Misra, N., Ko, S. H., Grigoropoulos, C. P., Miller, N., Haller, E. E., & Dubon, O., (2009). Melt-mediated coalescence of solution-deposited ZnO nanoparticles by excimer laser annealing for thin-film transistor fabrication. *Appl. Phys. A, 94*, 111–115.
12. Law, M., Greene, L. E., Johnson, J. C., Saykally, R., & Yang, P., (2005). Nanowire dye-sensitized solar cells. *Nat. Mater., 4*, 455–459.
13. Jiang, C. Y., Sun, X. W., Lo, G. Q., Kwong, D. L., & Wang, J. X., (2007). Improved dye-sensitized solar-cells with a ZnO-nanoflower photoanode. *Appl. Phys. Lett., 90*, 263501-1–263501-3.
14. Martinson, A. B. F., Elam, J. W., Hupp, J. T., & Pellin, M. J., (2007). ZnO nanotube based dye-sensitized, solar cells. *Nano Lett., 7*, 2183–2187.
15. Wang, X. D., Ding, Y., Summers, C. J., & Wang, Z. L., (2004). Large-scale synthesis of six-nanometer-wide ZnO nanobelts. *J. Phys. Chem. B, 108*, 8773–8777.
16. Kar, S., Dev, A., & Chaudhuri, S., (2006). Simple solvothermal route to synthesize ZnO nanosheets, nanonails, and well-aligned nanorod arrays. *J. Phys. Chem. B, 110*, 17848–17853.

17. Fu, M., Zhou, J., Xiao, Q. F., Li, B., Zong, R. L., Chen, W., & Zhang, J., (2006). ZnO nanosheets with ordered pore periodicity via colloidal crystal template assisted electrochemical deposition. *Adv. Mater., 18*, 1001–1004.
18. Ko, S. H., Lee, D., Kang, H. W., Nam, K. H., Yeo, J. Y., Hong, S. J., Grigoropoulos, C. P., & Sung, H. J., (2011). Nanoforest of hydrothermally grown hierarchical ZnO nanowires for a high efficiency dye-sensitized solar cell. *Nano Lett., 11*, 666–671.
19. Shin, S. S., Yeom, E. J., Yang, W. S., Hur, S., Kim, M. G., Im, J., Seo, J., Noh, J. H., & Seok, S., (2017). Colloidally prepared La-doped BaSnO$_3$ electrodes for efficient, photostableperovskite solar cells, *Science, 356*, 167–171.
20. D'Souza, F., Das, S. K., Zandler, M. E., Sandanayaka, A. S., & Ito, O., (2011). Bionano donor-acceptor hybrids of Porphyrin, ssDNA, and semiconductive single-wall carbon nanotubes for electron transfer via porphyrinexcitation. *J. Am. Chem. Soc., 133*, 19922–19930.
21. Lerner, M. B., D'Souza, J., Pazina, T., Dailey, J., Goldsmith, B. R., Robinson, M. K., & Johnson, A. T., (2012). Hybrids of a genetically engineered antibody and a carbon nanotube transistor for detection of prostate cancer biomarkers. *ACS Nano, 6*, 5143–5149.
22. Patwardhan, S. V., Mukherjee, N., Steintz-Kannan, M., & Clarson, S. J., (2003). Bioinspired synthesis of new silica structures. *Chem. Commun. (Camb.), 10*, 1122–1123.
23. Katz, E., (2006). Bioelectronics. *Electroanalysis, 18*, 1855–1857.
24. Walter, J. M., Greenfield, D., & Liphardt, J., (2010). Potential of light-harvesting proton pumps for bioenergy applications, *J. Curr. Opin. Biotechnol., 21*, 265–270.
25. Lu, Y., (2005). Photoelectric performance of bacteria photosynthetic proteins entrapped on tailored mesoporous WO$_3$-TiO$_2$ films. *Langmuir, 21*, 4071–4076.
26. Oesterhelt, D., & Stoeckenius, W., (1971). Rhodopsin-like protein from the purple membrane of halobacterium halobium, *Nature New Biology, 233*, 149–152.
27. Jap, B. K., Maestre, M. F., Hayward, S. B., & Glaeser, R. M., (1983). Peptide-chain secondary structure of bacteriorhodopsin. *Biophys J., 43*, 149–160.
28. Nonella, M., Windmuth, A., & Schulten, K., (1991). Structure of bacteriorhodopsin and in situ isomerization of retinal: A molecular dynamics study. *Photochem. Photobiol. 54*, 937–948.
29. Kuhlbrandt, W., (2000). Bacteriorhodopsin-the movie. *Nature, 406*, 569–570.
30. Govindjee, R., Balashov, S. P., & Ebrey, T. G., (1990). Quantum efficiency of the photochemical cycle of bacteriorhodopsin. *Biophys. J., 58*, 597–608.
31. Mak-Jurkauskas, M. L., Bajaj, V. S., Hornstein, M. K., Belenky, M., Griffin, R. G., & Herzfeld, J., (2008). Energy transformations early in the bacteriorhodopsin photocycle revealed by DNP-enhanced solid-state NMR. *Proc. Natl. Acad. Sci. U.S.A, 105*, 883–888.
32. Lanyi, J. K., & Pohorille, A., (2001). Proton pumps: Mechanism of action and applications. *Trends Biotechnol., 19*, 140–144.
33. Li, L. S., Xu, T., Zhang, Y. J., Jin, J., Li, T. J., Zou, B. S., & Wang, J. P., (2001). Photovoltaic characteristics of BR/p-silicon heterostructures using surfacephotovoltage spectroscopy. *J. Vac. Sci. Technol. A, 19*, 1037–1041.
34. Yen, C. W., Chu, L. K., & El-Sayed, M. A., (2010). Plasmonic field enhancement of the bacteriorhodopsin photocurrent during its proton pump photocycle. *J. Am. Chem. Soc., 132*, 7250, 7251.

35. Rehorek, M., & Heyn, M. P., (1979). Binding of all-trans retinal to the purple membrane. evidence for cooperativity and determination of the extinction coefficient. *Biochemistry, 18*, 4977–4983.
36. Birge, R. R., Gillespie, N. B., Izaguirre, E. W., Kusnetzow, A., Lawrence, A. F., Singh, D., Song, Q. W., Schmidt, E., Stuart, J. A., Seetharaman, S., & Wise, K., (1999). Biomolecular electronics: Protein-based associative processors and volumetric memories. *J. Phys. Chem. B, 103*, 10746–10766.
37. Spassov, V. Z., Luecke, H., Gerwert, K., & Bashford, D., (2001). pKa calculations suggest storage of an excess proton in a hydrogen-bonded water network in bacteriorhodopsin, *J. Mol. Biol., 312*, 203–219.
38. Braiman, M. S., Bousche´, O., & Rothschild, K. J., (1991). Protein dynamics in the bacteriorhodopsin photocycle: Submilli second Fourier transform infrared spectra of the L, M, and N photointermediates, *Proc. Natl. Acad. Sci. U.S.A., 88*, 2388–2392.
39. Chu, L., Yen, C., & El-Sayed, M. A., (2010). On the mechanism of the plasmonic field enhancement of the solar-to-electric energy conversion by the other photosynthetic system in nature (Bacteriorhodopsin): Kinetic and spectroscopic study. *J. Phys. Chem. C, 114*, 15358–15363.
40. Thavasi, V., Lazarova, T., Filipek, S., Kolinski, M., Querol, E., & Kumar, A., (2009). Study on the feasibility of bacteriorhodopsin as bio-photosensitizer in excitonic solar cell: A first report. *J. Nanosci. Nanotechnol., 9*, 1679–1687.
41. Allam, N. K., Yen, C. W., Near, R. D., & El-Sayed, M. A., (2011). Bacteriorhodopsin/ TiO_2 nanotube arrays hybrid system for enhanced photoelectrochemical water splitting. *Energy Environ. Sci., 4*, 2909–2914.
42. Balasubramanian, S., Wang, P., Schaller, R. D., Rajh, T., & Rozhkova, E. A., (2013). High-performance bioassisted nanophotocatalyst for hydrogen production. *Nano Lett., 13*, 3365–3371.
43. Naseri, N., Janfaza, S., & Irani, R., (2015). Visible light switchable BR/TiO_2 nanostructured photoanodes for bio-inspired solar energy conversion. *RSC Adv., 5*, 18642–18646.
44. Janfaza, S., Molaeirad, A., Mohamadpour, R., Khayati, M., & Mehrvand, J., (2014). Efficient bio-nano hybrid solar cells via purple membrane as sensitizer. *Bio. Nano. Sci., 4*, 71–77.
45. Roy, S., & Yadav, C., (2014). All-optical sub-ps switching and parallel logic gates with bacteriorhodopsin (BR) protein and BR-gold nanoparticles, *Laser Phys. Lett., 11*, 125901-1–125901-8.
46. Guo, Z. B., Liang, D. W., Rao, S. Y., & Xiang, Y., (2015). Heterogeneous bacteriorhodopsin/gold nanoparticle stacks as a photovoltaic system. *Nano Energy, 11*, 654–661.
47. Molaeirad, A., Janfaza, S., Karimi-Fard, A., & Mahyad, B., (2015). Photocurrent generation by adsorption of two main pigments of Halobacterium salinarum on TiO_2 nanostructured electrode. *Biotechnol. Appl. Biochem., 62*, 121–125.
48. Mohammadpour, R., & Janfaza, S., (2015). Efficient nanostructured biophotovoltaic cell based on bacteriorhodopsin as biophotosensitizer. *ACS Sustainable Chem. Eng., 3*, 809–813.
49. Lu, Z., Wang, J., Xiang, X., Li, R., QiaoY., & Li, C. M., (2015). Integration of bacteriorhodopsin with upconversion nanoparticles for NIR-triggered photoelectrical response. *Chem. Commun., 51*, 6373–6376.

50. Sharma, A., Prasad, E. S., & Chaturvedi, H., (2016). Photon induced separation of bio-nano hybrid complex based on carbon nanotubes and optically active bacteriorhodopsin, *Optical Materials Express*, *6*, 986–992.
51. Mochalova, K., Solovyevaa, D., Chistyakova, A., Zimkab, B., Lukashevc, E., Nabieva, I., & Oleinikova, V., (2016). Silver nanoparticles strongly affect the properties of bacteriorhodopsin, a photosensitive protein of *Halobacterium salinarium* purple membranes. *Materials Today: Proceedings*, *3*, 502–506.
52. Bondar, A. N., Elstner, M., Suhai, S., Smith, J. C., & Fischer, S., (2004). Mechanism of primary proton transfer in bacteriorhodopsin. *Structure*, *12*, 1281–1288.
53. Rakovich, A., Sukhanova, A., Bouchonville, N., Lukashev, E., Oleinikov, V., Artemyev, M., et al., (2010). Resonance energy transfer improves the biological function of bacteriorhodopsin within a hybrid material built from purple membranes and semiconductor quantum dots. *Nano Lett.*, *10*, 2640–2648.
54. Clapp, A. R., Medintz, I. L., & Mattoussi, H., (2006). Förster resonance energy transfer investigations using quantum-dot fluorophores. *Chem. Phys. Chem.*, *7*, 47–57.
55. Mitra, R. D., Silva, C. M., & Youvan, D. C., (1996). Fluorescence resonance energy transfer between blue-emitting and red-shifted excitation derivatives of the green fluorescent protein. *Gene*, *173*, 13–17.
56. Roy, P., Kantor-Uriel, N., Mishra, D., Dutta, S., Friedman, N., Sheves, M., & Naaman, R., (2016). Spin-controlled photoluminescence in hybrid nanoparticles purple membrane system. *ACS Nano*, *10*, 4525–4535.
57. Krivenkov, V. A., Samokhvalov, P. S., Bilan, R. S., Chistyakov, A. A., & Nabiev, I. R., (2017). *Optics and Spectroscopy*, *122*, 36–41.
58. Chen, Y., Wang, H., Dang, B., Xiong, Y., Yao, Q., Wang, C., Sun, Q., & Jin, C., (2017). Bio-Inspired nacre-like nanolignocellulose-poly (vinyl alcohol)-TiO$_2$ composite with superior mechanical and photocatalytic properties. *Sci. Rep.*, *7*, 1823.

CHAPTER 8

ZnS/ZnO Nanocomposite in Photovoltaics: A Computational Study on Energy Conversion

KEKA TALUKDAR

Department of Physics, Nadiha High School, Durgapur – 713211, West Bengal, India, E-mail: keka.talukdar@yahoo.co.in

ABSTRACT

Energy is the biggest need of today's world for development in any field. The natural sources of power are being used every day, and it is expected that within 30 years, the stock will come to an end. Renewable, clean sources of energy are one of the most important subject matter of study to both of the experimental and theoretical researchers. While experimental work is mostly to find new materials and methods, the theoretical studies are taken up for showing the way of gaining most optimum process and hence to reduce the cost of trial and error methods until some established method and material are developed. In photovoltaics (PVs) many nanomaterials (NMs) are being exploited for maximum conversion of light of which zinc sulfide is one. But due to the high bandgap of ZnS, ZnS/ZnO nanocomposite is modeled, and then it is subjected to the simulation by density functional approach, and many properties of this material related to energy conversion are discussed.

8.1 INTRODUCTION

Nanomaterials (NMs), according to the ISO definition, is a material with at least one external dimension in the nanoscale or having internal structure, or surface structure in the nanoscale, where nanoscale is defined as the size range from 1 nm to 100 nm. Nanoparticles (NPs) are minute pieces of matter with defined physical boundaries which can move as a unit having all three dimensions in the nanoscale. Nanoclusters have at least one dimension between 1 and 100 nanometers and a narrow size distribution.

8.1.1 NANOMATERIALS (NMS)

NMs may be of different types. Nanopowders are agglomerates of ultrafine particles, NPs, or nanoclusters. Nanometer-sized single crystals, or single-domain ultrafine particles, are often referred to as nanocrystals. NPs were used as dye materials in ceramics by ancient people [1]. Systematic experiments conducted on NMs have been started from the well known Faraday experiments [2] in 1857.

From the dimensional point of view, they may be classified as zero dimensional nanostructures such as metallic, semiconducting and ceramic NPs; one-dimensional nanostructures such as nanowires, nanotubes, and nanorods; two-dimensional nanostructures such as thin films. Besides this, individual nanostructures and ensembles of the nanostructures form high dimension arrays, assemblies, and superlattices. An interesting class of nanoparticles is known as fractal clusters of extremely small particles. Examples are most of the amorphous silica particles (known as white shoot) and amorphous Fe_2O_3 particles. A QD is defined as a crystalline NP that exhibits size-dependent properties due to quantum confinement effect on the electronic states. Table 8.1 gives the classification of NMs with their size and example.

TABLE 8.1 Classification of Nanomaterials

Types of NMs	Approximate Size	Materials
(a) Nanocrystals and clusters (zero-dimensional)	1–50 nm diameter	Nanopower, metal or semiconducting cluster
(b) Nanowires and nanotubes (one dimensional)	1–100 nm diameter	Nanotubes, nanowires, fibers, and rods
(c) Surfaces and thin films (two dimensional)	Several nm^2-μm^2	Nano film and nanosheets like graphene sheets
(d) Three-dimensional structures	Several nm in all dimensions	Metal, semiconductor, magnetic particles, buckyballs

Adapted and modified from Ref. [3]

8.1.2 DIFFERENCES IN PROPERTIES BETWEEN NANOMATERIALS (NMS) AND BULK MATERIALS

NMs have structural features in between atoms and the bulk materials. While most microstructured materials have properties similar to the corresponding bulk materials, the properties of materials with nanoscale dimensions are significantly different from those of atoms and bulks materials. Calculating the surface to volume ratio for a particle, the surface is found to increase with decreasing particle size. In thermodynamic applications, this property is of immense importance. Again, the area per mole greatly increases for an NM as it varies inversely in proportion to the particle diameter. So surface-related properties are greatly improved.

Also, solving the Fick's law for diffusion, we get diffusion scaling law which directly implies that when the grain size is in the nanometer range, the homogenization time is reduced to one of the milliseconds from several hours. Thus a sensor, made of an NM shows far better response compared to a sensor made with material of grain size in the micrometer range.

Thus, the nanometer size of the materials provides them: (i) a large fraction of surface atoms; (ii) high surface energy; (iii) spatial confinement; (iv) reduced imperfections, which do not exist in the corresponding bulk materials. The defects like dislocations, micro twins, and impurity precipitates, etc. may not influence the mechanical properties as they cannot multiply or such imperfections may not at all be present in the nanorange particles. Mechanical properties of the NMs are observed to be appreciably different from that of the bulk materials, particularly regarding their hardness, elastic modulus, fracture toughness, scratch resistance, fatigue strength, etc. Moreover, the external surfaces of NMs also have less or no defects, compared to the bulk materials, serving to enhance their mechanical properties [4, 5]. Large surface area of NPs also results in a lot of interactions between the intermixed materials in nanocomposites, leading to special properties such as increased strength and/or increased chemical/heat resistance.

8.1.3 NANOMATERIALS (NMS) IN PHOTOVOLTAICS (PVS)

Due to their special structure and properties, NMs are promising materials to improve the performance of the fabricated devices for energy conversion.

Their ability for light trapping and collecting photons of light are amazing compared to the materials of the macro world. Moreover, fabricating devices with NMs s is less costly, great life cycle and less power dissipation as heat. Polymer NPs [6–9], QD sensitized NPs [10–13], ZnO [14, 15], CdSe [16–18] crystals have huge potential for using in photovoltaic (PV) devices. Nanostructured materials are nowadays extensively used for designing p-n junction or p-i-n junction for PV use. Such NM based junctions have high light absorption quality and much-improved charge carrier generation efficiency [19]. How the emerging NMs are changing the world of energy and what is their life cycle and environmental impact, are well-reviewed by Kim et al., [20].

Generally, silicon is used in PVs for their low cost, availability in nature, non-toxic character. In thin-film solar cells (SCs) silicon is used, but light absorption is difficult in such cells compared to the wafer-based silicon SCs [21, 22]. Using metal NPs, surface plasmon resonance (SPR) may be increased in a particular frequency range. Thus absorption power may also be increased [24]. Silver NPs are found to be an effective means to enhance the SPR in the thin film so that light absorption efficiency is increased and at the same time, light scattering power in broadband also increases. 23% increase of the power conversion efficiency (PCE) was observed in such method by Chen et al., [25]. Silicon nanowire array SC with p-n junction in each wire shows improved PV activity [26–29]. Like silicon, silver NPs are used in clusters in heterojunction PVs [30]. Silicon NP is still an interesting material for their photoluminescence characteristics. Silicon particles lesser than 10nm size can act as a secondary light source which converts incident photon energy and converts them into visible light. Its efficiency increases by 1.64% more when conventional SC is coated with silicon NPs. [31]

SC fabricated with different materials show gradually improved efficiency up till now. SC made of blends of a branch of CdSe NPs, and the conjugated polymer is reported to be an upgraded PV device [32].

8.1.4 PLASMONICS IN PHOTOVOLTAIC (PV)

While the conventional SCs are based on silicon wafers, which are 180–300µm, the popularity of thin-film made SCs are much due to low production cost. For thin-film SCs, the thickness is 1–2 µm. Light trapping in such a narrow space is a challenging job indeed, mainly for the indirect bandgap semiconductor silicon, the near bandgap light absorbance is low [33]. With the increase of carrier collection length, the efficiency of the PV cell should not decrease for those new materials. But new methods are necessary. Some ly efficient materials like CdTe, copper indium gallium selenide (CIGS) or amorphous silicon can be used, but these have also failed to meet the global demand of large scale production according to the need. [34]. Up till now, by using a textured surface, the effective path length of light is increased by multiple reflections in the inner portion and hence increasing light trapping [35]. Plasmonic nanostructures, on the other hand, can increase the performance of SCs considerably by generating excitons at the interface of metal and dielectric for thin-film architecture [36].

Application of plasmonics is used for the improved absorption, Collection of photocarrier, and efficiency enhancement. Localized SPR (LSPR) is limited by the ohmic loss of the NMs. Changing the size and shape of the NPs; the form of the trapped light wave can be changed. Small particles initiate near field wave while the bigger particles scatter the incident photons in the form of absorption or re-emission. PV energy conversion efficiency can also be increased by hot-electron injection [37]. Enhancement of the performance of organic PVs (OPV) by plasmonic nanostructures is well-reviewed in many literature [38–40].

8.1.5 CNT AND GRAPHENE

In SC, CNTs and graphene are very fascinating material for their excellent mechanical, electrical, and optical properties. The fact that the huge available surface, CNTs, and graphene can be modified and interfaced with other materials chemically. This has helped the researchers to build CNT or graphene interfaced with other materials to prepare high-efficiency SC with minimum loss as heat energy. Often we get a better result with SWCNT and amorphous silicon hybrid with conversion efficiency increased by 1.5% [41]. Surface treatment of amorphous silicon and chemically doping of CNT further increase the efficiency. The PV conversion efficiency of 13.3% was achieved in the organo-lead iodide perovskite SC by creating a CNT/graphene buffer layer [42]. Electron mobility in CNT or graphene is very high with a lower percolation threshold. For a thicker device, the high carrier mobility helps to achieve high efficiency without losing internally like other organic SC. Graphene, compared to the CNTs exhibits more electron mobility that exceeds 60,000 cm^{2}V^{-1}s^{-1}. But the unreliability of the CVD process to grow uniform stacked structure and formation of lattice trapped states and air pockets in the graphene multilayer limit the conductivity of the device made of graphene. Introducing a self-assembled monolayer inside the graphene layers can increase the conductivity to a large amount [43–44].

8.1.6 EFFECT OF DOPING

Chemical doping may enhance the performance of the SCs. After doping graphene with bis(trifluromethanesulphony) amide, the conversion efficiency of a single layer graphene/n-Si Schottky junction SC improves by a factor 4.5 than that of an undoped graphene and reaches 8.6% [45]. Nitric acid infiltration of CNT networks combined with n-type silicon wafer reduces the internal resistance and enhances the conversion efficiency up to 13.8% by improving charge separation and transportation [46]. A conversion efficiency of 17.8% is normally obtained for an inverted perovskite SC with indium tin oxide (ITO). Whereas MoO$_3$ doped CNT and graphene-based calls reach efficiency up to 12.8% and 14.2% respectively. Due to lack of better graphene transfer method, the graphene driven SCs still suffer from lower reproducibility, and they are more susceptible to strain. The difference between the CNT and graphene-based SCs is due to their morphology too [47]. MoO$_x$ doping is however unstable and can be used for stable doping for CNT and graphene after thermal annealing of MoO$_x$-CNT composite. After annealing, they form a durable electrode thin film [48]. The main weak point the perovskite SC is thermal instability, which can be overcome by replacing the organic hole transport material by polymer functionalized SWCNTs. But graphene can only be used successfully after improving the transfer process [49]. Nitrogen doping in vertically aligned CNTs and graphene facilitates oxygen reduction reaction, which is helpful for electrocatalytic activity. Nitrogen doping of vertically aligned CNTs was successfully observed by Wei et al., [50]. In the same way, nitrogen doping was attempted by several researchers by different processes like CVD, plasma treatment of nitrogen, edge functionalization, or thermal treatment. Doping with boron, sulfur, phosphorous, and both boron and nitrogen is a very promising step towards the development of a metal-free catalyst for oxygen reduction reaction [51].

8.1.7 QUANTUM DOTS (QDS)

While the dye-sensitized SCs (DSSCs) got sufficient attention compared to the conventional energy sources due to their cost effectiveness, [52] the performance of SCs received special focus in respect of the conversion efficiency for the quantum-dot-sensitized SCs (QDSSCs). For the ease of preparation, bandgap tunability, excellent electrical and optical properties, high absorption coefficient, and ability to generate multiple excitons, DSSCs with graphene can be successfully designed. Almost all components are now being prepared with graphene as its component [53]. Different semiconductor dot SCs were discussed nicely in the review of Kamat [54]. Much lower PCE of 1.83% was obtained by with CdSe QD light absorber [55]. According to the authors, the possible reason might be the charge transfer of generated electrons by the photocatalytic reaction to the aqueous electrolyte. Compared to that a much better PCE of 4.13% and 6.11% was obtained by metal-free counter electrode made of TiNNPs on a carbon CNT-graphene hybrid QDSSCs with polysulphide electrolyte [56] and InP QD/N-doped MWCNT hybrid [57]. In the later work, the synergistic effect of InP QDs and N doped MWCNTs to exciton generation for charge transport and charge collection from the isolated electrons respectively improves the PCE to a much higher value. Further increase of PCE to 9.18% by the inclusion of graphene layers in hybrid colloidal QDSCs made of lead sulfide and single-layer graphene. Spontaneous charge transfer occurs to graphene by photoluminescence activity of QDs, and good transport properties are exhibited by this hybrid structure [58].

8.1.8 APPLICATIONS OF ZNS, ZNO, ZNS-ZNO COMPOSITES IN PHOTOVOLTAICS (PVS)

Both ZnO and ZnS nanostructures have a high bandgap of 3.44 Å and 3.72 Å. So they can hardly be used in PVs. But by making composites, the bandgap can be lowered. The high cost of silicon-based SCs and difficulties to prepare highly purified silicon; the motivation of researchers was to find new materials. CdTe, CdSe, though proved their efficiency as SCs, their low abundance in nature restricted their use commercially as Sc materials. In this context, the oxide materials and their composites are more abundant and less toxic. For example, ZnO is widely synthesized for definite purposes like nanobelts, particles used in sunscreen lotion, etc.

The only shortcoming of this material is in the nanosize is its relatively high bandgap. But due to different methods used in Nanotechnology, the bandgap can be engineered nowadays. Several composites [59, 60] and heterostructure, [61–63] hybrid formation [64, 65] method are reported till date to reduce the bandgap of ZnO. Fluorescence measurement of core-shell hexagonal structured ZnS/ZnO/CdS nanocomposites shows that the composite structure emits light in visible range [66]. Excellent photocatalytic activities are observed for ZnO-ZnS/C, prepared by simple heat treating method [67]. Coating ZnO with submonolayer ZnS thin film lowers the bandgap in the surface of the ZnO from 3.4 eV to 2.8 eV [68]. The energy to excite electrons from the surface is reduced due to layer formation and hence increases the energy conversion efficiency of the SC. Cauliflower like CdS/ZnS/ZnO, prepared by hydrothermal method, exhibits very improved absorption property along with optimum energy gap [69]. The optimal direction towards the proper bandgap, absorption, and carrier localization can be obtained by theoretical calculations by DFT [70]. Schrier et al., in their theoretical calculation showed how quantum-well-like and nanowire-based ZnO/ZnS heterostructure could be controlled to produce a bandgap of 2.07 eV. So this nanostructure can be used for photovoltaic applications as such nanostructure exhibits a conversion efficiency of 23%. Shuai and Shen also reported that core-shell nanostructure ZnO/ZnS, prepared by hydrothermal method shows very high optical and sensing properties [71]. Hereby DFT, different properties of ZnO/ZnS composite suitable for photovoltaic applications, are analyzed by three different methods.

8.2 THEORY AND MODELING

In the research of PVs, the main focus of this century to cut the cost, effort, and time to build a long-lasting highly efficient SC. This is to save the world from fossil fuel to burn and produce low cost and renewable energy conversion device. As discussed above, only the thin film technology can successfully produce low-cost SCs. In this context, it is necessary to understand the material properties deeply and relate nanoscale properties to the macro-scale device structure. For this, ab initio approaches are important. To reduce the cost of production, and thus the wastage of time in the trial and error methods, the DFT is finding its place for successful prediction of ground-state properties of materials.

In the following study, a p-i-n structure is simulated using Density Functional Theory (DFT) in Local Density Approximation (LDA). The SIESTA code given in nanohub.org is used for simulation. Transport property is modeled by the ADEPT model. For the success of DFT to calculate a property, the appropriate approximation is required. The main DFT theory [72, 73] is discussed in detail in chapter 10. Here Kohn-Sham equation and LDA is briefly discussed.

8.2.1 DENSITY FUNCTIONAL THEORY (DFT)

Wavefunction based methods are not suitable for a large system of molecules. DFT is based on the single quantity, density, and we don't need the exact wavefunction to know. It is a variable of three coordinates only. Hence the calculations need not be of high cost, but still very accurate results are obtained. In actual, DFT approximation is almost like the Hartree-Fock method.

According to Hohenberg-Kohn theorem [74], the Hamiltonian of a many-electron system can be written as:

$$H = T + U_{e-e} + V_{ext} \quad (1)$$

where T is the kinetic energy, U_{e-e}, the electron-electron interaction, and V_{ext}, the external potential. The system consists of interacting particles (electrons). For a nondegenerate ground state, the authors showed that only one external potential is possible that generates a ground-state electron density n(r). So electron density is given as:

$$n(r) = N \int |\Psi(r_1, r_2, r_3, \ldots r_N)|^2 dr_2 \ldots dr_N \quad (2)$$

where Ψ is the ground state wavefunction.

By simple approach, it can be proved that two different potentials cannot have the same charge density.

Ground state energy E can also be found in terms of ground-state charge density. It can be written as: $E[n(r)]$

$$\begin{aligned} E[n(r)] &= \langle \Psi | T + U_{e-e} + V_{ext} | \Psi \rangle \\ &= \langle \Psi | T + U_{e-e} | \Psi \rangle + \langle \Psi | V_{ext} | \Psi \rangle \\ &= F[n(r)] + \int n(r) V_{ext}(r) dr \end{aligned} \quad (3)$$

where $F[n(r)]$ is the universal functional, which is the functional of charge density but not of an external potential; $n(r)$ is the energy density; F is the universal functional of n, which is independent of an external potential. The correlation energy of a uniform electron gas can explain F completely.

In Kohn-Sham equation [75], F is expressed as the addition of three terms. In this equation, E can be written as:

$$E[n(r)] = T_s[n(r)] + \frac{1}{2}\iint \frac{n(r)n(r')}{|r-r'|}drdr' + E_{xc}[n(r)] + \int n(r)V_{ext}(r)dr$$

$$= -\frac{1}{2}\sum_i^N \langle \psi_i|\nabla^2|\psi_i\rangle + \frac{1}{2}\sum_i^N\sum_j^N \iint |\psi_i(\vec{r}_1)|^2 \frac{1}{r_{12}}|\psi_j(\vec{r}_2)|^2 d\vec{r}_1 d\vec{r}_2 + E_{xc}[n] - \sum_i^N\sum_A^M \int \frac{Z_A}{r_{1A}}|\psi_i(\vec{r}_1)|^2 d\vec{r}_1 \quad (4)$$

where $T_s[n(r)]$ is the kinetic energy of the non-interacting electron gas with density $n(r)$. The second term is the electrostatic Hartree energy, and the third term is the exchange-correlation energy.

$$T_s[n(r)] = -\frac{1}{2}\sum_{i=1}^N \int \psi_i^*(r)\nabla^2 \psi_i(r)dr \quad (5)$$

If an effective potential

$$V_{eff}(r) = V_{ext}(r) + \int \frac{n(r')}{|r-r'|}dr' + V_{xc}(r) \quad (6)$$

then Kohn-Sham equation is rewritten as simply

$$\frac{\delta T_s[n(r)]}{\delta n(r)} + V_{eff}(r) = \mu \quad (7)$$

The effective potential is called from now V_{KS}.

$$2\sum_{i=1}^{N/2}\varepsilon_i = T_s[n] + \int drn(r)V_{KS}(r)$$

$$= T_s[n] + \int drdr'\frac{n(r)n(r')}{|r-r'|} + \int drn(r)V_{xc}(r) + \int drn(r)V_{ext}(r) \quad (8)$$

For interacting system

$$E = 2\sum_{i=1}^{N/2}\varepsilon_i - \frac{1}{2}\int drdr'\frac{n(r)n(r')}{|r-r'|} - \int drn(r)V_{xc}(r) + E_{xc}[n] \quad (9)$$

Eigenvalues thus obtained, is not so significant as this is a sum of single-particle energy.

However, if the form of exchange potential is known, the results will be exact. But it's extremely difficult to know the exact form of the exchange potential. Some approximations on the potential can efficiently solve the problem. Of them, LDA is a very simple one.

8.2.2 LOCAL DENSITY APPROXIMATION (LDA)

It is assumed that for uniform electron gas exchange electron correlation energy density at every point in space is the same. The two terms are exchange energy, and the second is correlation energy.

$$E_{xc}^{LDA}[n] = E_{x}^{LDA}[n] + E_{c}^{LDA}[n] \qquad (10)$$

The expression of exchange energy is

$$E_{x}^{LDA}[n] = C \int n^{\frac{4}{3}}(r)dr \qquad (11)$$

The correlation energy can be calculated from Quantum Monte Carlo calculations [76] in the electron gas model. By LDA model, many physical properties of the materials can be found with a satisfactory level of accuracy.

8.3 RESULTS AND DISCUSSION

ZnO/ZnS composite structure shows very interesting results concerned to various properties of its PV property. Three methods are used differently to calculate the efficiency of SC.

8.3.1 SIESTA RESULTS

In SIESTA software, ab initio density functional method is used with Kohn-Sham self-consistent equations in LDA. Here pseudopotentials are used with atomic orbital as a basis set. Before stating and explaining the results, the important parameters used for simulation are given below:

- Thickness of p-type and n-type semiconductor region–400 nm.
- Thickness of i region–200nm.
- Hole mobility–500 $cm^2V^{-1}S^{-1}$, electron mobility–1500 $cm^2V^{-1}S^{-1}$.
- The intrinsic part is flooded with carriers of p and n region.

Figure 8.1 shows the photon flux per energy level vs. energy graph while in the same figure, the variation of the absorption coefficient is plotted in the blue line. Very high values of absorption coefficient prove that the material is very good light absorber. The absorption coefficient decreases as the photon energy increases. However, the slope of the curve is high after energy 3–4 eV.

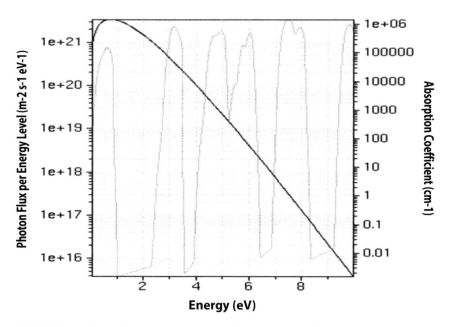

FIGURE 8.1 Photon flux per energy level and absorption coefficient vs. energy.

In Figure 8.2, the incident flux against photon energy is plotted along with the absorption coefficient up to 9 eV. The absorption coefficient is close to 0.001 cm^{-1} at 7eV energy, after which it remains almost the same. Up to this range of energy flux and photon flux fluctuates and reaches a high value. Maximum energy flux obtained is 600 Wm^{-2} eV^{-1}.

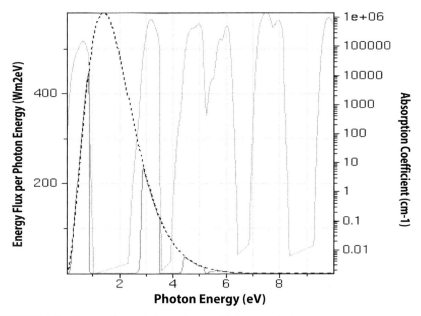

FIGURE 8.2 Energy flux and absorption coefficient vs. photon energy.

Absorption is the property of the material which is related to the method of transferring energy from the incident photon to the molecules of the material. Absorptance is the ratio of the energy transferred to the material when exposed to an incident photon flux and the total incident energy. So absorptance A is defined as

$$A = 1 - T - R \tag{12}$$

where T and R denote the transmission and reflection, respectively.

Absorbance Abs is defined as

$$\begin{aligned}\text{Abs} &= -\log_{10}(T) \\ &= -\log_{10}(R)\end{aligned} \tag{13}$$

Coefficient of absorption (a) is defined as the rate at which light is absorbed while traveling through a material.

$$a = \frac{Abs}{l} = -\frac{\log_{10}(T)}{l} \tag{14}$$

The estimation of the absorption power can be obtained from Figure 8.3. The percentage of incident power and d/dx of the incident power are plotted gainst thickness. How much of incident power is absorbed, is obtained from the d/dx of the incident power per nanometer of the material. It shows a high absorption power of the composite material. However, here, the plots are showing results in percentage.

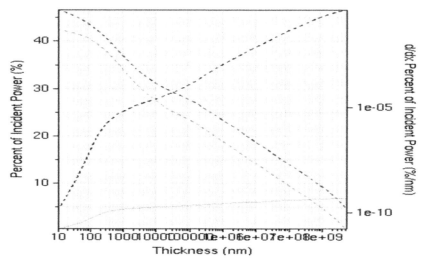

FIGURE 8.3 Plot of power absorbed as a function of thickness.

The efficiency in terms of power density is calculated and is plotted in Figure 8.4. It can be said that the thermodynamic I-V curve too. Current is plotted in a unit of Am^{-2}. The efficiency was increasing in a straight line and reached a very high value, which is shown as a dotted line. The efficiency is calculated as 53.3 Wm^{-2} at 0.57 V. The current density is almost constant up to 0.47 V and then current density starts to decrease, which comes down to zero at 0.57V. A very high value of current density is reported in this study. The enhanced performance of the composite is the combined effect of ZnS-ZnO composite as when the performance of ZnO is observed, it gives much less efficiency compared to the composite structure. Figure 8.5 shows the current density and efficiency of the ZnO NPs. The efficiency shows the strange nature of attaining a maximum at 0.002 V and then reduces gradually to 0. However, the nature of the curves resembles with the experimental results [77, 78].

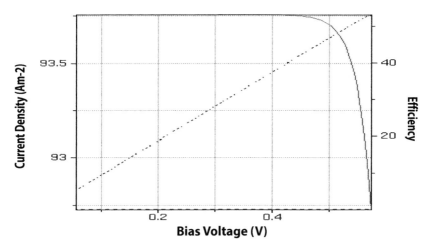

FIGURE 8.4 Current density and efficiency with voltage.

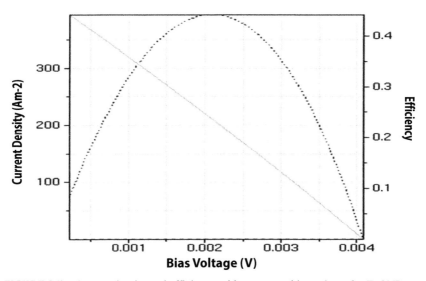

FIGURE 8.5 Current density and efficiency with respect to bias voltage for ZnONP.

The maximum current per photon energy is depicted in Figure 8.6. It rises up to 539.87 A/m^2 eV at a photon energy of 0.54eV. Clear current pulse is obtained peaks of which decreasing gradually.

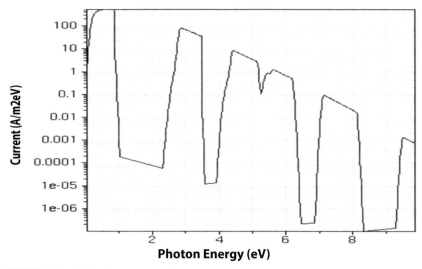

FIGURE 8.6 Induced current per photon energy.

The variation of epsilon 2 with energy is given in Figure 8.7. Epsilon 2 is the imaginary part of the dielectric constant and has a significant interest in the optical properties of materials. The refractive index varies directly as the square root of the dielectric constant of the medium. The variation of dielectric constant hence implies the change in refractive index. High refractive index usually means that the light moves more slowly in the material that allows a very thin material.

8.3.2 ABSORPTION COEFFICIENT METHOD

By this method, epsilon 2 is generated from the absorption matrix input. In the thickness of the device (1000 nm), the photon energy is absorbed according to the absorption power of the material. Absorption matrix is generated, and epsilon 2 is plotted in Figure 8.8.

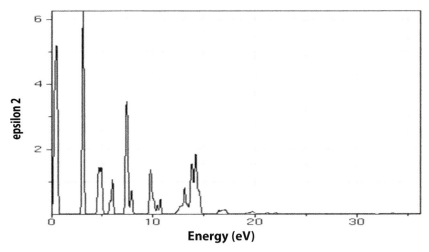

FIGURE 8.7 Variation of epsilon 2 with energy.

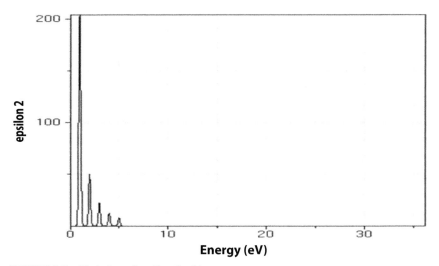

FIGURE 8.8 Variation of epsilon 2 with respect to energy.

When light falls on the front surface of the p-i-n device, it suffers from single or multiple reflections, diffraction or scattering. The power redistribution takes place due to all other surface-related phenomena, and this is a continuous function of polar and azimuthal angle. Depending on the properties of the material and its surface, power distribution takes place inside the material. The distribution is thus a vector with discrete elements. According to Lambert-Beer's law, the power inside a homogeneous absorbing medium varies exponentially decreasing manner. Here in this study, an absorption matrix is generated from which surface-related phenomena are calculated. During multiple reflections, the absorption vector after reflection in the first layer of the material is taken for a simulation starting point. The propagation of the light the task is repeated via multiplication of the matrix. In this method, the photon flux per energy level is given in Figure 8.9. This gives a curve which is like a wave, but the falling apart is not so sharp like that obtained in the previous method (Figure 8.10).

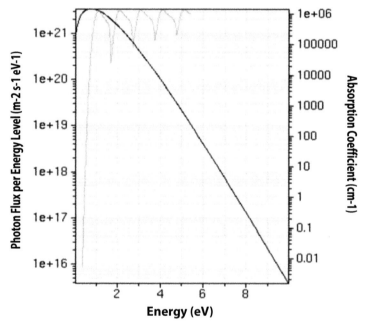

FIGURE 8.9 Photon flux per energy level and absorption coefficient vs. energy in the absorption coefficient method.

The percentage of the incident photon and photon energy absorbed as a function of thickness both are plotted in Figure 8.11. High current density compared to the SIESTA method is obtained in this process. The efficiency in terms of power density is also showing high value (Figure 8.12).

ZnS/ZnO Nanocomposite in Photovoltaics

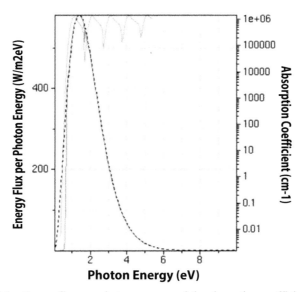

FIGURE 8.10 Energy flux per photon energy and the absorption coefficient vs. photon energy in the absorption coefficient method.

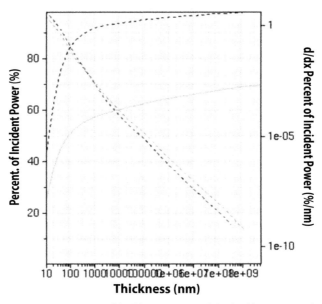

FIGURE 8.11 The percentage of incident power and the incident power absorbed as a function of thickness in the absorption matrix method.

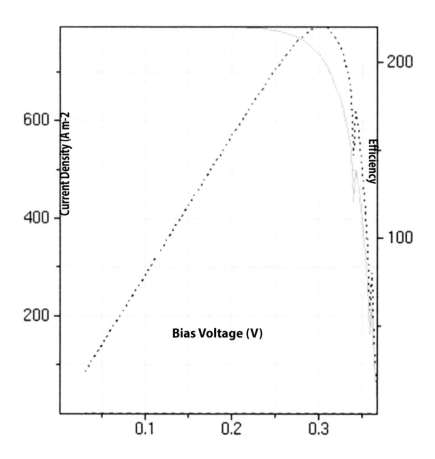

FIGURE 8.12 The current density and efficiency vs. bias voltage in the absorption matrix method.

8.3.3 BANDGAP ASSUMPTION METHOD

In this method, a bandgap measured from the previous method is assumed as initial bandgap, and constant absorption is assumed above the initial bandgap. Here Poisson's equation is solved by the ADEPT model. Mobility of electrons and hole in the p-i-n structure is set. The speed of the charge carriers governs the traps and recombination of carriers inside the layers. The dopant density modulates the Fermi level and a number of free carriers in the layer. Traps are modeled by Shockley, Read, and Hall model. The Auger coefficients for recombination and the lifetime of the holes and electrons for modeling the traps are changed. The values are given below.

Assumed bandgap – 2.5eV
Donor dopant density – 1.5E16cm^{-3}
Hole lifetime – 1e6s
Electron lifetime – 1e6s
Trap level with respect to Fermi level – 0.3eV
Relative recombination coefficient – 1.1e–14 cm^{3}s^{-1}
Hole Auger coefficient – 0.3e–30cm^{6}s^{-1}
Electron Auger coefficient – 1.1e–30cm^{6}s^{-1}

In this method, the spectrum type is blackbody type. The solar spectrum, frequency-specific absorption, and absorption coefficient along with the absorption spectrum are analyzed. The display thickness is chosen as 1000 nm. The results obtained are stated below:

Optimum bias: 0.873V
Optimum current: 454.892 Am^{-2}
Optimum power: 397.236 Wm^{-2}
Optimum efficiency: 30.129%
Total spectrum power: 1318.443 W m^{-2}
Power after thermalization: 562.031 W m^{-2}
Percent power after thermalization: 42.628%
Percent power absorbed: 72.069%
Short circuit current: 468.359A m^{-2}

The induced current with respect to photon energy is obtained as Figure 8.13. Current density and efficiency in terms of power density are shown in Figure 8.14. Very high current density, as well as power density, is showing the performance of a highly efficient photovoltaic cell made of ZnS/ZnO composite material.

Figures 8.15 and 8.16 depict the absorption coefficient and absorbed power in this method for the taken sample of ZnS/ZnO composite. The same trend of high values persists in this method too. Energy flux per photon energy in Figure 8.17 is, however, somewhat lesser than that obtained in the bandgap method.

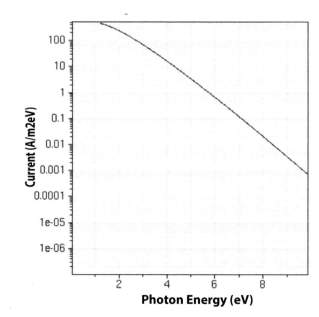

FIGURE 8.13 Current vs. photon energy in the bandgap assumption method.

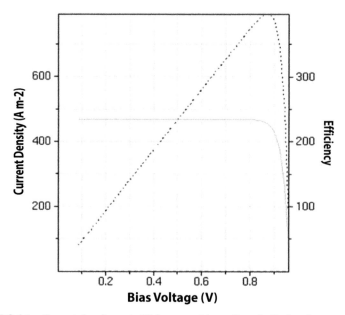

FIGURE 8.14 Current density and efficiency vs. bias voltage in the bandgap assumption method.

ZnS/ZnO Nanocomposite in Photovoltaics

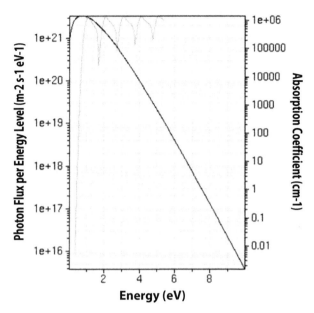

FIGURE 8.15 Photon flux per energy level and absorption coefficient in the bandgap method.

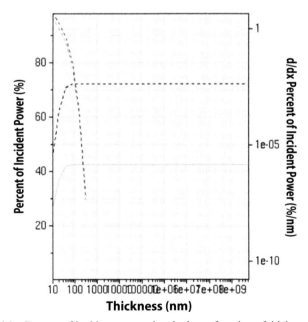

FIGURE 8.16 Percent of incident power absorbed as a function of thickness.

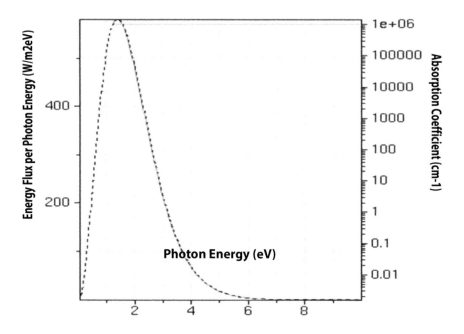

FIGURE 8.17 Energy flux and absorption coefficient vs. photon energy.

8.4 CONCLUSIONS

In this theoretical study using DFT and LDA, the PV properties of ZnO/ZnS composite are investigated in three different ways: (i) SIESTA method, (ii) absorption method, and (iii) Bandgap assumption method. Three methods point towards high PV activity of the material. PCE of the cell is calculated as 53.53 Wm^{-2} in SIESTA method. High absorption power is reported in all methods. A current density of 539.87 A/m^2 eV at photon energy of 0.54 eV is recorded in SIESTA. Absorption method reports an efficiency of 220 Wm^{-2} with a current density of 800 A/m^2. The PV cell made of only ZnO shows efficiency much lesser than the composite structure. Assuming a fixed bandgap of 2.5 eV, the efficiency can be increased up to 300 Wm^{-2}. Following today's bandgap engineering, we can tailor the bandgap of our sample composite to achieve highly efficient composite material to meet new challenges.

KEYWORDS

- nanomaterials
- nanoparticles
- photovoltaics
- power conversion efficiency
- quantum dot
- surface plasmon resonance

REFERENCES

1. Rao, C. N. R., & Cheetham, A. K., (2001). Science and technology of NMs: Current status and future prospects. *J. Mater. Chem., 11*, 2887–2894.
2. Zhao, H., & Ning, Y., (2000). Techniques used for the preparation and application of gold powder in ancient China. *Gold Bull., 33*, 103–105.
3. Aldama, T. F., (2018). *Online Presentation*, Website: https://en.ppt-online.org/226443 last accessed 9/9/ (Accessed on 25 July 2019).
4. Kulkarni, S. K., (2007). *"Nanotechnology: Principles and Practices."* Capital Publishing Company, Pune.
5. Dresselhaus, M. S., Dresselhaus, G., & Eklund, P. C., (1995). *"Science of Fullerenes and Carbon Nanotubes."* San Diego, Academic Press.
6. Balazs, A. C., Emrick, T., & Russell, T. P., (2006). NP polymer composites: Where two small worlds meet. *Science, 314*, 1107–1110.
7. Kong, E. S. W., Tromholt, T., & Krebs, F. C., (2015). *NMs Polymers, and Devices: Materials Functionalization and Device Fabrication* (pp. 319–340). Wiley & Sons, Hoboken, New Jersey.
8. Dayal, S., Kopidakis, N., Olson, D. C., Ginley, D. S., & Rumbles, G., (2010). Photovoltaic devices with a low band gap polymer and CdSe nanostructures exceeding 3% efficiency. *Nano Lett.*, 239–242.
9. Colberts, F. J. M., Wienk, M. M., & Janssen, R. A. J., (2017). Aqueous NP polymer SCs: Effects of surfactant concentration and processing on device performance. *ACS Appl. Mater. Interfaces, 9*, 13380–13389.
10. Shen, Q., Kobayashi, J., Diguna, L. J., & Toyoda, T., (2008). Effect of ZnS coating on the photovoltaic properties of CdSe quantum dot - sensitized solar cells. *Journal of Applied Physics, 103*, 084304.
11. Jun, H. K., Careem, M. A., & Arof, A. K., (2013). QD-sensitized SCs—perspective and recent developments: A review of Cd chalcogenide QDs as sensitizers. *Renewable and Sustainable Energy Reviews, 22*, 148–167.

12. Voznyy, O., Lan, X., Liu, M., Walters, G., Quintero-Bermudez, R., et al., (2017). Mixed quantum-dot SCs. *Nature Commun., 8,* 1325.
13. Sharma, D., Jha, R., & Kumar, S., (2016). QD sensitized SC: Recent advances and future perspectives in photoanode. *Solar Energy Materials and SCs, 155,* 294–322.
14. Vittal, R., & Ho, K. C., (2017). Zinc oxide based dye-sensitized solar cells: A review. *Renewable and Sustainable Energy Reviews. 207*(70), 920–935.
15. Anta, J. A., Guillén, E., & Tena-Zaera, R., (2012). ZnO-based dye-sensitized SCs. *J. Phys. Chem. C., 116,* 11413–11425.
16. Lee, H., Wang, M., Chen, P., Gamelin, D. R., Zakeeruddin, S. M., et al., (2009). Efficient CdSeQD-sensitized SCs prepared by an improved successive ionic layer adsorption and reaction process. *Nano Lett., 9,* 4221–4227.
17. Huang, F., Zhang, L., Zhang, Q., Hou, J., Wang, H., et al., (2016). High efficiency CdS/CdSeQD sensitized SCs with two ZnSe layers. *ACS Appl. Mater. Interfaces, 8,* 34482–34489.
18. Lu, Y. B., Li, L., Su, S. C., Chen, Y. J., Song, Y. L., et al., (2017). A novel TiO_2 nanostructure as photoanode for highly efficient CdSeQD-sensitized SC. *RSC Adv., 7,* 9795–9802.
19. Husain, M., & Khan, J. H., (2016). *Advances in NMs.* Springer.
20. Kim, J., Rivera, J. L., Meng, T. Y., Laratte, B., & Chen, S., (2016). Review of life cycle assessment of NMs in photovoltaics. *Solar Energy, 133,* 249–258.
21. Stuart, H. R., & Hall, D. G., (1998). *Appl. Phys. Lett., 73,* 38159.
22. Köntges, M., Kajari-Schröder, S., & Kunze, I., (2013). Crack statistic for wafer-based silicon SC modules in the field measured by UV fluorescence. *IEEE Journal of Photovoltaics, 3,* 95–101.
23. Gardelis, S., Nassiopoulou, A. G., Manousiadis, P., Vouroutzis, N., & Frangis, N., (2013). A silicon-wafer based p-n junction SC by aluminum-induced recrystallization and doping. *Appl. Phys. Lett., 103,* 241114.
24. Pillai, S., Catchpole, K. R., Trupke, T., & Green, M. A., (2007). Surface plasmon enhanced silicon SCs. *J. Appl. Phys., 101,* 093105.
25. Chen, X., Jia, B., Saha, J. K., Cai, B., Stokes, N., et al., (2012). *Nano Lett., 12,* 2187–2192.
26. Kelzenberg, M. D., Turner-Evans, D. B., Kayes, B. M., Filler, M. A., & Putnam, M. C., (2008). Photovoltaic measurements in single-nanowire silicon SCs. *Nano Lett., 8,* 710–714.
27. Tsakalakos, L., Balch, J., Fronheiser, J., Korevaar, B. A., & Sulima, O., (2007). Silicon nanowire SCs. *Appl. Phys. Lett., 91,* 233117.
28. Stelzner, T., Pietsch, M., Andr, G., Falk, F., Ose, E., et al., (2008). Silicon nanowire-based SCs. *Nanotechnol., 19,* 295203.
29. Adachi, M. M., Anantram, M. P., & Karim, K. S., (2013). Core-shell silicon nanowire SCs. *Sci. Reports., 3,* 1546.
30. Wang, D. H., Park, K. H., Seo, J. H., Seifter, J., Jeon, J. H., et al., (2011). Enhanced power conversion efficiency in PCDTBT/PC 70 BM bulk heterojunction photovoltaic devices with embedded silver NP clusters. *Adv. Energy Mater., 1,* 766–770.
31. Rasouli, H. R., Ghobadi, A., Ghobadi, T. G. U., Ates, H., & Topalli, K., (2017). Nanosecond pulsed laser ablated sub-10 nm silicon NPs for improving photovoltaic conversion efficiency of commercial SCs. *J. Opt., 19,* 105902.

32. Sun, B., Marx, E., & Greenham, N. C., (2003). Photovoltaic devices using blends of branched CdSeNPs and conjugated polymers. *Nano Lett., 3,* 961–963.
33. Wadia, C., Alivisatos, A. P., & Kamme, D. M., (2009). Materials availability expands the opportunity for large-scale photovoltaics deployment. *Environ. Sci. Technol., 43,* 2072.
34. Catchpole, K. R., & Polman, A., (2008). Plasmonic SC. *Opt. Exp., 16,* 21793.
35. Stuart, H. R., & Hall, D. G., (1997). Thermodynamic limit to light trapping in thin planar structures. *J. Opt. Soc. Am. A, 14,* 3001.
36. Atwater, H. A., & Polman, A., (2010). Plasmonics for improved photovoltaic devices. *Nature Mater., 9,* 205–213.
37. Fan, W., & Leung, M. K. H., (2016). Recent development of plasmonic resonance-based photocatalysis and photovoltaics for solar utilization. *Molecules, 21,* 180.
38. Yang, X., Liu, W., & Chen, H., (2015). *Recent Advances in Plasmonic Organic Photovoltaics, 58,* 210–220.
39. Ahn, S., Rourke, D., & Park, W., (2016). Plasmonic nanostructures for organic photovoltaic devices. *J. Optics, 18,* 033001.
40. Chan, K., Wright, M., Elumalai, N., Uddin, A., & Pillai, S., (2017). Plasmonics in organic and perovskite SCs: Optical and electrical effects. *Adv. Opt. Mater., 5,* 1600698.
41. Funde, A. M., Nasibulin, A. G., Syed, H. G., Anisimov, A. S., Tsapenko, A., et al., (2016). Carbon nanotube-amorphous silicon hybrid SC with improved conversion efficiency. *Nanotechnol., 27,* 185401.
42. Wang, F., Endo, M., Mouri, S., Ohno, Y., Wakamiya, A., et al., (2016). Highly stable perovskite SC with an all-carbon hole transport layer. **Nanoscale, 8,** 11882–11888.
43. Loh, K. P., Tong, S. W., & Wu, J., (2016). Graphene and graphene-like molecules: Prospects in SCs. *JACS, 138,* 1095–1102.
44. Liu, Y., Yuan, L., Yang, M., Zheng, Y., Li, L., Gao, L., et al., (2014). *Giant Enhancement in Vertical Conductivity of Stacked CVD Graphene Sheets by Self-Assembled Molecular Layers, 5,* 5461.
45. Miao, X., Tongay, S., Petterson, M. K., Berke, K., Rinzlert, A. G., et al., (2012). High efficiency graphene SCs by chemical doping. *Nano Lett., 12,* 2745–2750.
46. Jia, Y., Cao, A., Bai, X., Li, Z., Zhang, L., et al., (2011). Achieving high efficiency silicon-carbon nanotube heterojunction SCs by acid doping. *Nano Lett., 11,* 1901–1905.
47. Habisreutinger, S. N., Leijtens, T., Eperon, G. E., Stranks, S. D., & Nicholas, R. J., (2014). Carbon nanotube/polymer composites as a highly stable hole collection layer in perovskite SCs. *Nano Lett., 14,* 5561–5568.
48. Hellstrom, S. L., Vosgueritchian, M., Stoltenberg, R. M., Irfan, I., Hammock, M., et al., (2012). Stron and stable doping of carbon nanotubes and graphene by MoO_x for transperant electrodes. *Nano Lett., 12,* 3574–3580.
49. Jeon, I., Yoon, J., Ahn, N., Atwa, M., Delacou, C., et al., (2017). Carbon nanotube versus graphene as flexible transparent electrodes in inverted perovskite SCs. *J. Phys. Chem. Lett., 8,* 5395–5401.
50. Wei, Q., Tong, X., Zhang, G., Qiao, J., Gong, Q., et al., (2015). Nitrogen-doped carbon nanotube and graphene materials for oxygen reduction reactions. *Catalysts, 5,* 1574–1602.

51. Liu, J., Xue, Y., Zhang, M., & Dai, L., (2012). Graphene-based materials for energy applications. *MRS Bulletin, 37*, 1265–1272.
52. Ye, Meidan., Wen, X., Wang, M., Icozzia, J., & Zhang, N., (2015). Recenet advances in dye-synthesized SCs: From photoanodes, sensitizers and electrolytes to counter electrodes. *Materialstoday., 18*, 155–162.
53. Liu, J., Xue, Y., Zhang, M., & Dai, L., (2012). Graphene-based materials for energy applications. *MRS Bulletin, 37*, 1265–1272.
54. Kamat, P. V., (2008). Quantum SCs semiconductor nanocrystals as light harvesters. *J. Phys. Chem. C., 112*, 18737–18753.
55. Giménez, S., Mora-Seró, I., Macor, L., Guijarro, N., Lana, V., et al., (2009). Improving the performance of colloidal quantum-dot-sensitized SCs. *Nanotechnol., 20*, 295204.
56. Youn, D. H., Seol, M., Kim, J. Y., Jang, J. W., & Choi, Y., (2013). TiNNPs on CNT-graphene hybrid support as noble-metal-free counter electrode for quantum-dot-sensitized SCs. *Chem. Sus. Chem., 6*, 261–267.
57. Lee, J. M., Kwon, B. H., Park, H. II., Kim, H., Kim, M. G., et al., (2013). Exciton dissociation and charge-transport enhancement in organic SC with quantum dot/N-doped CNT hybrid NMs. *Adv. Mater., 25*, 2011–2017.
58. Kim, B. S., Neo, D. C. J., Hou, B., Park, J. B., & Cho, Y., (2016). High performance PbSQD/graphene hybrid SC with efficient charge extraction. *ACS Appl. Mater. Interfaces, 8*, 13902–13908.
59. Liu, Y., Gu, Y., Yan, X., Kang, Z., Lu, S., et al., (2015). *Design of ZnO/ZnS/Au Sandwich Structure Photoanode for Enhanced Efficiency of Photoelectrochemical Water Splitting, 8*, 2891–2900.
60. Hao, J, Wang, X., Liu, F., Han, S., Lian, J., et al., (2017). Facile synthesis ZnS/ZnO/Ni(OH)$_2$ composites grown on Ni foam: A bifunctional materials for photocatalysts and supercapacitors. *Scientific Reports, 7*, 3021.
61. Yu, X. L., Song, J. G., Fu, Y. S., Xie, Y., Song, X., et al., (2010). *ZnS/ZnO Heteronanostructure as Photoanode to Enhance the Conversion Efficiency of Dye-Sensitized SCs, 114*, 2380–2384.
62. Schrier, J., Denis, O. D., Wang, W. L., & Paul, A. A., (2007). Optical properties of ZnO/ZnS and ZnO/ZnTe heterostructures for photovoltaic applications. *Nano Lett., 7*, 2377–2382.
63. Luo, S., He, X., Shen, H., Li, J., & Yin, X., (2017). Vertically aligned ZnO/ZnTe core/shell heterostructures on an AZO substrate for improved photovoltaic performance. *RSC. Adv., 7*, 14837–14845.
64. Maloney, F. S., Poudyal, U., Chen, W., & Wang, W., (2016). Influence of QD concentration on carrier transport in ZnO: TiO2 Nano-hybrid photoanodes for QD-sensitized solar cells. *NMs, 6*, 191.
65. Srivastava, S. K., & Mittal, V., (2017). *Hybrid NMs, Advances in Energy, Environment and Polymer Nanocomposites*. Wiley & Sons, USA.
66. Ramasamy, V., Anandan, C., & Murugadoss, G., (2013). Synthesis and characterization of ZnS/ZnO/CdS nanocomposites. *Materials Science in Semiconductor Processing, 16*, 1759–1764.
67. Ma, H., Han, J., Fu, Y., Song, Y., Yu, C., et al., (2011). Synthesis of visible light responsive ZnO–ZnS/C photocatalyst by simple carbothermal reduction. *Appl. Catalysis B. Environmental, 102*, 417–423.

68. Lahiri, J., & Batzill, M., (2008). Surface functionalization of ZnO photocatalysts with monolayer ZnS. *J. Phys. Chem. C., 112*, 4304–4307.
69. Shuai, X. M., & Shen, W. Z., (2011). A facile chemical conversion synthesis of ZnO/ZnS core/shell nanorods and diverse metal sulfide nanotubes. *J. Phys. Chem., 115*, 6415–6422.
70. Schrier, J., Demchenko, D. O., Wang, L., & Alivisatos, A. P., (2007). Optical properties of ZnO/ZnS and ZnO/ZnTe heterostructures for photovoltaic applications. *Nano Lett., 7*, 2377–2382.
71. Zhifeng Liu., Guo, K., Wang, Y., Zheng, X., Ya, J., Li, J., Han, L., et al., (2014). Preparation of cauliflower-like CdS/ZnS/ZnO nanostructure and its photoelectric properties. *J. Nanopart. Res., 16*, 2446.
72. Drwizler, R. M., & Gross, E. K. U., (1990). *Density Functional Theory*. Springer.
73. Parr, R. G., & Weitao, Y., (1994). *Density Functional Theory of Atoms and Molecules*. Oxford University Press.
74. Hohenberg, P., & Kohn, W., (1964). Inhomogeneous electron gas. *Phys. Rev. B., 76*, 6062.
75. Kohn, W., & Sham, L. J., (1965). Self-consistent equations including exchange and correlation effects. *Phys. Rev., 140*, A1133.
76. Ceperley, D. M., & Alder, B. J., (1980). Ground state of the electron gas by a stochastic method. *Phys. Rev. Lett., 45*, 566.
77. Shen, P. C., Lin, M. S., & Lin, C. F., (2014). Environmentally benign technology for efficient warm-white light emission. *Sci. Rep., 4*, 5307.
78. Rouhi, J., Mamat, M. H., Ooi, C. H. R., Mahmud, S., & Mahmood, M. R., (2015). High-performance dye-sensitized solar cells based on morphology-controllable synthesis of ZnO–ZnS heterostructure nanocone photoanodes. *Plos One., 4*, 0123433.

CHAPTER 9

Bio-Nanometal-Cluster Composites for Renewable Energy Storage Applications

R. GOVINDHAN and B. KARTHIKEYAN

Department of Chemistry, Annamalai University, Annamalainagar – 608002, Tamil Nadu, India

ABSTRACT

The innovation on the nanocomposites brings rapid growth in the nano field. It is assumed that the repopulation will be needed to improve on the energy supply for the coming generations. Synthesizing the controllable composites and its societal applications for sustainable energy is highly appreciated. Nanotechnology has widened up new frontiers in materials science and engineering to meet this challenge by developing new materials, particularly bio-nanometal-cluster composites useful for efficient energy conversion and clean storage. Compared with the existing energy materials, bio-nanometal-cluster composites possess unique size-/surface-dependent (e.g., morphological, electrical, optical, and mechanical) properties useful for enhancing the energy-conversion and storage performances. Bio-nano-metal-cluster composites as energy materials are at tremendous progress, for example, solar cells (SCs) and fuel cells supercapacitors and batteries. In this chapter, we reported such bio-physical renewable energy storage applications particularly by metal clusters. The bio-nanometal-cluster composites have been characterized by transmission electron microscopy (TEM), scanning electron microscopy (SEM) and computational studies. The bio-nanometal-cluster composites were also analyzed using UV-Vis spectroscopy. Bio-physical kinetics was used to examine and to calculate the stabilization energy of the formed bio-nanometal-cluster composites. These bio-nanometal-cluster composite gave an excellent renewable solar energy conversion as well as storage capacity applications.

9.1 INTRODUCTION

Noble metal nanoparticles (NPs) found many applications in different fields such as different types of catalytic activity, photonics, and electronics, because of their unique optical, electronic, sensing, mechanical, magnetic, and chemical properties. The most common method employed for the synthesis of metal NPs is the reduction of metal ions in the solution phase. Metals with free electrons possess surface plasmon resonances (SPRs) in the visible spectrum, which give rise to such intense colors. These properties are mainly observed in Au, Ag, and Cu because of the presence of free conduction electrons. As copper NPs are widely used in the areas of electrochemical response, human contact, the necessity to develop eco-friendly methods for nanoparticle synthesis that do not use toxic chemicals has been constantly growing. Development of green nanotechnology is generating interest of researchers toward eco-friendly biosynthesis of NPs [1–5]. Gold (Au), Platinum (Pt), Silver (Ag), Palladium (Pd) and Rhodium (Rh) are important noble metals [6–8]. Although they are admired primarily for their ability to reflect light; their applications have become far more sophisticated with increased understanding and control of the size. Today, these metals are widely used in various fields, but when they are fashioned into structures with nanometer-sized dimensions, they also become enablers for a completely different set of applications that involve light. These new applications go far beyond merely reflecting light and have renewed interest in guiding the interactions between metals and light in the field known as plasmonics [9–13].

In this chapter, synthesis, characterization of trimetallic Au/Pt/Ag colloidal nanocomposites (Au/Pt/Ag TNCs) and its surface-enhanced Raman scattering (SERS), sensing of 7-azaindole (7-AI) and methanol oxidation activity were studied and discussed. Trimetallic nanoparticles (TNPs) usually possess higher catalytic activity and selectivity than the corresponding monometallic ones when they are used as colloidal dispersion catalysts. The presence of one or more elements may provide a way to improve the activity of Au-based multimetallic colloidal catalysts [6, 7]. The structure of TNPs is strongly related to their catalytic performance. Among various morphologies formed by the TNP, the core-shell and random alloy structures were most interesting, since such composites usually exhibit improved properties like catalysis, sensing, and SERS. It is well-known that the smaller the size of the NPs, the higher the catalytic activity [7, 8]. Addition to that is TNPs, metal atoms at the active sites are adjacent to foreign metal atoms; it can show higher catalytic activity than the individual monometallic nanoparticles (MNPs). TNPs composed of 3d-transition and noble metals with specific structures will provide a great number of new catalysts for various chemical reactions, since the catalytic properties of TNPs can be potentially tailored by both ligand (electronic effect) and ensemble effect (geometrical effect) [9–12]. In the case of the electronic effect, the added metal atom can electronically affect the electron density at the catalytic site, which activates or deactivates the catalytic strength and selectivity. The reaction substrates interact only with the catalytic metal atoms without direct interaction with the added metal atoms. But, in the case of the geometrical effect, the substrate should interact not only with the catalytic metal atoms but also with the added metal atoms. Thus, both metal atoms can directly interact with the substrate and affect the activity and selectivity of the reaction [14]. SERS is a unique phenomenon in which the scattering cross-section of molecules adsorbed on particular monometallic surfaces is enhanced [14–16]. In recent years, it has been reported that even single-molecule identification is possible by the study of SERS, suggesting that the enhancement factor can reach as much as 10^{14} to 10^{18}. SERS has been used in many areas of science and technology, including catalysis, sensors, chemical analysis, molecular electronics, corrosion, and lubrication, etc. [14, 17–19].

Higher catalytic activity for the TNCs than the bimetallic nanoparticles (BNPs) and MNPs are expected because; a possible mechanism of electronic charge transfer effect between the various kinds of adjacent elements was reported as the reason for the high catalytic activities of BNPs and TNPs [20]. This kind of consideration could be applied to the selected Au/Pt/Ag TNC system. In Au/Pt/Ag TNCs, Ag (ionization energy is 7.576 eV) is most electronegative among all elements (*Note:* ionization energy of Pt and Au are 8.62 and 9.225 eV, respectively), therefore, Ag atoms would donate electrons to the neighboring Au and Pt atoms through the electronic charge transfer effect. There are at least two possible modes of charge transfer; one is, from Ag to Au atom and another from Ag to Pt atom. Therefore, the synergistic effect of these two charge-transfer modes might play an important role for the higher catalytic activity of the Au/Pt/Ag TNCs compared with that of corresponding mono- and bimetallic counterparts. Order of the charge transfer may be consistent with the order of the ionization potential of Au, Pt and Ag metal. The possible way of electronic charge transfers in Au/Pt/Ag TNCs and ionization energy of the corresponding bulk metals is shown in Figure 9.1.

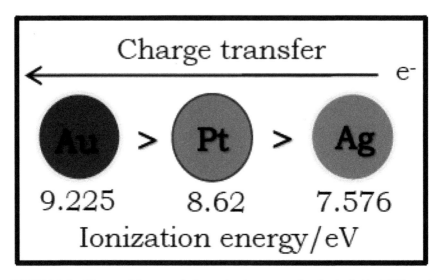

FIGURE 9.1 The possible way of electronic charge transfers in Au/Pt/Ag TNCs and ionization energy of the corresponding bulk metals.

9.2 EXPERIMENTAL METHODS

9.2.1 SYNTHESIS OF TRIMETALLIC AU/PT/AG COLLOIDAL NANOCOMPOSITES (AU/PT/AG TNCS)

Trisodium citrate reduced Au NPs were synthesized according to the reported method [6–8]. Briefly, 10 mL of aqueous 0.1% metal salt (HAuCl$_4$·3H$_2$O) was heated to boil, and 2 mL of 1% trisodium citrate was then added with stirring. The reaction mixture was heated for 4 min, and cooled down to room temperature (RT). The color change indicated the formation of Au NPs. Then, 10 mL of 0.1% metal salt (H$_2$PtCl$_6$·xH$_2$O) was added to the Au NPs followed by the addition of 2 mL of 1% trisodium citrate with stirring. Finally, 10 mL of 0.1% metal salt (AgNO$_3$) were added into the Au/Pt NPs.

9.2.2 UV-VISIBLE SPECTROSCOPY

UV-Vis absorbance spectrum of the TNCs was recorded over a range of 800–200 nm with a Shimadzu UV-1650 PC spectrophotometer, operated at a resolution of 0.5 nm. The samples were filled in a quartz cuvette of 1 cm light-path length, and the light absorption spectra were given with reference to deionized water and distilled ethanol. Fourier transform infrared (FT-IR) spectrum and surface-enhanced infrared spectrum (SEIRS) were recorded with a thermo-scientific spectrophotometer, model no. iS5 equipped with attenuated total reflectance (ATR) facility which is implemented with Zn-Se crystal detector. Each spectrum was recorded with an acquisition time of 18 s. FT-Raman and surface-enhanced Raman spectrum (SERS) was recorded with an integral microscope BRUKER RFS 27 spectrometer equipped with 1024 × 256 pixels liquefied nitrogen-cooled germanium detector. A 1064 nm line of the neodymium-doped yttrium aluminum garnet (Nd: YAG) laser was used as an excitation wavelength. The laser power was 100 mW, and the data acquisition time was 30 s. The TNCs were activated by the addition of 2 M sodium chloride to promote the aggregation prior to SERS measurement.

Dynamic light scattering (DLS) technique, DLS was used to measure the hydrodynamic diameter and particle size distribution of the TNCs. Measurements were carried out using a Malvern Zetasizer Ver. 6.32 working at a fixed angle of 173°. TNCs were properly diluted with water and poured in polystyrene (PS) cuvette before each measurement. Data obtained were analyzed using the Zetasizer software.

9.2.3 HIGH-RESOLUTION TRANSMISSION ELECTRON MICROSCOPY (HR-TEM)

HR-TEM images were recorded using a JEOL 3010 high-resolution transmission electron microscopy (TEM) with an ultra-high-resolution (UHR) pole piece operated at an accelerating voltage of 300 kV with an energy dispersive X-ray (EDX) accessory operated at 30 kV. Samples for HR-TEM images were prepared by depositing a drop of the synthesized sample on a carbon-coated Cu grid and allowing it to evaporate. The morphology of the TNCs were examined by FEI Quanta FEG 200 3010 high-resolution scanning electron microscopy (HR-SEM), with an EDX accessory, operates at an accelerating voltage of 30 kV. Apart from giving the high-resolution surface morphological images, the Quanta 200 FEG also has the analytical capabilities such as detecting the presence of elements on any solid materials through the EDX it providing crystalline information up to few nano-meter depths of the material surface *via* electron backscattered detection (BSD). Samples for HR-SEM characterization were prepared by spreading out few drops of TNCs on a silica glass plate. The plate was then dried at 100°C for 5 min. The surface morphology of the material was visualized by using a SUPRA 55-CARL ZEISS (Germany) field emission scanning electron microscopy (FESEM) with field emission filament operating at 20 kV. It has a resolution of 1.5 nm and equipped with a through-the-lens detector (TLD) working in UHR mode. For particle size determination, the average size was calculated from at least 25 randomly chosen features visible in an image. A sample for atomic force microscopy (AFM) characterization were prepared by adding a small quantity of TNCs or powder sample with carbon tetrachloride and was sonicated for 30 min in a fast-clean ultrasonic cleaner. Then this solution was spread out on the silica glass plate. The plate was then dried at 100°C for 5 min.

9.3 RESULTS AND DISCUSSION

9.3.1 UV-VISIBLE SPECTROSCOPY

The optical properties of the multi-component metal NPs may be used to compare and get the structural insight of Au/Pt/Ag TNCs and its corresponding MNPs. UV-Vis absorption spectra of Au, Pt, Ag MNPs, and Au/Pt/Ag TNCs are shown in Figure 9.2. The UV-Vis absorption spectrum of Au NPs shows a maximum peak at 531 nm, which is a typical surface Plasmon band of Au NPs. It can also be seen that the Pt NPs do not show any absorption in the selected wavelength range. The absorption peak of Ag NPs appears at 430 nm. For Au/Pt/Ag TNCs, a clear single absorbance peak at 528 nm was observed. The absence of multiple bands in the spectrum ruled out the presence of individual Au, Pt, and Ag NPs in TNCs [21].

9.3.2 DYNAMIC LIGHT SCATTERING (DLS) TECHNIQUES

The hydrodynamic diameter of the as-synthesized colloidal nanocomposites is measured by DLS technique. A representative DLS curve of the Au/

Pt/Ag TNCs is shown in Figure 9.3. The Z-average value of the Au/Pt/Ag TNCs is 34.73 d. nm.

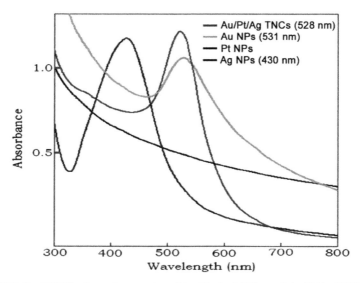

FIGURE 9.2 UV-Vis absorption spectra of Au, Pt, Ag MNPs, and Au/Pt/Ag TNCs.

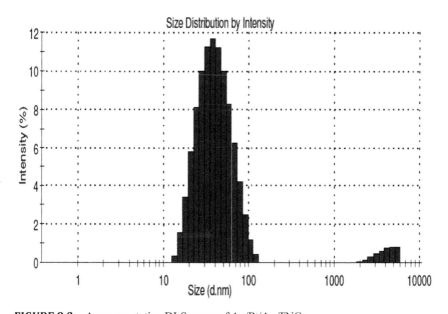

FIGURE 9.3 A representative DLS curve of Au/Pt/Ag TNCs.

9.3.3 ZETA POTENTIAL (ZP) AND ZETA DEVIATION ANALYSIS

The magnitude of the ZP gives an indication of the potential stability and surface charge of the synthesized colloidal nanocomposites [20, 22]. The ZP curve of the Au/Pt/Ag TNCs is shown in Figure 9.4. The ZP value is about –31.9 mV, which indicated the kinetic stability of the TNCs. This negative ZP value discloses that the Au/Pt/Ag TNCs has good stability. The zeta deviation value of the TNCs is found to be 10.6 mV.

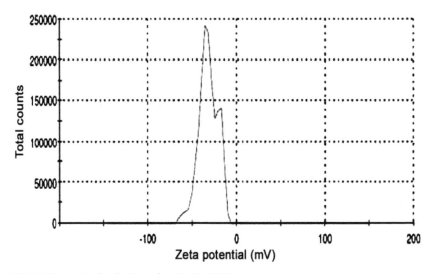

FIGURE 9.4 ZP distribution of Au/Pt/Ag TNCs.

9.3.4 TRANSMISSION ELECTRON MICROSCOPY (TEM) ANALYSIS

TEM is utilized to characterize the size, morphology, and structure of the Au/Pt/Ag TNCs. Figure 9.5 depicts the TEM images of the Au/Pt/Ag TNCs. Figure 9.5 clearly shows that TNCs are formed in spherical shaped structure (selected particles are highlighted). Each NPs exhibits inhomogeneous electron density (electron richer and lesser dense) with a dark core surrounded by a lighter shell (black in color core and light color shell are clearly seen) [23–27]. The length of the individual selected particle (highlighted in Figure 9.5b) is examined by the line plot profile of "Image J Viewer software," and found that the average particle size (respectively with their width) of the selected particle was about ca. ~ 20 nm.

Bio-Nanometal-Cluster Composites for Renewable Energy 223

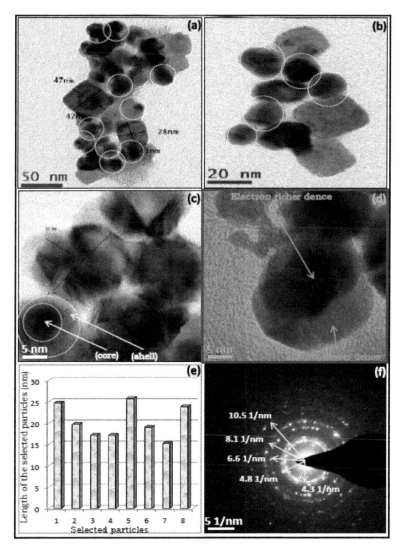

FIGURE 9.5 TEM images of Au/Pt/Ag TNCs, *scale bars* (a) 50, (b) 20 and (c, d) 5 nm, (e) particle size distribution of the selected particles (particles are highlighted in Figure 9.5b) and (f) SAED pattern of Au/Pt/Ag TNCs.

Core and shell forms are highlighted in Figure 9.5c to support the formation of core-shell. A typical TEM image is shown in Figure 9.5c and showing spherical particles with contrast difference. The selected NP is highlighted in Figure 9.5d; the core-shell form is inferred through the difference in density of electrons [23–28]. Particle size histogram of Au/Pt/Ag TNCs is shown in Figure 9.5e. The particle sizes were roughly distributed from 15 to 26 nm. The selected area electron diffraction (SAED) pattern of Au/Pt/Ag TNCs is presented in Figure 9.5f. The SAED pattern of the TNCs suggesting that they have polycrystalline features.

9.3.5 FIELD EMISSION SCANNING ELECTRON MICROSCOPY (FESEM) AND EDX ANALYSIS

Figure 9.6 depicts FE-SEM images of the Au/Pt/Ag TNCs with different magnification. This analysis clearly reveals the Au/Pt/Ag TNCs are having hetero-structured morphology. Figure 9.6c and 9.6d show the aggregated behavior of TNCs. Aggregations are due to the experimental process involved in the characterization of Au/Pt/Ag TNCs. Further, surface plot and line profile of the marked area (shown in Figure 9.6f and 9.6g) in Figure 9.6c confirmed the aggregation of smaller NPs which forms microcube like structure.

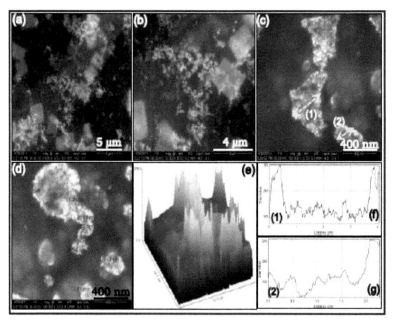

FIGURE 9.6 FE-SEM images of Au/Pt/Ag TNCs, *magnification*: (a) 6K, (b) 12K and (c, d) 100K, *scale bars* (a) 5, (b) 4 μm and (c, d) 400 nm, (e) surface plot and (f, g) line profile of Figure 9.6c (selected particles are highlighted in Figure 9.6c).

9.3.6 X-RAY PHOTOELECTRON SPECTROSCOPY (XPS) ANALYSIS

The surface composition and structure are indispensable information in multimetallic nanocomposites. Figure 9.7 depicts the results of XPS analysis of the Au/Pt/Ag TNCs with their corresponding binding energy values *versus* intensity. Figure 9.7a represents the survey spectrum of the Au/Pt/Ag TNCs. XPS pattern of the TNCs showed a significant Au4f, Pt4f, and Ag3d signals corresponding to the binding energy of Au, Pt, and Ag, respectively (shown in Figure 9.7b–9.7d). A significant signal corresponding to the binding energy of C is also found (Figure 9.7e). Au, Pt, and Ag showed two obvious peaks of Au4f, Pt4f, and Ag3d (shown in Figure 9.7b–9.7d), respectively. Peaks at 83.45 and 87.25 eV are attributed to Au ($4f_{7/2}$, $4f_{5/2}$, respectively). Spin-orbit doublet peaks indicating that Au is present in Au (0) state, whereas Pt (4f) and Ag (3d) peaks could be resolved into two sets of spin-orbit doublets. Pt ($4f_{7/2}$) peaks at 72.5 and Pt ($4f_{5/2}$) at 75.4 eV indicates that Pt (0) oxidation state and Ag ($3d_{5/2}$) peaks appeared at 367.15 and Ag ($3d_{3/2}$) at 373.25 eV indicated the Ag forms Ag (0) oxidation state [29].

9.3.7 ELECTROCHEMICAL STUDIES

9.3.7.1 ELECTRO CATALYSIS OF METHANOL OXIDATION REACTION (MOR) BY AU/PT/AG TNCS

Density functional theory (DFT) basis set of (B3LYP/Lanl2DZ) optimized the lowest energy structure of Pt *n* clusters (*n* = 3 to 6), are shown in Figure 9.8.

The detail of the optimized clusters and occupied the highest molecular-lowest unoccupied molecular orbital (*HOMO-LUMO*) energy gap (*Eg*) values are compiled in Table 9.1.

Determination of the energies of the *HOMO* and *LUMO* is important in quantum chemical calculations. The *HOMO* is the orbital that primarily acts as an electron donor, and the *LUMO* is the orbital that largely acts as electron acceptor [20]. The E_g values ($E_g = E_{LUMO} - E_{HOMO}$) were calculated from the total density of state (DOS) plot. Electronic DOS plot of the DFT optimized Pt*n* clusters using B3LYP/Lanl2DZ level of theory is shown in Figure 9.9. TNCs are important in catalysis because of the synergistic interaction between three different metallic elements contained in one particle, which also leads to enhance the electrocatalytic activity towards MOR. Thus Pt metallic clusters are focused and optimized in Au/Pt/Ag TNCs.

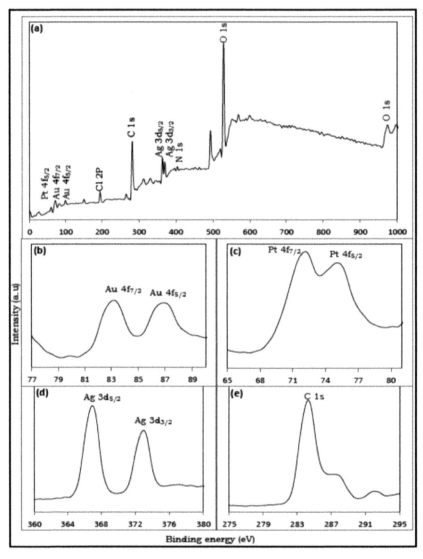

FIGURE 9.7 XPS of Au/Pt/Ag TNCs, (a) XPS survey spectrum, (b-e) XPS of Au (4f), Pt (4f), Ag (3d) and C (1s) respectively with its binding energy (in eV).

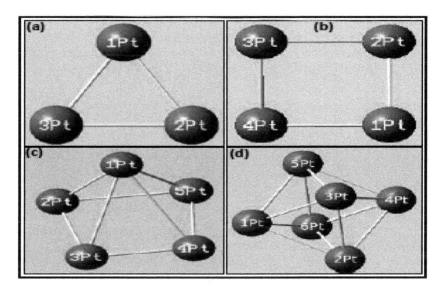

FIGURE 9.8 DFT-optimized structures of Pt*n* clusters ((a-d) Pt*3*, Pt*4*, Pt*5*, and Pt*6* clusters).

TABLE 9.1 DFT Optimized Structures and *HOMO-LUMO* Energy Gap (*Eg*) Values

SL. No.	Optimized Structures	E_{LUMO} (eV)	E_{HOMO} (eV)	E_g (eV)
1	Pt_3	4.03	−5.88	1.85
2	Pt_4	−3.92	−5.88	1.96
3	Pt_5	−4.17	−5.68	1.51
4	Pt_6	−3.81	−5.23	1.42

Pt-based materials are being mostly used as a catalyst in fuel cells for the oxidation reactions. Electrocatalytic activity of the Pt-containing TNCs by using a modified GCE in the electrochemical oxidation of methanol was investigated. The electrocatalytic performance was evaluated by cyclic voltammetry (CV) and chronoamperometry (CA) investigations. The recorded CV of trimetallic Au/Pt/Ag colloidal nanocomposite modified GCE (Au/Pt/Ag-GCE) in 0.5 M H_2SO_4 in 1 M methanol at different scan rates are presented in Figures 9.10a and 9.10b. One observes that the potential and peak current for the Au/Pt/Ag-GCE oxidation processes showed a strong correlation to the scan rates. The oxidation peak current (*ip*) for methanol oxidation became larger with the increase of the scan rate, with the scan rate ranging from 0.01 to 0.5 V/s, the anodic peak current (*ipa*) increases linearly. Considering the CV behavior of Au/Pt/Ag-GCE, an *ipa* appeared at about 0.6 V. So the *ipa* at potential of 0.6 is selected, and a plot of it *versus* the scan rate is shown as an inset in Figure 9.10b. It was found that it is linearly proportional to the scan rate with the regression equation *ipa* = 11.849 × –0.3035 with R^2 = 0.9959 (*ipa*: 1×10^{-4} A, scan rate: V/s). It was inferred from that, at a sufficiently positive potential, the electrocatalytic methanol oxidation is controlled by diffusion of methanol. The peak height for the *ipa*, is shown as an inset in Figure 9.10c. The *ipa* is found to be a linear function of the square root of the scan rates, and the regression equation is *ipa*= 9.7076× –1.5849 with R^2 = 0.9426. The methanol oxidation reaction (MOR) at Au/Pt/Ag-GCE and bare GCE in 0.5 M H_2SO_4 of 1 M methanol at 0.5 V/s is shown in Figure 9.10c. Electrochemical oxidation of the methanol is not observed at bare-GCE, whereas remarkable improvement in the *ipa* (*ipa* = 5.66 × e^{-4} A) is observed at Au/Pt/Ag-GCE. It justifies the advantage of MORs with Au/Pt/Ag-GCE. The well-known hydrogen adsorption/desorption peak related to Pt are observed in the range between –0.2 to 0.5 V (Figure 9.10c). This characteristic peak indicated that the role of Pt in the MOR process. The appearance of a cathodic peak at 0.16 V (Figure 9.10c) is related to the redox reaction of Pt.

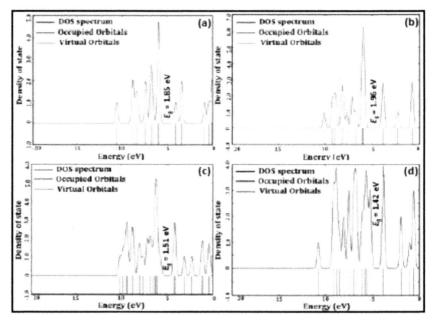

FIGURE 9.9. DOS plot of the DFT optimized Pt*n* clusters ((a-d) Pt*3*, Pt*4*, Pt*5*, and Pt*6* clusters). *Inset*: Energy gap values in eV.

Bio-Nanometal-Cluster Composites for Renewable Energy 229

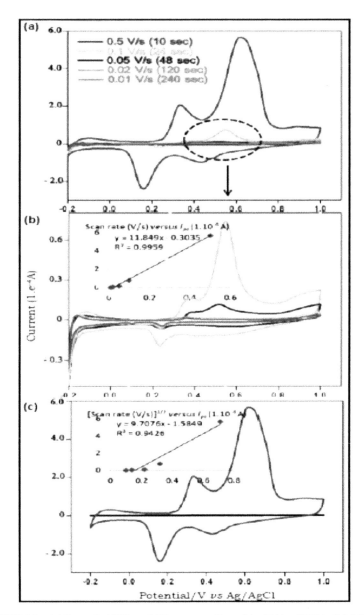

FIGURE 9.10 (a) Recorded CVs and (b) zoom portion of Au/Pt/Ag-GCE in 0.5 M H_2SO_4 + 1 M methanol at different scan rates (c) CV of Au/Pt/Ag-GCE along with bare-GCE for the methanol oxidation in 0.5 M H_2SO_4 + 1 M methanol. *Scan rate* is 0.5 V/s. Inset figure in 9.10b and 9.10c Anodic peak current (*ipa*) of the CV up to 0.6 V/s obtained with Au/Pt/Ag-GCE *versus* the scan rate and anodic peak height *versus* the square root of the scan rate.

9.3.7.2 CHRONOAMPEROMETRY (CA) STUDY OF AU/PT/AG-GCE

The CA curve of the Au/Pt/Ag-GCE is shown in Figure 9.11. The CA curve exhibited a slower current attenuation with high retention of current after 1000 s, revealing that the Au/Pt/Ag-GCE is promoting better catalytic activity in the electro-oxidation of methanol (MOR).

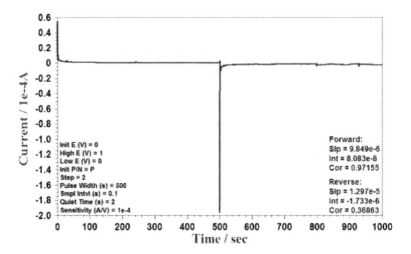

FIGURE 9.11 CA curve of Au/Pt/Ag-GCE in 0.5 M H_2SO_4 with 1 M methanol.

9.3.8 SOLAR CELL APPLICATION

The prepared silicate catalyst is used for the photocatalysis of some dye molecules.

It is found that the functionalized catalyst acts as a good catalyst by producing desired radicals, which will be useful in the degradation of the dyes. The schematic mechanism of the action is given in Figure 9.12.

9.4 CONCLUSION

In the present work, the bio-nanometal-cluster composites were prepared and characterized by TEM, SEM, and computational studies. These bio-nanometal-cluster composites were also analyzed using UV-Vis spectroscopy. Bio-physical kinetics was used to examine and to calculate the stabilization energy of the formed bio-nanometal-cluster composites. These bio-nanometal-cluster composite gave an excellent renewable solar energy conversion as well as storage capacity applications. Electrocatalytic activity of the Pt-containing TNCs by using a modified GCE in the electrochemical oxidation of methanol was reported by the CV and CA techniques. So it is demonstrated that tri-metal noble nanometal catalysts are excellent candidates for future applications in sustainable energy solutions like fuel cells and solar energy conversion techniques.

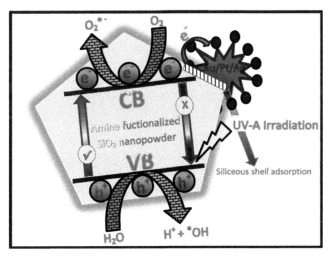

FIGURE 9.12 Schematic mechanism of photocatalysis of TNCs doped amine-functionalized silicate nanocomposites.

KEYWORDS

- **bio-nanometal-cluster composites**
- **chronoamperometry**
- **green catalyst**
- **photophysical kinetics**
- **renewable solar energy**
- **TEM and sensing**
- **theoretical applications**

REFERENCES

1. Duan, Y. N., Zhang, J. M., & Xu, K. W., (2014). *Sci. China-Phys. Mech. Astrom.*, *57*, 644–651.
2. Wang, Y., Wanzhi, W., Zeng, J., Liu, X., & Zeng, X., (2008). *Microchim. Acta*, *160*, 253–260.
3. Cai, R., Zhu, J., Chen, J., Xu, C., Zhang, W., Zhang, C., Shi, W., Tan, H., Dan, D., Hng, H. H., Lim, T. M., & Yan, Q., (2013). *J. Phys. Chem. C.*, *116*, 12468–12474.
4. Mittu, R., (2016). *Int. Adv. Res. J. Sci., Eng. Tech.*, 3, 5. doi: 10.17148/IARJSET. 2016.3508.
5. Jana, N. R., Wang, Z. L., Sau, T. K., & Pal, T., (2000). *Curr. Sci.*, *79*, 1367–1370.
6. Wertime, T. A., (1973). *Science*, *182*, 875–887.
7. Wertime, T. A., (1964). *Science*, *146*, 1257–1267.
8. Branigan, K., (1982). *Nature*, *296*, 701.
9. Shalaev, V. M., (2008). *Science*, *322*, 384–386.
10. Brongersma, M. L., & Shalaev, V. M., (2010). *Science*, *328*, 440, 441.
11. Gramotnev, D. K., & Bozhevolnyi, S. I., (2010). *Nat. Photonics*, *4*, 83–91.
12. Lal, S., Link, S., & Halas, N. J., (2007). *Nat. Photonics*, *1*, 641–648.
13. Schuller, J. A., Barnard, E. S., Cai, W., Jun, Y. C., White, J. S., & Brongersma, M. L., (2010). *Nat. Mater.*, *9*, 193–204.
14. Fang, N., Lee, H., & Zhang, X., (2005). *Science*, *308*, 534–537.
15. Chang, D. E., Sorensen, A. S., Hemmer, P. R., & Lukin, M. D., (2006). *Phys. Rev. Lett.*, *97*, 053002.
16. Chang, D. E., Sorensen, A. S., Demler, E. A., & Lukin, M. D., (2007). *Nat. Phys.*, *3*, 807–812.
17. Joachim, C., Gimzewski, J. K., & Aviram, A., (2000). *Nature*, *408*, 541–548.
18. Ozbay, E., (2006). *Science*, *311*, 189–193.
19. Atwater, H. A., & Polman, A., (2010). *Nat. Mater.*, *9*, 205–213.
20. Rang, M., Jones, A. C., Zhou, F., Li, Z. Y., Wiley, B. J., Xia, Y., & Raschke, M. B., (2008). *Nano Lett.*, *8*, 3357–3363.
21. Yan, R., Gargas, D., & Yang, P., (2009). *Nat. Photonics*, *3*, 569–576.
22. Sanders, A. W., Routenberg, D. A., Wiley, B. J., Xia, Y., Dufresne, E. R., & Reed, M. A., (2006). *Nano Lett.*, *6*, 1822–1826.
23. Sapsford, K. E., Algar, W. R., Berti, L., Gemmill, K. B., Casey, B. J., Oh, E., Stewart, M. H., & Medintz, I. L., (2013). *Chem. Rev.*, *113*, 1904–2074.
24. Daniel, M. C., & Astruc, D., (2004). *Chem. Rev.*, *104*, 293–346.
25. Graham, D., Thompson, D. G., Smith, W. E., & Faulds, K., (2008). *Nat. Nanotechnol.*, *3*, 548.
26. Zhang, G. R., Zhao, D., Feng, Y. Y., Zhang, B., Su, D. S., Liu, G., & Xu, B. Q., (2012). *ACS Nano*, *6*, 2226–2236.
27. Cheong, S., Watt, J. D., & Tilley, R. D., (2010). *Nanoscale*, *2*, 2045–2053.
28. Chen, A., & Holt-Hindle, P., (2010). *Chem. Rev.*, *110*, 3767–3804.
29. Chen, J., Lim, B., Lee, E. P., & Xia, Y., (2009). *Nano Today*, *4*, 81–95.

CHAPTER 10

AlNi and AuZn Nanohybrids for Capacitors: A Computational Study

KEKA TALUKDAR

Department of Physics, Nadiha High School, Durgapur – 713211, West Bengal, India, E-mail: keka.talukdar@yahoo.co.in

ABSTRACT

Motivated by the energy- related applications of nanomaterials and nanohybrids, the energy storage capacity of AlNi and AuZn nanohybrids is thoroughly investigated by density functional theory. Devices formed with the crystals, separated by some distance in the atomic range, are simulated in local density approximation. The capacitance of the capacitors is found by two different methods. The capacitance of the two nanohybrids shows difference in their value and nature in both the two methods. Negative capacitance is observed for AuZn nanohybrid. In this chapter, transmission spectra are also calculated by extended Huckel method that reveals very low transmission of the hybrid structures. Much improved transmission is obtained for the interface formed by the two structures. The high electrical conductivity of the interface is another remarkable result of the investigation.

10.1 INTRODUCTION

After the discovery of electricity, the advancement of the human race is now totally dependent on electric power. Due to portability and small size, the battery is the way to energy miniaturization. Starting from the ancient times, now the lithium-ion battery has gained sufficient enlightenment such as high energy density, good cycle life, and high power capacity. Capacitors with electrochemical double layers [1, 2] are basically high power capacitors compared to conventional capacitors. Compared to normal batteries, they also have high power density, where the charge is stored in between the layers. Other classes of capacitors are pseudocapacitors [3] that store charges by a redox reaction. They have very high charge and discharge rate, very high power density, excellent cycle stability. Due to these features along with their eco-friendly nature, pseudocapacitors have caught maximum attention for preparing energy storage devices. These are usually made of pseudocapacitive materials including graphene composites. However, the first choice is preferred when a large amount of power is required to store and deliver repeatedly with performance failure. The conventional batteries fail to do so consistently for their large resistance and frequent failure. These type of capacitors are built with carbon nanotubes (CNTs) [4] graphene [5, 6] carbon cloth [7] activated carbon [8], etc.

10.2 NANOPARTICLES (NPS) FOR ENERGY APPLICATIONS

Extensive research is reported on metal nanohybrids, which can serve as building blocks of smaller energy storage devices. But these materials are not abandoned in the earth, and hence their application is limited. By the application of different strategies, like improving their catalytic power and reducing their usage amount, these materials can be used in solar cell, photocatalysts, drug delivery, etc. [9]. Nonvolatile memory can be fabricated by nanohybrids by using them as a floating gate. Of them, different types of nanohybrids, including hetero-nanohybrids, are embedded in between two oxide layers. Metal nanohybrids are most effective in expanding the charge retention time, and hence give better performance of such memory devices [10]. Metal with high dielectric constant further improves the performance.

A high capacitance of 17.3 $\mu F/cm^2$ was obtained by Genc et al., [11] by using reduced graphene oxide as cathode and octyl-bis(3-methylimidazolium) dioxide as the electrolyte. The cathode material, electrolyte, and the synthesis method of capacitor materials influence the capacitance of the supercapacitors. The vacancy defects introduced into the nanosheet layers increase the pseudocapacitance of the capacitor [12]. High value of capacitance of highly crystalline zinc cobaltite nanohybrids was reported by Davis et al., [13] Epoxide driven sol-gel method was found to be very much effective, and by changing the solvent and epoxide, considerable changes were noticed. Gold nanohybrids are very much promising candidates in fabricating capacitor. Several remarkable results are reported by different studies. Double-layered gold nanohybrids, embedded in SiO_2 matrix, show high charge storage capacity when the nanohybrids are made slightly large shaped. In that case, the dielectric in between is charged quickly in the presence of gold particles [14]. The energy conversion efficiency of dye-synthesized solar cell increases in the presence of Au nanoparticles (NPs) [15]. The carrier transport property is increased by the modification of defects and oxygen vacancy density in the device.

For the smooth running of the human race, the quest for energy is a non-ending process. The hunt for energy in this decade is increased many folds for our civilization is at risk due to rapidly decreasing quantity of fossil fuel. Worldwide concept of energy materials is changing fast, and now after the discovery of CNTs [16] and graphene [17], received a new dimension of thinking. Due to the remarkably exclusive properties of these two materials and due to the progress of nanotechnology, the scientists and researchers have paid maximum concentration to the energy conversion and energy storage applications of these comparatively new materials. Researches on the individual properties of CNTs and graphene tell about their highly optimistic mechanical, electrical, thermal and optical properties which lead them to use these materials in energy storage and energy conversion materials [18–21].

The performance of supercapacitors mainly depends on the electrochemical performances of electrode material like high power density, long life, and low cost, which all are possible with CNTs. CNTs have a high aspect ratio, high mechanical strength, porous structure, high electrical conductivity which are being exploited in supercapacitor electrode. Nitrogen-doped CNT-graphite felt acts as advanced electrode materials for vanadium redox flow batteries. Nitrogen doping increases the electronic properties of the CNTs, and chemisorption properties are also improved due to the formation of defect sites. [22]. Due to the toughness and remarkable electronic properties, CNT fibers can act as an electrode material as well as in electronic devices [23]. Electricity generation in a microbial fuel cell from wastewater can also be facilitated by choosing MWCNTs as electrode materials, which shows much better performance than normal carbon cloth [24]. These electrodes have fast charging and discharging rate and excellent cycle stability and high power density.

10.3 NEGATIVE CAPACITANCE (NC)

Negative capacitance (NC) is a phenomenon observed in different substances like conducting polymer embedded with nanohybrid [25], solid-state device structures [26], light-emitting diodes (LEDs) [27, 28], interfaces of metal and semiconductor [26], photodetectors [29, 30], etc. This is an intrinsic property of Winger Crystal [31], which means electrons in the solid phase, and it is deviated from the b.c.c. or triangular lattice structure. Winger crystal-forming polymer nanowires also exhibit NC [32]. NC is very important and is originated due to several reasons. In an organic light-emitting diode, the recombination process of the injection carrier introduces NC in them [33]. Electron-hole recombination in some heterostructure leads to negative capacity. In p-i-n photodetectors, NC is resulted from carrier confinement. Where there is the existence of NC, an increase in the bias voltage decreases the charge in the parallel plates of the capacitor. The capacitor exhibits an inductive behavior.

The lattice trap states at the interface of the heterostructure [34] are responsible for such difference. In low forward bias condition, sometimes, the electrons near the Schottky barrier reside near the interface, and their excess energy helps to knock out some electrons from the trapped state near the interface. This energy of taking out the electrons, being lesser than the energy to produce electron-hole pair, the charge accumulation at the surfaces varies a lot [34]. The strong temperature dependency of the NC is studied and explained by Arslan et al., in forward bias condition, the NC decreases with increasing temperature [35]. However, the NC generated at the interface of organic solar cell made by poly(3-hexyl thiophene)/1-(3-methoxycarbonyl)propyl-1-phenyl[6,6]C 61 (P3HT:PCBM) blends is inhibited with the increase of light intensity [36]. The possible reason of which may be explained by the increase of electron-hole recombination rate due to highly intense light. NC has considerable use in increasing the optical receiver bandwidth in photodetectors [37], developing high-quality field-effect transistor (FET) with low power dissipation [38, 39],

10.4 COMPUTATIONAL STUDY ON ENERGY APPLICATIONS OF NANOMATERIALS (NMS)

Computational study on the energy applications of nanomaterials (NMs) is mainly based on density functional theory (DFT). Some combined experimental and computational analysis is also reported. Mohammad et al., [40] obtained maximum power conversion efficiency (PCE) among different ruthenium sensitizers as 11.18%. Computational reporting was also in agreement with experiment. Calculations were done with time-dependent DFT. The study of Oguchi et al., [41] proved experimentally and computationally the stabilized high-temperature phase of $LiBH_4$, which is a potential candidate for solid-state electronics. Here also, both studies are in well agreement.

In this chapter, the energy storage applications of AlNi and AuZn nanocrystals are investigated by Density functional approach. Parallel plate nano capacitors with separating distance in nano range are subjected to atomic-scale simulation and capacitance of the nanodevice is found by finding the electrostatic energy as a function of bias voltage. Calculations are done by ATK-DFT engine offered by the commercial software ATK-quantum wise. Electrostatic difference potential and electrostatic difference density are found at a bias voltage of 1 volt and plotted against distance. Future study of this work includes the investigation of change of capacitance of the crystals in various conditions and also the interface properties of the nanohybrids.

10.5 THEORY

The theory behind this analysis is simple physics of capacitor that is made up of two parallel plates called electrodes separated by a distance, and a bias voltage is applied in between them. Positive and negative charges are built up in the opposite plates, i.e., an electric field is generated.

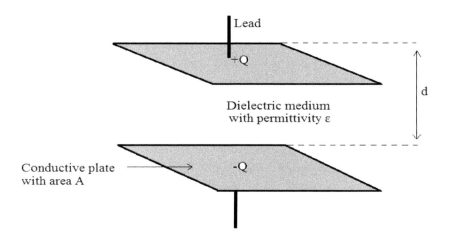

If a bias V is applied across the leads opposite charges, ±Q is built-up in different plates. Electric field E is generated in between the plates. The capacitance of the capacitor

$$C = \frac{Q}{V} = \frac{Q}{Ed} = \frac{A\varepsilon}{d} \quad (1)$$

Here A is the area of each plate, ε is the permittivity of the space between the plates, and d is the distance between the plates.

The electrostatic energy is found out by first principle calculation, and from that capacitance is calculated. Capacitance is also calculated from the Mulliken charge populations of the atoms. The first principle calculations are done by commercial software Quantum Wise [42]. Firstly, the Mulliken charge population, electrostatic difference potential, and electrostatic difference density are found out. On the application of finite bias voltage, net charges are accumulated on the plates. The net charge if plotted against bias voltage, the slope of the curve gives the capacitance.

The second method of finding the capacitance is to find the electrostatic energy E as a function of the applied bias V.

$$E(V) = \frac{1}{2} \int \delta v(r,V) \delta n(r,V) dr \quad (2)$$

Here δv and δn are the induced electrostatic potential and the induced density. The induced density and electrostatic potential as a function of the distance between the plates. The capacitance is found by fitting a parabola to the relation.

$$E(V) = C \frac{V^2}{2} \quad (3)$$

In this approach, we assume that some part of the electron density is distributed in some finite region where the electric field is present. Whereas in the former assumption the electrons are only distributed over the two surfaces of the plates.

The equations are valid for the ideal capacitor, in an atomic scale the properties may be quite different. However, evaluating the electrostatic energy and then capacitance, the main process may be understood. Also from the Mulliken charge populations of the atoms, the capacitance can be found. These two processes are described, and results are obtained after calculation. These two results are compared.

AlNi and AuZn crystals are cleaved along the plane with Miller indices [100]. Each structure is repeated six times and starting from the middle portion, the two parts are separated by a distance, and a nanodevice is made with the whole structure. Here separation of two parts is done by considering two ghost atoms in the middle and then shifting the right portion to more right with the ghost atom. As the distance between the two portions is still too small, hence the right part is shifted more up to 2 Å. after forming the device, it now represents a parallel plate capacitor.

The device is subjected to simulation with DFT-LDA method with double zeta polarized set. Grid points are set as 9×9×100 k point. The Mulliken population (MP), Electron difference potential (EDP), and Electron Difference density (EDD) are calculated first. The change in surface charge with the change in bias voltage, i.e., $\frac{dq}{dV}$ is calculated. Capacitance is calculated in two different ways.

10.5.1 DENSITY FUNCTIONAL THEORY (DFT)

For periodic system, DFT is a popular method. It helps to find the energy surfaces of molecules in quantum mechanics. Also, it can be used to find dipole moment, charge distribution, etc. Here simulation is done where we adopt the linear combination of atomic orbitals (LACO).

In quantum mechanics, the Hamiltonian of a system can be written simply as:

$$H = T + U + V$$

where the mutual interaction term is given as U and interaction term with an external field V is defined as:

$$V = \sum_{i=1}^{n} v(r_i) \qquad (4)$$

10.5.2 BORN-OPPENHEIMER APPROXIMATION

Born-Oppenheimer approximation [43] is:

$$\hat{H}\Psi(r_1, r_2, ...r_{N)}) = E\Psi(r_1, r_2, ...r_{N)}) \qquad (5)$$

In the material simulation, the third term of equation (1) is the interaction of the electrons and nuclei only.

Here wavefunction Ψ is very complicated, and thus we need a physics observable which form the Hamiltonian a priori. Pierre Hohenberg and Walter Kohn [44] developed a theory, known as DFT which states that to find the energy of the system, only the two-particle probability density is important and that is a function of three coordinates. One doesn't have to find all 3N coordinates associated with the wave function. For their work, Walter Kohn was awarded the noble prize in 1998 [45].

10.5.3 KOHN SHYAM EQUATION

According to Hohenbergy Kohn theory, different densities are chosen of which the lower one represents the more accurate one. So choosing an improved density and the solution of the question of applying that density to the variational equation without the help of wave function detail is important. Up to this point, Hamiltonian of the system can be determined by the electron density. This theory, along with the variational principle gives DFT. The theory points towards the existence of a universal functional $F(\rho)$ minimization of which gives ground state density and energy. So the basic DFT can be written as:

$$\delta\left[F[\rho] - \mu \int \rho(r)dr - N\right] = 0 \quad (6)$$

Here $\rho(r)$ is the electronic density. The ground state energy and density is the minimum of the functional $F(\rho)$ when the density represents the exact number of electrons. μ is the chemical potential. The universal functional term does not require the detail of the external potential field.

Now according to equation (1), energy functional contains three terms. These are the kinetic term, i.e., the term arising from external potential and the term originated from electron-electron interaction. So writing the total energy, we get the three terms in the energy expression.

$$E[\rho] = T_e[\rho] + V_{e-e}[\rho] + V_{ext}[\rho] \quad (7)$$

$V_{ext}[\rho]$ is trivial. The other two terms are to be approximated.

In the Kohn-Sham approach [46], a system of N non-interacting electrons is described by a single determinant wavefunction in N orbitals. Electronic energy of a molecule is now:

$$E = T_e + V_{ne} + V_{e-e} + V_{xc} \quad (8)$$

The last term comes from exchange-correlation. V_{ne} is the Coulomb interaction with nuclei. The non-interacting kinetic energy T_e can be found by writing the density as a product of orbital densities and hence T from orbitals.

The energy functional is:

$$E[\rho(r)] = \sum_{i}^{N} \left(\langle \chi_i | -\frac{1}{2}\nabla_i^2 | \chi_i \rangle - \langle \chi_i | \sum_{k}^{nuclei} \frac{Z_k}{|r_i - r_k|} | \chi_i \rangle \right)$$
$$+ \sum_{i}^{N} \langle \chi_i | \frac{1}{2} \int \frac{\rho(r')}{|r_i - r'|} dr' | \chi_i \rangle + E_{xc}[\rho(r)] \quad (9)$$

For open and bound systems are respectively found by Non-equilibrium Green's functions (NEGF) and diagonalized Kohn-Sham approach.

While calculating energy, we assume one electron Kohn-Sham Hamiltonian

$$H = -\frac{h^2}{2m}\nabla^2 + V^{eff}[n](r) \quad (10)$$

Where the first term is the kinetic energy of the electron and other is the effective potential of all interaction due to other electrons and that with the external electric field. Certainly, in DFT, electrons are described in terms of energy density.

The total wavefunction is assumed to be a linear combination of orbitals. Transferring the differential equation to a matrix equation, two matrices, Hamiltonian matrix, and overlap matrix are to be solved.

Electron density in terms of the density matrix is written as:

$$\rho(r) = \sum_{ij} D_{ij} \phi_i(r) \phi_j(r) \quad (11)$$

and the *density matrix* D_{ij}

$$D_{ij} = \sum_{\alpha} f_\alpha c_{\alpha i}^* c_{\alpha j} \quad (12)$$

$c_{\alpha i}$ is expansion coefficients.

EDD of a many-body system:

$$\Delta \rho(r) = \rho(r) - \sum_{\mu} \rho^{atom}(r - R_j) \quad (13)$$

Here R_j is the position of the jth atom. The expression gives the difference in the electron density of a many-body system and the individual atom-based electron densities.

10.6 RESULTS AND DISCUSSION

10.6.1 ENERGY STORAGE

The device structure of AlNi and AuZn nanohybrids are shown in Figure 10.1(a) and (b). C axis actually indicates the z-axis.

242   *Nanomaterials-Based Composites for Energy Applications*

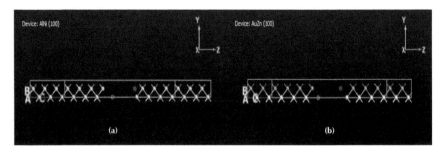

FIGURE 10.1 Device structure of atomic-scale capacitor made of (a) AlNi (b) AuZn nanohybrids.

The induced charge on applying bias voltage is calculated as Mulliken charges, accumulation of which along the two sides of the parallel plates of AlNi hybrid is shown in Figure 10.2. Figure 10.3 exhibits the same for AuZn hybrid.

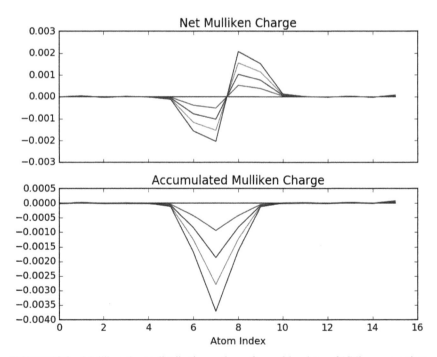

FIGURE 10.2 Mulliken charge distribution on the surfaces of the plates of AlNi nanocapacitor.

AlNi and AuZn Nanohybrids for Capacitors

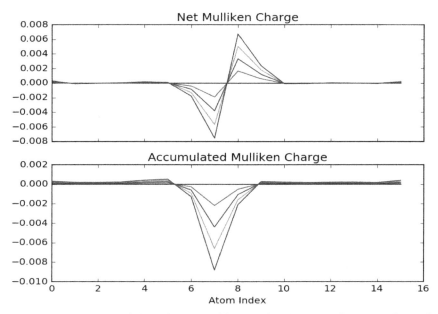

FIGURE 10.3 Accumulation of a net positive/negative charge on the two surfaces of AuZn nanohybrid in the atomic level.

MP analysis [47] is based on the linear combination of an atomic orbital. The two structures offer two different charge distributions; that means, it depends on structure, mass, density, everything. The atomic masses of AlNi and AuZn are 85.675 Da and 262.376 Da, respectively. Their densities are 6.21 Mg·m^{-3} and 13.96 Mg·m^{-3}, respectively. Both of them have b.c.c. structure of prototype CsCl. The results of this calculation can next be used for convergence in the self-consistent-field calculation.

$\frac{dq}{dV}$, i.e., the capacitance can now be calculated from by calculation MP and charge distribution for four different bias voltages 0.25 V, 0.5 V, 0.75 V and 1.0 V. Induced charge vs. bias voltage is plotted in Figure 10.4 and 10.5 for AlNi and AuZn, respectively.

For an ideal parallel plate capacitor, capacitance:

$$C = \frac{Q}{V} = \frac{Q}{Ed} = \frac{A\varepsilon}{d}$$

where $\varepsilon=1$ is the permittivity of vacuum (Figure 10.3). Calculating the area of each plate, we can find:

Capacitance = 5.9392e–22 Farad.

Parallel plate d = 10.7759 Angstrom

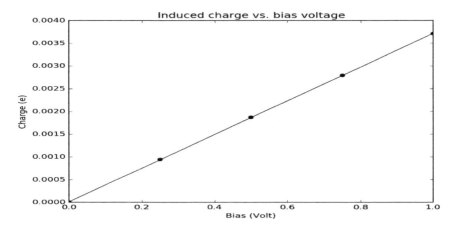

FIGURE 10.4 Charge vs. bias voltage for AlNi nanohybrid parallel plate capacitor.

From equation (2) electrostatic energy as a function of the applied bias V. In the second process we assume that some part of the electron density is distributed in a region where an electric field exists other than on the plate surfaces (Figure 10.5).

The dependence of capacitance on the bias is analyzed too. The dependence is shown in Figure 10.4. The electrostatic potential and electrostatic densities are calculated at a voltage of 1 V. The density and potential are given in Figure 10.6 for AuZn nanohybrid. For AlNi, the calculation is done in the same way, and capacitance is found close to that obtained by Mulliken analysis. For AuZn nanohybrid, the calculated value of capacitance and distance between the parallel plates in both the methods are C = –1.28683e–21 Farad and distance d = 0.70018 Angstrom (Figure 10.6).

AlNi and AuZn Nanohybrids for Capacitors

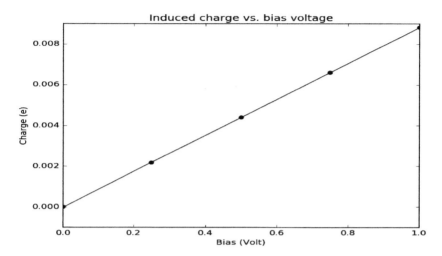

FIGURE 10.5 Charge vs. bias voltage of the parallel plate capacitor made of AuZn nanohybrid.

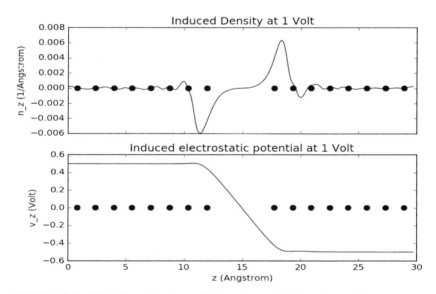

FIGURE 10.6 The induced density and induced potential along the z-axis.

Fitting a parabola to the relation given in equation (3), we can calculate capacitance. Energy-bias curve is exhibited in Figure 10.7. Interesting result is obtained for AuZn hybrid. NC is reported in this sample. The evidence of NC was obtained in different literature where the phenomenon was reported in solid-state device structures, polymer-based nanocomposites, conducting polymer, metal-semiconductor interface, solar cells (SCs), and many more. The matter is discussed in the introduction part.

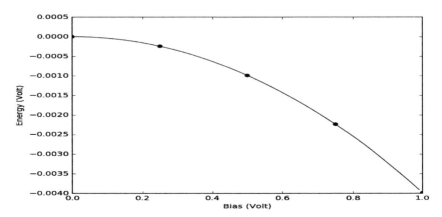

FIGURE 10.7 Energy vs. voltage of a AuZn nanohybrid parallel plate capacitor.

The capacitance vs. voltage is given in Figure 10.8, which compares the results of two processes for AuZn nanohybrid.

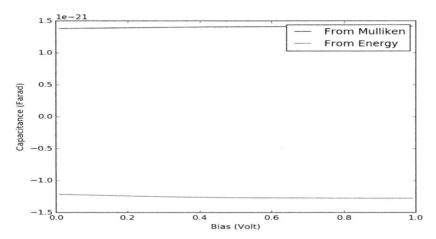

FIGURE 10.8 Change of capacitance with bias voltage in two different processes-by Mulliken analysis and from energy for AuZn nanohybrid.

10.6.2 TRANSMISSION SPECTRA

The transmission spectra of the nanohybrids are then analyzed by Extended-Huckel method. Figure 10.9 depicts the transmission spectra of the AlNi nanohybrid capacitor structure. The red line the sum of up and down spin. The Fermi level is shown as a dotted line. The transmission is very low with the range −1 to +1 eV energy range. The same k points are used here as the above calculation. Here Self-consistent function calculation is not done. Neumann boundary condition is employed. It assumes the constraint that the derivative of the electrostatic potential is zero on the boundary. The asymmetric position of the band edge relative to the Fermi level is reported in an unbiased condition.

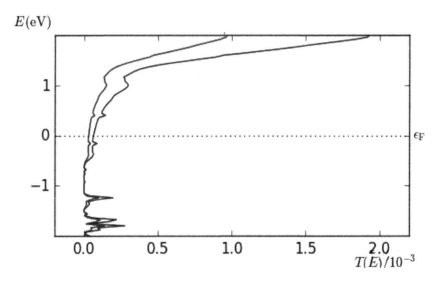

FIGURE 10.9 Transmission spectra of AlNi nanohybrid capacitor.

The transmission zone is depicted in Figure 10.10. An interpolated contour plot is shown in Figure 10.11. It is a plot of transmission coefficients against the reciprocal vectors k_A and k_B. The resolution of the contour plot can be increased by the increase of the number of k-points.

The transmission spectra of AuZn nanohybrid is given below. The contour plot for this spectra is shown in Figure 10.12.

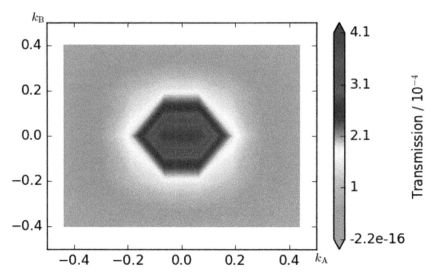

FIGURE 10.10 **(See color insert.)** Contour plot of transmission coefficients along the reciprocal vectors k_A and k_B for AlNi nanohybrid.

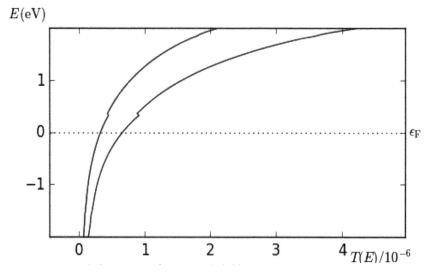

FIGURE 10.11 **(See color insert.)** Transmission spectra of AuZn nanohybrid.

Now an interface is modeled with AuZn and AlNi, and transmission spectra is observed (Figure 10.13). An abrupt change in the spectra is

AlNi and AuZn Nanohybrids for Capacitors

observed at the interface at zero bias condition. Very low transmission results. The interface is formed for the minimum strain of this configuration. The transmission spectra of the interface are given in Figure 10.14. The band edge distribution is changed, and almost symmetric distribution of transmission bands are obtained on both sides of the Fermi level.

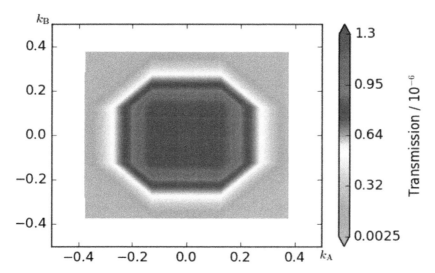

FIGURE 10.12 (See color insert.) Contour plot of transmission coefficients along the reciprocal vectors k_A and k_B for AuZn nanohybrid.

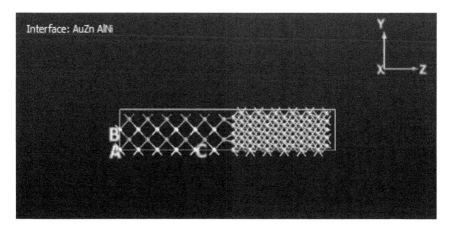

FIGURE 10.13 Interface formed with AuZn and AlNi nanohybrids.

250   Nanomaterials-Based Composites for Energy Applications

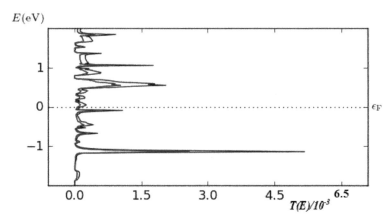

FIGURE 10.14   Transmission spectra of AuZn/AlNi interface.

10.6.3   ELECTRICAL CONDUCTIVITY AT THE INTERFACE

To find how the properties at the interface change, the thermoelectric conductivity is calculated for the interface. This can be obtained from the transmission spectra. The conductivity of the AlNi and AuZn nanohybrids is very low, but remarkable conductivity is observed for the interface of the two materials, which is plotted in Figure 10.15.

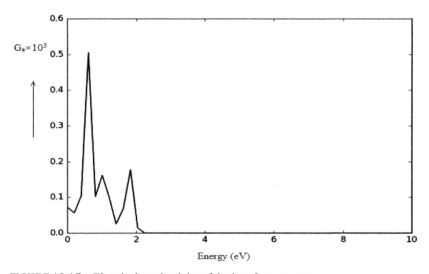

FIGURE 10.15   Electrical conductivity of the interface vs. energy.

10.7 CONCLUSIONS

The accumulation of charges as Mulliken charge is plotted for the two types of nanohybrids. Capacitance is found by the MP and charge distribution. The capacitance of AlNi and AuZn nanohybrids is found as 5.9392e–22 Farad and –1.28683e–21 Farad, respectively. The induced charge increases with the bias voltage. The capacitance found by calculating induced density and induced potential is in close resemblance with that calculated by the MP method. Change of electrostatic energy with bias voltage is observed here, which shows different nature in the two nanohybrids. By extended Huckel method, the transmission spectra of the hybrid structures are calculated. Though very low transmission is observed, the transmission spectra are much improved when the interface is formed with the two nanohybrids. The transmission spectra are used to find the electrical conductivity at the interface, which is calculated as very high. Maximum conductivity reaches to 0.5×10^3 S/m.

KEYWORDS

- carbon nanotubes
- field-effect transistor
- light-emitting diodes
- nanoparticles
- negative capacitance
- power conversion efficiency

REFERENCES

1. Dar, R. A., Giri, L., Karna, S. P., & Srivastava, A. K., (2016). Performance of palladium nanoparticle-graphene composite as an efficient electrode material for electrochemical double layer capacitors. *Electrochim. Acta., 196*, 547–557.
2. Nanoparticle Technologies: From lab to market, (2013). *Farid Bensebaa* (1st edn., pp. 347–348). Acaemic Press: UK.
3. Wang, W., Guo, S., Lee, I., Ahamed, K., Zhong, J., et al., (2014). Hydrous ruthenium oxide nanoparticles anchored to graphene and carbon nanotube hybrid foam for supercapacitors. *Scientific Reports., 4*, 4452.

4. Hu, L., Chen, W., Xie, X., Liu, N., Yang, Y., et al. (2011). Symmetrical MnO_2-carbon nanotube-textile nanostructures for wearable pseudocapacitors with high mass loading. *ACS Nano, 5*, 8904–8913.
5. Wang, H., Casalongue, S. H., Liang, Y., & Dai, H., (2010). $Ni(OH)_2$ nanoplates grown on graphene as advanced electrochemical pseudocapacitor materials. *J. Am. Chem. Soc., 132*, 7472–7477.
6. Zhao, Y., Li, M. P., Liu, S., & Islam, M. F., (2017). Superelasticpseudocapacitors from freestanding MnO_2 -decorated graphene-coated carbon nanotube aerogels. *ACS Appl. Mater. Interfaces, 9*, 23810–23819.
7. Gao, S., Liao, F., Ma, S., Zhu, L., & Shao, M., (2015). Network-like mesoporous $NiCo_2O_4$ grown on carbon cloth for high-performance pseudocapacitors. *J. Mat. Chem. A, 3*, 16520–16527) activated carbon.
8. Teng, Y., Liu, K., Liu, R., Yang, Z., Wang, L., et al., (2017). *A Novel Copper Nanoparticles/Bean Dregs-Based Activated Carbon Composite as Pseudocapacitors, 89*, 33–41.
9. Davis, M., Gumeci, C., Black, B., Korzeniewski, C., & Hope-Weeks, L., (2012). *Tailoring Cobalt-Doped Zinc Oxide Nanohybrids With High Capacitance Activity: Factors Affecting Structure and Surface Morphology, 2*, 2061–2066.
10. Liu, Z., Lee, C., Narayanan, V., Pei, G., & Kan, E. C., (2002). Metal nanohybrid memories-part II: Electrical characteristics. *IEEE Trans. Electron Devices, 49*, 1606–1613.
11. Genc, R., Alas, M. O., Harputlu, E., Repp, S., Kremer, N., et al., (2017). *Scientific Reports, 7*, 11222.
12. Gao, P., Metz, P., Hey, T., Gong, Y., Liu, D., et al., (2017). The critical role of point defects in improving the specific capacitance of δ-MnO_2 nanosheets. *Nature Communications, 8*, 14559.
13. Fan, Z., & Zhang, H., (2016). Template synthesis of noble metal nanohybrids with unusual crystal structures and their catalytic applications. *Acc. Chem. Res., 49*, 2841–2850.
14. Shah, A. A., Umar, A. A., & Salleh, M. M., (2016). Efficient quantum capacitance enhancement in DSSC by gold nanoparticles plasmonic effect. *Electrochimica Acta, 195*, 134–142.
15. Chakraborty, C., & Bose, C., (2016). Effect of size and position of gold nanohybrids embedded in gate oxide of SiO_2/Si MOS structures. *J. Adv. Dielect., 6*, 1650001.
16. Iijima, (1991). *Nature, 354*, 56.
17. Novoselov, K. S., Geiml, A. K., Morozov, S. V., Jiangl, D., Zhang, Y., Dubonos, S. V., Grigorieva, I. V., & Firsov, A. A., (2004). Electric field effect in atomically thin carbon films. *Science, 306*(5696), pp. 666–669.
18. Yiran, W., Huige, W., Yang, L., Suying, W., Evan, K. W., & Zhanhu, G., (2015). Multifunctional carbon nanostructures for advanced energy storage applications. *Nanomaterials, 5*, 755–777.
19. Shun, M., Ganhua, L., & Junhong, C., (2015). Three-dimensional graphene-based composites for energy applications. *Nanoscale, 7*, 6924.
20. Liming, D., Dong, W. C., Jong, B. B., & Wen, L., (2012). Carbon nanomaterials for advanced energy conversion and storage. *Small, 8*(8), 1130–1166.

21. Guan, W., Pengfeng, T., Dongxing, W., Zhe, L., Lu, P., Ying, H., Caifeng, W., Wei, Z., Su, C., & Wei, C., (2017). High-performance supercapacitors based on electrochemical-induced vertical-aligned carbon nanotubes and polyaniline nanocomposite electrodes. *Scientific Reports, 7*, Article Number: 43676, doi: 10.1038/srep43676.
22. Shuangyin, W., Xinsheng, Z., Thomas, C., & Arumugam, M., (2012). Nitrogen-doped carbon nanotube/graphite felts as advanced electrode materials for vanadium redox flow batteries. *J. Phys. Chem. Lett., 3*(16), pp. 2164–2167.
23. Juan, J. V., & Rebeca, M., (2015). Tough electrodes: Carbon nanotube fibers as the ultimate current collectors/active material for energy management devices. *Chem. Mater., 27*(20), pp. 6901–6917.
24. Thepsuparungsikul, N., Phonthamachai, N., & Ng, H. Y. (2012). Multi-walled carbon nanotubes as electrode material for microbial fuel cells. *Water Sci. Technol., 65*(7), 1208–1214.
25. Bakueva, L., Konstantatos, G., Musikhin, S., Ruda, H. E., & Shik, A., (2004). Negative capacitance in polyer-nanohybridcomposites. *Appl. Phys. Lett., 85*, 3567–3569.
26. Wu, X., Yang, E. S., & Evans, H. L., (1990). Negative capacitance at metal-semiconductor interfaces. *J. Appl. Phys., 68*, 2845.
27. Pingree, L. S. C., Scott, B. J., Russell, M. T., Marks, T. J., & Hersam, M. C., (2005). Negative capacitance in organic light-emitting diodes. *Appl. Phys. Lett., 86*, 073509.
28. Guan, M., Niu, L., Zhang, Y., Liu, X., Li, Y., et al., (2017). Space charges and negative capacitance effect in organic light-emitting diodes by transient current response analysis. *RSC Adv., 7*, 50598–50602.
29. Ershov, M., Liu, H. C., Buchanan, M., Wasilewski, Z. R., et al., (1997). Usual capacitance behavior of quantum well infrared photodetectors. *Appl. Phys. Lett., 70*, 1828.
30. Gharbi, R., Abdelkrim, M., Fathallah, M., Tresso, E., & Ferrero, S., (2006). Observation of negative capacitance in a-Sic:H/a-Si:H UV photodetectors. *Solis State Electronics., 50*, 367–371.
31. Wigner, E., (1934). On the interaction of electrons in metals. *Phys. Rev., 46*, 1002–1011.
32. Rahman, A., & Sanyal, M. K., (2009). Negative capacitance in Winger crystal forming polymer nanowires. *Appl. Phys. Lett., 94*, 242102.
33. Zhu, C., Feng, L., Cong, H., Zhang, G., & Chen, Z., (2009). Negative capacitance in light-emitting devices. *Solid State Electronics., 53*, 324–328.
34. Byrum, L. E., Ariyawansa, G., Jayasinghe, R. C., Dietz, N., Perera, A. G. U., et al., (2009). Capacitance hysteresis in GaN/AlGaN heterostructures. *J. Appl. Phys., 105*, 023709.
35. Arslan, E., Şafak, Y., Altındal, Ş., Kelekçi, Ö., & Özbay, E., (2010). Temperature dependent negative capacitance behavior in (Ni/Au)/AlGaN/AlN/GaN heterostructures. *Journal of Non-Crystalline Solids, 356*, 1006–1011.
36. Lungenschmied, C., Ehrenfreund, E., & Sariciftci, N. S., (2009). Negative capacitance and its photo-inhibition in organic bulk heterojunction devices. *Organic Electronics, 10*, 115–118.
37. Youn, J. S., Kang, H. S., Lee, M. J., Park, K. Y., & Choi, W. Y., (2009). High-speed CMOS integrated optical receiver with an avalanche photodetector. *IEEE Photonics Technology Letters, 21*, 1553–1555.

38. Wang, X., Chen, Y., Wu, G., Li, D., & Tu, L., (2017). Two-dimensional negative capacitance transistor with polyvinylidene fluoride-based ferroelectric polymer gating. *NJP 2D Materials and Applications*, 38.
39. Vural, Ö., Şafak, Y., Türüt, A., & Altındal, Ş., (2012). Temperature dependent negative capacitance behavior of Al/ahodamine-101/n-GaAs Schottky barrier diodes and R_s effects on the C-V and G/w-V characteristics. *Journal of Alloys Compounds, 513*, 107–111.
40. Nizamuddin, M. K., Angelis, F. D., Fantacci, S., Selloni, A., Viscardi, G., et al., (2005). Combined experimental and DFT-TDDFT computational study of photoelectrochemical cell Ruthenium sensitizers. *JACS, 127*, 16835–16847.
41. Oguchi, H., Matsuo, M., Hummelshøj, J. S., Vegge, T., Nørskov, J. K., et al., (2009). Experimental and computational studies on structural transitions in the $LiBH_4$-LiI pseudobinary system. *Appl. Phys. Lett., 94*, 141912.
42. Brandbyge, M., Mozos, J. L., Taylor, J., & Stokbro, K., (2002). Atomistix tool kit version 2016.2. Quantum wise A/S (www.quantumwise.com). *Density-Functional Method for Nonequilibrium Electron Transport Phys. Rev. B, 65*, 165401.
43. Kohn, W., (1998). Electronic structure of matter-wave functions and density functionals. *Noble Lecture, 71*, 1253.
44. Hohenberg, P., & Kohn, W., (1964). Inhomogeneous electron gas. *Phys. Rev., 136*, B864.
45. Born, M., & Oppenheimer, J. R., (1927). On the quantum theory of molecules. *Annalen der Physik., 389*, 457–484.
46. Kohn, W., & Sham, L. J., (1995). Self consistent equations including exchange and correlation effects. *Phys. Rev., 140*, A1133.
47. Mulliken, R. S., (1955). Electronic population analysis on LCAO-MO molecular wave functions. *I. J. Chem. Phys., 23*, 1833–1840.

CHAPTER 11

Recycling of Polymer Nanocomposites of Carbon Fiber Reinforced Compressed Natural Gas Reservoirs

GILBERTO JOÃO PAVANI,[1] SÉRGIO ADALBETO PAVANI,[2] and CARLOS ARTHUR FERREIRA[3]

[1]*Instituto Federal do Rio Grande do Sul, Av. São Vicente 785, Farroupilha, Brazil, E-mail: gilberto.pavani@farroupilha.ifrs.edu.br*

[2]*Universidade Federal de Santa Maria, Av. Roraima 1000, Santa Maria, Brazil*

[3]*Universidade Federal do Rio Grande do Sul, Av. Bento Gonçalves 9500, Porto Alegre, Brazil*

ABSTRACT

This research work deals with the application of polymeric nanocomposites for the production of compressed natural gas (CNG) reservoirs reinforced with carbon fiber preimpregnated with epoxy resin through the filament winding process. It offers technical solutions adopted for the use of nanoclay to reduce gas permeability and reinforcement with carbon fiber to achieve the pressure resistance required for the vehicular storage of CNG. The obtained results indicate that the used methodology enables the production of pressure vessels for CNG storage for vehicular use in polymeric composites, according to the guidelines of ISO 11439: 2000 for CNG-4 cylinders which states that the constituent materials must be recycled as a part of the product's life cycle and used to replace virgin raw material in the production of injected polymers, for example, aiming for economy and the preservation of natural resources.

11.1 INTRODUCTION

The need to reduce fuel consumption and atmospheric emissions has led the automotive and aerospace industry to increasingly use composite materials such as carbon fiber [1], creating the need to recycle waste from components that are disused or which have reached the end of their useful life as raw material for the manufacturing of new industrial products employing environmentally sustainable and economically viable solutions.

Environmentally friendly technologies are a growing demand from society and government bodies for the purpose of preserving natural resources through innovation, involving the entire product life cycle from design to final disposal [2].

Based on the recycling procedures used by the *Aircraft Fleet Recycling Association* – AFRA [3], the carbon fiber used in the coating of reservoirs for compressed natural gas (CNG) was reused, with the goal to use it in the injection process of thermoplastics such as polyethylene, replacing glass fiber, which is widely used in the industry [4].

However, the recycling of nanopolymers such as the one used in reservoir's liner is not done with procedures as solidly defined as those determined by AFRA, requiring a particular analysis for the preservation of the polymer's mechanical and physicochemical properties.

Natural gas is considered the fossil fuel with the lowest emission of carbon monoxide [5], presenting low emissions of pollutants and residue from the combustion process, and can be used efficiently in internal combustion engines, helping to preserve the environment.

Since natural gas has a low energy density compared to liquid fuels, it is usually stored under pressure in steel tanks whose own weight is proportional to their storage capacity and the maximum permissible pressure of the steel from which it is manufactured [6].

The use of polymer reservoirs instead of metallic reservoirs aims to reduce the weight of inert load present in vehicles fueled with CNG, allowing fuel economy and increased life of the engine and the braking and suspension systems, as well as having ecological and environmental benefits, such as the reduction of the emission of atmospheric pollutants and the recyclability of its components.

The polymer liner offers characteristics such as lightness, pressure resistance, corrosion protection, and durability to the reservoir, being produced in polyethylene by rotomolding, as it allows large-scale production of seamless liners using low-cost molds.

As the polyethylene used to make the liner is permeable to methane [7], main component of CNG [8], the addition of nanoclay is necessary to ensure its efficient and safe storage, reducing the risk of fires or explosions.

The high pressure of vehicular CNG storage requires the polymer liner to be reinforced with carbon fiber, using the filament winding (FW) process which has low production cost, high productivity, high reliability and can provide almost isotropic properties to the final product through the superposition of fiber layers at different winding angles [9].

11.2 POLYMER RESERVOIRS FOR COMPRESSED NATURAL GAS (CNG)

The main disadvantages presented by natural gas as an automotive fuel are the high cost of compression and the weight of the steel reservoir for transport and storage [10], making it difficult for the automobile fleet to use on a large scale.

The reservoir for CNG made of composite material aims to reduce the inert load present in the vehicles supplied with this type of fuel, allowing economy and increasing the useful life of the engine, as well as reducing the emission of pollutants, contributing to the preservation of the environment.

The polymer reservoir for CNG can be divided into the liner, whose main function is to act as a gas barrier, and coating, whose main function is to resist the high storage pressure, which is verified in the hydrostatic test.

The polymer liner offers characteristics such as lightness, resistance to pressure and impacts, corrosion protection and durability to the reservoir, being produced in polyethylene by rotomolding, due to the low cost of the raw materials and of the production process.

Currently, polyethylene is the most widely used resin in rotomolding due to its low cost, chemical inertia, ease of processing and mechanical properties that can be improved by crosslinking or adding nanoclays to polyethylene [11].

Rotomolding is indicated for the production of hollow and dimensionally stable pieces of large size such as the liner of the CNG reservoir, presenting surface finish comparable to that of injection molding. In addition, it has a lower mold cost and greater hollow part production capacity than injection molding [12].

The high storage pressure of CNG requires the polymer liner to be coated with carbon fiber because of its high mechanical resistance, using the filament winding process due to the low cost of production, high productivity, high reliability and its capacity to provide almost isotropic properties to the product [13].

Carbon fiber presents high rigidity, but the values presented in Table 11.1 are for comparison, since they depend on the production process [14].

TABLE 11.1 Tensile Strength of Commercial Fibers

Material	MPa
Kevlar	2757
E-Glass	3450
Carbon fiber	4127

The liner was used as a mandrel in the FW process, allowing the deposition of the fiber for structural reinforcement at different angles and patterns in the direction of the main stresses, by superposition of layers composed of angular and circumferential windings.

Pressure vessels for CNG are manufactured in accordance with ISO 11439:2013–Gas cylinders–High-pressure cylinders for on-board storage of natural gas as fuel for automotive vehicles [15], which specifies the requirements for the serial production of Reservoirs for the storage of CNG for automotive vehicles, including buses and trucks.

This standard classifies as CNG-4, the cylinders produced with non-metallic materials, using manufacturing methods adapted to the service conditions of the vehicular transportation of natural gas, such as the reservoir in development.

Each polymer reservoir was coated with three layers of T700SC–12k carbon fiber pre-impregnated with UF3360 TCR epoxy resin [16] composed of a non-geodesic winding with a 10° angle and a circumferential winding, being subjected to hydrostatic testing for its service pressure [17], the result of which is shown in Figure 11.1.

Three-layer reservoirs reached an average pressure of 350 bar ± 5% when subjected to hydrostatic testing but did not meet the requirements of ISO 11439:2013 [15] which determines the minimum test pressure for a CNG-4 reservoir as 2.25 higher than the service pressure, set at 220 bar by the National Petroleum Agency [18], resulting in a test pressure of 495 bar.

The analysis of the results indicated that a fourth layer will meet the requirements of said norm, according to a calculation by linear regression and the proposed equation for the definition of the number of layers for any internal pressure applied in the CNG reservoir.

FIGURE 11.1 Reservoir destroyed with fracture detail (author's own).

The results indicate that the methodology used enables the production of reservoirs for vehicular use classified as CNG-4, according to ISO 11439:2013 [15], allowing a reduction of at least 60% in weight in relation to an equivalent steel cylinder.

Therefore, the use of the reservoir under development is a viable alternative for CNG fueled vehicles by reducing inert load, associating fuel economy with increasing engine life, as well as adding ecological and economic benefits to the society and optimizing the useful volume of the trunk by making available new fuel tank formats to meet the growing variety of models due to the internationalization of the automotive market.

However, the large-scale production of the polymer reservoir for CNG will result in a heavy environmental liability due to its estimated useful life being five years making it necessary to be recycled in accordance with current environmental standards such as ISO 14000 – Environmental Management [19].

A reservoir, under several cycles of loading, tends to break under stresses less than those obtained in the hydrostatic test. The magnitude of these stresses decreases with the number of cycles due to the mechanical degradation of the material, known as fatigue, whose cracks usually start at points of maximum stress, irregularities, or structural defects.

Since the number of cycles is proportional to the life of the reservoir, it is important to select the right materials, the liner manufacturing process and the fiber-winding angle to avoid stress concentration, as well as the additives used in the polymer liner.

In this sense, after the hydrostatic test, the polyethylene and the carbon fiber composite removed from the liner were sent to the Materials Laboratory of the Federal University of Santa Maria to investigate suitable recycling techniques.

Aiming to recover the carbon fiber from the scraps of CNG polymer reservoirs to reduce environmental impact, the recycling technologies used by the aerospace industry such as pyrolysis [20] and the catalytic conversion of the resin were investigated [21].

Pyrolysis recycling results in the decomposition of the resin of composite materials by heating the carbon fiber in a high vacuum oven, above the curing temperature of the resin, avoiding to burn the carbon fiber to maintain its mechanical properties similar to the original material [22] for use in other industrial applications.

The high vacuum pressure significantly reduces the residual air in the furnace or autoclave, producing components free of surface oxidation and nitriding, due to the absence of oxygen and nitrogen during heating, preserving the carbon fiber.

Epoxy resins are the most widely used polymer composites in the automotive and aerospace industries [23], because their formulation and the stoichiometric ratio between resin and hardener have a decisive effect on the mechanical properties of the final product [24].

As polyethylene allows the passage of methane under pressure through the walls of the liner, it needs the addition of nanoclay to reduce its gas permeability. In this research, to quantify the permeability of the nanopolymer liner to the methane, oxygen was used, due to it having polarity and molecular diameter compatible with this hydrocarbon.

In the polymeric nanocomposite, the maleic acid compatibilized montmorillonite nanoclay (Cloisite 15A) produced by the company Southern Clay was used, according to the manufacturer's instructions.

The literature recommends a weight percentage of nanoclay between 1% and 5% [25–27] for the gas barrier effect, but the limit value of 5% was chosen [28–30] in order to verify the influence of the type of nanoclay in the barrier and mechanical behavior of the reservoirs in order to later find the optimized value of the amount of nanoclay deemed appropriate for the purposes of this search.

The carried out research shows that the use of nanoclay to reduce the gas permeability of polyethylene and carbon fiber as reinforcement to support the storage pressure make possible the use of polymer reservoirs for vehicular storage of CNG, as well as the feasibility of adequate recycling of the components after the end of their useful life.

11.3 CHARACTERIZATION OF THE POLYMER RESERVOIR

11.3.1 DESCRIPTION

The real-scale polymer liner (Ø225 x 725 mm), with a capacity of 22 liters of water, was manufactured by rotomolding with a PEBD/HDPE blend, developed in an earlier study carried out at the Polymeric Materials Laboratory of the Federal University of Rio Grande do Sul South [31].

For reinforcement, carbon fiber pre- impregnated with epoxy resin was used because of its high mechanical properties that are presented in Table 11.2.

TABLE 11.2 Typical Properties of Pre-Impregnated Carbon/Epoxy Fiber

Property	Carbon Fiber T700SC-12k	Epoxy Resin UF3360
Density (g/cm^3)	1.80	1.20
Tensile strength (MPa)	4,900.00	68.85
Tensile modulus (GPa)	230.00	3.17

Source: http://www.toraycfa.com/pdfs/T700SDataSheet.pdf.

The simulation of the application of the reinforcement layers by FW was carried out with the CADWIND 2007 software, version 8271, of the company Material SA.

The liner was coated in successive layers composed of a non-geodesic winding at 10° and a circumferential winding applied by a Kuka Roboter Gmbh robot, type KR 140 L 100-2 ensuring reliability and repeatability, characteristics necessary to the reservoir under development.

The carbon fiber was applied to the liner at room temperature (RT), with a heating cycle of <5°F (2.8°C) per minute up to 210°F (90°C), permanence for 24 hours and cooling at <5°F per minute until 150°F (66°C) prior to removal from the drying oven, with ambient cooling.

11.3.2 ULTRASOUND ASSAY

Ultrasound is a non-destructive test technology (NDT) that aims to verify the integrity of different materials by detecting intrinsic defects or those caused during the manufacturing process, including polymer composites [32].

Ultrasonic inspection, widely used in the petrochemical industry [33, 34], especially in polyethylene pipes [35, 36], reduces the degree of uncertainty in the inspection of important pieces such as the liner used in the CNG reservoir, because the ultrasonic wave, when traversing the polymer, is reflected as it reaches a discontinuity, allowing its identification, in addition to making it possible to measure the thickness of the wall of said liner.

Ultrasonic inspection has high sensitivity in the detection of internal discontinuities, dispensing intermediary processes for its interpretation, as well as expediting the inspection by dispensing safety measures necessary for radiographic tests, for example [37].

In ultrasonic inspection the acoustic waves make the particles that make up the material oscillate around their equilibrium position, transferring the kinetic energy to the next planes until all the particles of the material oscillate in the direction of propagation of the wave.

However, the sound wave suffers the effects of dispersion and absorption as it travels through the material, resulting in sonic attenuation that can be visualized on the ultrasound device as background reflection echoes from a piece with parallel surfaces [38].

The dispersion is due to the fact that the material is heterogeneous, generating different interfaces, because the sound waves give energy so that the particles of the material oscillate. Part of the energy of the sonic wave is also lost by divergences that manifest as it moves away from its emitting source, that is, the more the edge of the sonic beam affects the discontinuity, the smaller the echo amplitude [39].

As the sound propagates at a determined speed, depending on the propagation medium, the transducer emits a signal with a defined frequency that passes through the material and returns when reflected in a different medium, recording the response time. It is possible to identify faults and calculate the thickness of the sample.

Thus, the ultrasonic wave that crosses the liner wall is reflected in the interface of the bottom of the piece, originating echoes that return to the transducer, generating the electrical signal that is shown on the screen of the device as an echo, indicating its thickness.

However, ultrasonic inspection is not indicated to verify the distribution of montmorillonite clay particles because of their small size, whose mean diameter is 300 nm [40], whereas a transducer with a frequency of 4 Mhz can detect particles with a diameter equivalent to half the wavelength, that is, 1,000 times larger than the diameter of the nanoclay [41].

Ultrasonic inspection allows a quick identification of the integrity of the material, as the porosity of the material prevents any readings on the device, as well as allowing the verification of the wall thickness of the liner in the case of intact material, according to Figure 11.2.

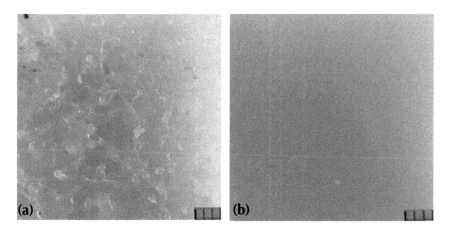

FIGURE 11.2 Porous material (a) and whole material (b).

In the ultrasound inspection, three liners were selected, and a sample did not allow readings in the ultrasound device, because the internal surface was not integrated, the sound that crossed the liner wall was diffracted when the surface was found to be rough, not returning to the transducer.

Ultrasonic inspection was performed with an SIUI device, model CTS-9005, calibrated with a rotomolded polyethylene plate with ultrasonic propagation velocity initially defined for 2.460 m/s [42], but as the liner is produced with a 95% high-density polyethylene (HDPE) and 5% linear low-density polyethylene (LLDPE) blend, the velocity was adjusted to 2.445 m/s, obtaining the accuracy with one decimal place in relation to the measurements made with a calibrated Mitutoyo caliper with accuracy of 0.05 mm.

The inspection was performed at RT with a standard single – crystal (zero degrees) transducer with 10 mm diameter, SIUI brand; model RP4-10L and frequency of 4 Mhz.

The normal transducer is used for inspection of parts with parallel surfaces or when it is desired to detect discontinuities in the direction perpendicular to the surface of the piece, being applicable for checking objects with slightly curved surfaces such as the liner.

To allow the comparison of the thickness of the samples through ultrasound and with the use of the caliper, each liner was divided into five longitudinal segments with 100 mm length while the cross-section was divided into six sectors of 60°, totaling thirty measurement points.

The segments have a length of 100 mm, which corresponds to ten times the diameter of the transducer to avoid incorrect readings due to the edge effect of the sample.

11.3.3 HYDROSTATIC TESTING

The hydrostatic test of the reservoirs in development was performed on four samples with one layer, four samples with two layers and four samples with three layers of carbon fibers, pressurized at 0.1 MPa/s using the Flutrol equipment (150 psi) with Load cell HBM 50 MPa, with Spindler interface 8, 60 Hz, the data being processed with Catman 4.0 Professional software.

The fracture of the liner, in 80% of the cases, regardless of the number of layers, occurred in the region near the tops, due to the action of radial tension and longitudinal tension, as well as the lower wall thickness due to failures in the roto-moulding process, which corresponds to the finite element simulation (FEM) performed by Velosa [43].

From the results of the hydrostatic test, with the occurrence of most of the fractures near the tops, it was decided to adjust the winding design to create carbon fiber reinforcing rings in these regions instead of increasing the number of layers to reduce the costs and the environmental liability that requires recycling.

Calorimetry analysis by differential scanning calorimetry indicated that the curing cycle for the pre-impregnated epoxy resin carbon fiber used in the tested reservoirs was completed.

11.3.4 NUMBER OF RESERVOIR LAYERS

Carbon fiber offers lightness, rigidity and mechanical strength, being more resistant than fiberglass because the tensile strength of the carbon fiber reaches 70,000 kg/cm^2, almost double the 45,000 kg/cm^2 that can be reached by the Fiberglass [44].

Despite the success of the experiment, the three-layer reservoir with an approximate average weight of 3.430 kg reached an average pressure of 350 bar, not meeting the requirements of ISO 11439:2013 [15] that defines the minimum pressure of 450 bar for hydrostatic testing of a CNG-4 type reservoir.

However, according to linear regression calculation and the proposed equation for the number of layers for any internal pressure, a fourth layer will meet the requirements of said standard, maintaining the comparative advantage in relation to the weight.

From the results of the hydrostatic test, the feasibility of applying fiber bands near the tops to increase the resistance of the reservoir with a saving of material was verified, generating less waste to be sent for recycling.

If the composite to be recycled was not known, it would be necessary to identify the curing agent used through processes such as FT-IR spectroscopy [45].

In a cylindrical reservoir, the following two main stresses occur:

- longitudinal stress (σl) that tends to break the tops, being supported by non-geodesic winding; and
- circumferential stress (σc) that tends to break the lateral wall, being supported by the circumferential winding.

For the dimensioning of a cylindrical reservoir, the circumferential tension is used, because the longitudinal tension corresponds to the half of this tension.

The tensile test resulted in a tensile strength of 56.65 kgf/mm² ± 15.04% for a fiber filament pre-impregnated with epoxy resin, corresponding to the circumferential tensile strength.

The tensile strength of the non-geodesic winding is calculated by multiplying the value obtained in the tensile test by sin (10°), resulting in a mean tensile strength of 9.84 kgf/mm² for a filament.

According to a previous study [46], the number of layers of a cylindrical reservoir made of composite material is calculated by the following equation:

$$t = p.r/k.\sigma c \qquad (1)$$

where t: layer thickness (mm) = ?; p: internal pressure obtained in the hydrostatic test for the three-layer reservoir = 3.57 kgf/mm²; r: cylinder radius = 110 mm; k: material factor, where k = 2 for pre-impregnated carbon fiber, obtained by experimental data; σc: circumferential tension per layer considering the value obtained in the tensile test = 66.46 kgf/mm². Therefore, $t \sim 5.9$ mm = 6 layers.

The value coincided with the linear regression calculation and the proposed equation for calculating the number of layers for any internal pressure, maintaining the advantage over the weight of the steel cylinders.

11.4 POLYMER NANOCOMPOSITES

Nanocomposites are the result of the combination of two or more materials with different properties, and the dispersed phase has at least one of its dimensions at the nanometer scale, being used in a small amount, typically 5% of the weight [47].

Montmorillonite clay is the most widely used inorganic phase in the preparation of polymer nanocomposites because of its high aspect ratio, good delamination capacity, high solvent resistance and thermal stability suitable for polymerization and extrusion processes [48].

Polymer nanocomposites have superior properties to conventional composites, as they have higher tensile and modulus resistance, as well as lower gas permeability [49].

The small size and the adequate dispersion of the nanoparticles (NPs) create a large interfacial area and a high aspect ratio that lead to the improvement of nanocomposite properties [50].

The small amount of NPs does not affect the density and processability of the polymer compared to traditional composites, which outweighs the cost of nanoclay [51].

As the polymer matrix maintains the characteristics of the pure polymer and the volumetric fraction of the clay is small, the reduction in permeability is due to the tortuous pathway to the diffusion of the gaseous molecules, created by the nano-tiles due to its shape and dispersion [52].

The silicates are not miscible in polymers, and their hydrophilic surface must be converted to organophilic by ion-exchange reactions [53].

The dispersion of clay nanolayers in a polymer optimizes the number of reinforcing elements available for the transportation of applied load, facilitating the transfer of stress, and improving its mechanical properties [54].

In the CNG reservoir, nanoclay can aid in the absorption of impact energy and reduce the propagation of microcracks [55], but the excess of nanoclay tends to weaken the composite, impairing its resistance to fatigue [56].

The processing conditions are fundamental, because the nanocomposites produced in co-rotating twin-screw extruders have higher exfoliation and dispersion than those processed in mono threads, since the resin is subjected to the same level of shear [57].

11.4.1 SELECTION OF THE NANOCLAY

The tensile test of the injected and rotomolded samples was carried out in a universal test machine of the brand Instron to select the best raw materials for the manufacturing of the liner for mechanical resistance.

The samples tested were as follows:

- Polyethylene blend;
- Polyethylene blend with 5% weight of national nanoclay (montmorillonite clay with a granulometry of 74 microns produced by the company Bentonisa S/A);
- Polyethylene blender with 5% by weight of American nanoclay (Cloisite 30B montmorillonite clay produced by Southern Clay Products).

The products were manually blended to the polyethylene blend and extruded into a single screw machine while the samples were taken directly from rotomolded parts that were machined in an MTC Robotics bench milling machine.

The use of machined samples sought to validate the rotomolding process while the injected samples were produced, with the same material, using a mini-injector of the Thermo Scientific brand, model Haake MiniJet II, with a normalized mold, being injected at 185°C.

Five samples of each raw material, both machined and injected, were subjected to a tensile test at 5,000 mm/min.

The tensile test of the carbon fiber samples was performed on a universal test machine Emic DL20000 with TRD26 load cell, TRD15 strain gauge, claws for polymer films and MTest version 3.0 programs to find the maximum tensile strength of a single filament of this fiber.

11.4.2 GAS PERMEABILITY IN POLYMERS

The molecular structure is important in the gas permeability, because if it is amorphous, the molecules can move more easily and the material will have higher permeability, but the polarity and the molecular diameter of the permanent must also be considered [58].

The environment can affect the diffusibility, being the temperature a critical factor, because its increase makes the polymer structure more open, allowing the gas to permeate more easily.

The presence of one gas or liquid may also affect the permeability of another gas. In the case of CNG, methane permeability increases due to the presence of carbon dioxide, plasticizing the polymeric structure, which tends to swell, facilitating the passage of methane [59].

The permeability of a given gas depends on the diffusion coefficient of the barrier material, its thickness, and the concentration difference of the permanent gas through the barrier.

Thus, if the vehicle is off, there will be stable equilibrium conditions, such as the steady-state concentration on both sides of the barrier, which can be expressed by Fick's First Law of Diffusibility:

$$J = - D.\delta C/\delta x \qquad (2)$$

where J is the gas flow through the barrier material in a unit of area, D is the diffusion coefficient of the material, δC is the change in gas concentration of one face of the barrier material to the measurement point, and δx is a distance from one side of the material to the measuring point. The negative sign is introduced because δx usually opposes the direction of flow.

If there is fuel consumption, the permeability of the material will vary over time, being expressed by Fick's Second Law of Diffusibility:

$$\delta C/\delta t = D.\delta^2 C/\delta x^2 \tag{3}$$

where C is the gas concentration, t is the permeation time, D is the diffusion coefficient, and x is the distance from one face of the material to the measurement point. Partial differentials indicate that the rate of change in concentration depends on the thickness of the barrier material (x) and time (t).

Fick's equations allow the barrier properties of different polymers to be compared quantitatively as well as to calculate the amount of a permanent gas or liquid through a given thickness of known barrier material at a given time.

The permeation of gases occurs through the intermolecular spaces of the barrier material in the following steps:

i. Sorption and solubilization of the permeant on the surface of the material;
ii. Diffusion through the material due to the concentration gradient;
iii. Desorption and evaporation of the permeant on the other side of the material.

When a face of the material is exposed to a gas at a certain partial pressure, the first (i) and last (iii) steps of the permeation process are faster than the diffusion. Thus, the rate of diffusion starts to control the gas permeation.

The size of the spaces between the polymer chains depends on the type and the physical state of the polymer, because seeing as the crystalline structure is practically impermeable due to molecular packing; the diffusion takes place in the amorphous zone.

For example, the oxygen permeability rate (TPO_2) is defined as the amount of oxygen passing through a unit of area parallel to the surface of a barrier per unit of time under the test conditions.

The incorporation of nanoclay tends to reduce the permeation of gases in polymers [60], because the distribution of the NPs creates tortuous paths that delay the movement of the molecules of the gas through the polymer [61].

For example, the permeability of the CO_2/CH_4 mixture in an HDPE nanocomposite with 5 wt. % nanoclay at 50°C and 100 bar was reduced by 47% and the diffusion coefficient by 35% compared to pure HDPE [62].

The transport properties in nanocomposites are related to the free volume of the polymer matrix. For example, the permeability of pure CH_4 and the CH_4/CO_2 mixture in HDPE increases with temperature due to increased mobility of HDPE and gas molecules, enhancing diffusion [63].

As for the effect of the internal pressure on the transport coefficients, the increase leads to an increase in the density of the polymer due to compaction, reducing the free volume available to the molecule to permeate, but also increases the gas concentration, thus increasing the gas permeability [64].

The diffusion coefficient is also influenced by the tortuosity that depends on the aspect ratio, shape, and orientation of the clay nano tiles [65]. For example, in injection-molded HDPE nanocomposites, the permeability reduced fivefold to the aspect ratio of 1000, but for particles parallel to the direction of gas diffusion, there is no barrier effect [66].

The gas permeability test aimed to quantify the permeability of the liner to methane, but the permeability to oxygen, which has polarity and molecular diameter compatible with this combustible gas was verified.

The test was performed with two groups of samples according to ASTM F 1927 (plates) and ASTM D 3985 (films) standards.

The first group of samples consisted of hot-pressed die-cut sheets, cut to a diameter of 12 cm and machined up to a thickness of 1 mm in the MTC milling, and three specimens were tested as follows:

- Polyethylene blend;
- Polyethylene blend with 5% weight of Brazilian nanoclay; and
- Polyethylene blend with 5% weight of American nanoclay.

The permeability result on the tested plates is shown below:

- Polyethylene blend: 13,660 mL (NTP/m²/day) at 1 atm and 23°C;
- Polyethylene blends with Brazilian nanoclay: 32,829 mL (NTP/ m²/day) at 1 atm and 23°C;

- Polyethylene blend with American nanoclay: 150,000 mL (NTP/m²/day) at 1 atm and 23°C (equipment detection limit).

The high permeability exhibited by rotomolded specimens with nancolay is due to the monoscrew extruder processing, which did not allow adequate distribution and shear of the clay particles to obtain the barrier effect on the polyethylene.

The second group of samples, made up of films of the polyethylene blend with nanoclay, had the pellets extruded in a corotating twin-screw extruder at 250 rpm, after heating in an oven at 80°C for 24 hours for drying, being produced in a Seibt brand extruder, model ES 35.

The following samples were analyzed in films:

- Sample A: HDPE (95%) blends with LDPE (5%) and sodium nanoclay (without compatibilization);
- Sample B: HDPE Blend (95%) with LDPE (5%) and Cloisite 15A nanoclay;
- Sample C: HDPE Blend (95%) with LDPE (5%) and Cloisite 30B nanoclay.

The oxygen permeability rates (TPO_2) were determined by coulometric method, according to ASTM D3985-05 [67], in Oxtran, model 2/20, from Mocon, operating with pure oxygen as a permanent gas.

The tests were performed at 23°C, dry, with the conditioning of the specimens at 23°C and dry for 48 hours in an effective permeation area of 100 cm² in each specimen.

The results obtained were corrected to 1 atm of oxygen partial pressure gradient between the two surfaces of the film, because this gradient corresponds to the driving force for the permeation of oxygen through the polyethylene film.

The material was processed in a twin-screw extruder, but the speed of 250 rpm did not allow adequate distribution and shear of the clay particles.

Once the TPO_2 was determined, the oxygen permeability coefficient (P) was calculated from the permeability rate, as follows:

$$P = TPO_2 \cdot e/p \qquad (4)$$

where P: oxygen permeability coefficient (mL (NTP).μm.m−2. day−1. Atm−1); TPO_2: rate of oxygen permeability (mL (NTP). m−2. day−1); e: mean thickness of the specimen (μm); p: partial pressure of oxygen in the permeating gas chamber of the diffusion cell, because the partial pressure of O_2 in the chamber of the drag gas is considered null.

The nanoclay was incorporated to the polyethylene to decrease the gas permeability and to increase the mechanical resistance of the liner for the reduction of the wall thickness and the final weight of the reservoir, being the samples submitted to tensile and gas permeability tests to select the most adequate one.

However, the aspect ratio of the particles and their orientation also influence the barrier properties of HDPE nanocomposites in injection-molded recipients, seeing as the permeability reduces fivefold when the particles are perpendicular to the direction of the diffusion.

The sample with Cloisite 15A nanoclay has a lower oxygen permeability coefficient, confirming the manufacturer's recommendation that indicates its compatibility with maleic acid for polyethylene due to the hydrophobicity of this polymer.

11.4.3 SELECTING MATERIALS FOR THE LINER

In order to increase the mechanical strength of the liner and to reduce the nominal thickness of its wall, reducing the final weight of the CNG tank, new materials were obtained through tensile tests.

The tensile test of these materials showed that the injected specimens presented higher tensile strength than the rotomolded ones, that the addition of nanoclay increased the tensile strength and that the American nanoclay (Cloisite 30B) presented better results than the Brazilian one, as seen in Table 11.3.

TABLE 11.3 Results of the Traction Test of Candidate Materials (N)

Material	Rotomolded	Injected
PE Blend	143.84 ± 32.03	238.70 ± 18.37
PE Blend with 5% weight of Brazilian nanoclay	137.80 ± 11.38	205.12 ± 8.24
PE Blend with 5% weight of American nanoclay	140.74 ± 6.25	220.27 ± 3.79

The von Mises criterion indicates that the cylinder rupture pressure is proportional to the yield tension, whose value is close to the tensile stress (σrt) obtained in the tensile test. Thus, the higher the value of σrt, the greater the pressure supported by the cylinder in the hydrostatic test.

For Walters [68], Equation (6) should not be used to predict burst pressure, since σy differs from the conventional yield tension (σe) obtained in the tensile test.

Therefore, in qualitative terms, it is possible to use the tensile test to select candidate materials for the manufacture of CNG reservoirs.

11.5 RECYCLING OF POLYMER NANOCOMPOSITES

The lack of adequate management of polymer waste results in improper discarding, generating serious environmental problems due to its low degradability, requiring technical and economic feasibility for its recycling [69, 70].

In addition, virgin resins are manufactured with specific properties according to the needs of the final application, while the recycled material exhibits variations in its properties due to the different grades of the same resin.

Recycling of polymers is fundamental in environmental, social and economic terms, as it preserves non-renewable sources of raw materials, contributes to job creation and reduces waste disposal costs, as well as contributing to the reuse of waste [71].

Although recycled HDPE is widely used in industry in general, few studies have been carried out to analyze its rheological properties [72].

The polymers can receive primary, secondary, tertiary, or quaternary recycling according to ASTM D7209-06–Standard Guide for Waste Reduction, Resource Recovery, and Use of Recycled Polymeric Materials and Products [73].

Primary and secondary recycling are known as mechanical recycling consisting of the physical transformation of polymeric residues into granules.

The primary recycling consists of transforming polymeric industrial waste into a single type of resin, being the most used in Brazil, absorbing 5% of the plastic in Brazil while the secondary one uses waste after consumption.

In Brazil, plastic products are identified for recycling according to ABNT NBR 13230: 2008 – Packaging and storage of recyclable plastics – Identification and symbology [74], which standardizes the identification of different types of plastic resins in order to facilitate the Step of sorting the plastic waste sent to recycling.

The resin identification system was introduced in 1988 by the Society of the Plastics Industry [75], which has been adopted worldwide by manufacturers of plastic packaging because it allows the quick identification of the type of resin, facilitating the separation of discarded packaging [76].

In the case of lamination or coextrusion of several materials for the manufacture of a package, the two main structural components must be identified to assist in the mechanical recycling by processes that do not require the separation of the structural layers [77].

However, this standard does not specify symbology for polymeric composites, aiming at their separation to avoid compromising the quality of the recycling chain that is, they require tests to identify their fillers and additives.

Although pyrolysis recycling preserves the mechanical properties of carbon fiber for use in new industrial applications, the most feasible option is the use of CNG reservoir coatings as reinforcement of polymer matrices in substitution of fiberglass in the process of injection transformation for the production of various industrial products due to the lower abrasiveness.

Fiberglass presents high performance, being the most used in polymeric composites due to the combination of its properties as high resistance and rigidity, as well as low cost in relation to carbon and aramid fibers.

The addition of fibers to the polymer allows the stress applied to the matrix to be transferred to the fibers, significantly increasing the mechanical strength of the composite [78], but this mechanism can be improved with additives and coupling agents.

As stress transfer occurs through shear at the interface, the long fibers provide greater strength gain and stiffness than the short and particulate fibers, presenting a certain degree of anisotropy, that is, many properties of the composite depend on the direction of the fibers in relation to the stress to which the composite is subjected.

11.6 CONCLUSIONS

Despite the success of the destructive hydrostatic test, the three-layer reservoir did not meet the requirements of ISO 11439:2013 [15] which sets the minimum pressure of 450 bar for the test of a CNG-4 reservoir, i.e., those totally produced with non-metallic materials, seamlessly.

It is estimated that the fourth layer of carbon fiber will meet the requirements of ISO 11439:2013 [15], supporting the minimum pressure of 495 bar for the hydrostatic test, maintaining the advantage over the weight of the steel cylinders, according to the calculation by linear regression and Eq. (6).

In the hydrostatic test, it was verified that in 80% of the tested tanks, the fracture of the liner, regardless of the number of layers, occurred in the region near the tops, due to the lower wall thickness, which is as a result of the rotomolding manufacturing process.

Despite the difference in the shape between the reservoir under development and that presented by Velosa [43], there is similarity in the location of fractures resulting from the hydrostatic test and those found in the simulation, due to the concentration of stresses close to the tops.

In the case of the reservoir under development, the region near the tops also has the smallest wall thickness, as measured by the liner, requiring adjustments in the design to increase its final resistance.

In order to increase the mechanical strength of the liner and to reduce the nominal thickness of its wall from 10 mm to 7 mm, the maximum thick-ness allowed by the rotomolding process, new materials were obtained with rotomolded and injected samples.

It is verified that the injected specimens show higher tensile strength than the rotomolded ones, it was also verified that the addition of nanoclay increases the tensile strength and that the Cloisite 30B nanoclay presents better results than the national one due to the smaller granulometry.

The low results obtained in the rotomolded specimens are due to the monoroscrew extruder processing that did not allow adequate distribution and shear of the clay, unlike the injected specimens that had better processing conditions.

The method of qualitative selection, through the tensile test, to evaluate the mechanical strength of the specimens is justified by the application of the von Mises criterion for the yield of a plastic material in a triaxial state of tensions adapted to the uniaxial state of tensions, According to Walters [68].

Gas permeability is an essential requirement for the selection of a material for the production of CNG reservoirs, but since it was impracticable to perform the test with methane, the main component of CNG, oxygen, which is compatible with it in terms of molecular diameter and polarity, was used.

The high permeability presented by the rotomolded samples with nanoclay is due to the monoroscrew extrusion process, which did not allow good distribution and shear of the clay particles, inferred by the permeability results that increased with the addition of the clay.

As for the film, the permeability coefficient was used, which characterizes the composite, in order to minimize the effect of the variation of thickness in the comparison of samples. The permeability coefficient characterizes the barrier of homogeneous materials, not being totally correct in this case.

The material was processed in a twin-screw extruder, but the speed of 250 rpm did not allow adequate distribution and shear of the clay particles, preventing the samples from being homogeneous.

From the obtained data, it is verified that the sample with Cloisite 15A nanoclay presents lower coefficient of permeability to oxygen.

There is a possibility of having the volume of the compressed natural gas (CNG) reservoir, maintaining equivalent autonomy to liquid fuels, by increasing the storage pressure to 636 bar, implying a hydrostatic test pressure of 1,431 bar according To ISO 11439:2013 [15], requiring 18 layers of composite to be supported.

Therefore, the production of nanocomposite reservoirs is a viable alternative for vehicles supplied with CNG, associating fuel economy with ecological and economic benefits.

The recovered carbon fiber can be used as reinforcement in polymers in substitution of the fiberglass that, due to its abrasive effect, damages the screw in the process of injection molding, besides being ecologically advantageous, as it avoids that it gets deposited in landfills, allied to the economic advantage, because the cost of recovered carbon fiber tends to be lower than virgin fiberglass.

The epoxy resin has a varied chemical composition, but can also be recycled by reusing its components by separating anhydride-type curing agents through suitable for its specific formulation.

The recycling of polyethylene with nanoclay involves grinding which tends to fracture the exfoliation of the nanoplates, making their orientation random, and the extrusion in single screw machine produced samples that have similar permeability values to the polymer without nanoclay.

The results indicate that the production of reservoirs in polymer composites is a viable alternative for vehicles fueled with natural gas, associating fuel economy with an increased life of the engine and of the suspension and braking systems.

Therefore, it is possible to produce pressure vessels for the storage of CNG for vehicular use in polymer nanocomposites, according to the guidelines of standard ISO 11439:2013 [15] for CNG-4 cylinders whose constituent materials can be recycled and used as a substitute for virgin raw material in the production of injected polymers, for example.

The higher the resistance and the useful life of the reservoir, the lower the need for raw material and the need for periodic replacements, reducing the need for recycling of its components and the consumption of natural resources, reducing the environmental impact in its large scale implantation.

KEYWORDS

- **carbon fiber**
- **compressed natural gas**
- **gas permeability**
- **nanocomposites**
- **recycling**
- **ultrasound**

REFERENCES

1. Donnet, J. B., et al., (1998). *Carbon Fibers* (3rd edn.). Marcel Dekker, New York, USA, p. 71.
2. Stak, J., (2011). *Product Lifecycle Management: 21st Century Paradigm for Product Realization.* Springer, p. 369.
3. AFRA – Aircraft Fleet Recycling Association. http://www.afraassociation.org (Accessed on 25 July 2019).
4. Lee, S. M., (1993). *Handbook of Composite Reinforcements.* Wiley – VCH, USA, p. 330.
5. Faiz, A., et al., (1996). *Air Pollution from Motor Vehicles: Standards and Technologies for Controlling Emissions.* World Bank Publications, USA.
6. Chandra, V., (2006). *Fundamentals of Natural Gas: An International Perspective.* Penn Well, USA, p. 125.
7. Termeer, C., (2013). *Fundamentals of Investing Oil and Gas,* USA, p. 202.
8. Adewole, J. K. et al. (2011). *J Chem Eng, 9999*, 1–13.
9. Sih, G. C., et al, (1995). *Advanced Technology for Design and Fabrication of Composite Materials and Structures: Applications to the Automotive, Marine, Aerospace and Construction Industry.* Kluwer Academic Publisher, Netherlands, p. 344.

10. Rojey, A., et al., (1997). *Natural Gas: Production, Processing and Transport.* Institut Français du Pétrole Publications, Éditions Technip, Paris, France.
11. Visakh, P. M., & Morlanes, M. J. M., (2015). *Polyethylene-Based Blends, Composites and Nanocomposites.* John Wiley, USA, p. 142.
12. Crawford, R. J., (1993). *Rotational Molding.* New York: Rapra Technology, p. 14.
13. Shen, F. C., (1995). A filament-wound structure technology overview. *Materials Chemistry and Physics, 42.*
14. Corum, J. M., et al., (2000). *Basic Properties of Reference Crossply Carbon-Fiber Composite.* OAK Ridge National Laboratory, U.S. Department of Energy.
15. ISO 11439:2013 – Gas Cylinders – High Pressure Cylinders for the On-Board Storage of Natural Gas as a Fuel for Automotive Vehicles. http://www.iso.org/iso/home/store/catalogue_ics.htm (Accessed on 25 July 2019).
16. TCR – Resin Data Sheet (2016). http://tcrcomposites.com (Accessed on 25 July 2019).
17. Baker, R. O., et al., (2015). *Practical Reservoir Engineering and Characterization.* Elsevier, USA, p. 100.
18. ANP–Agência Nacional do Petróleo, Gás Natural e Biocombustíveis, (2016). http://www.anp.gov.br/wwwanp/ (Accessed on 25 July 2019).
19. ISO 14000–Environmental Management. http://www.iso.org/iso/home/standards/management-standards/iso14000.htm (Accessed on 25 July 2019).
20. Romão, B. M. V., et al., (2003). Characterization by FT-IR of curing agents used in epoxy resins. *Polímeros, 13*(3), São Carlos, Brazil.
21. Boeing Environmental Technotes, (2007). *Environmental Assurance, Aircraft & Composite Recycling, 12*(1), USA.
22. Pompidou, S., et al., (2012). Recycling of carbon fiber: Identification of bases for a synergy between recyclers and designers. *11th Biennial Conference on Engineering Systems Design and Analysis. ASME.*
23. Jiayu, G., et al., (2000). Termochimica, *Acta, 352,* 153.
24. Varley, R. J., et al., (1995). *Polymer, 36*(7), 1347.
25. Brody, A. L., (2006). Nano and food packaging technologies converge. *Food Technology, 60*(3), 92–94.
26. Khalaf, M. N., (2012). Control the discontinuity of the flow curve of the polyethylene by nanoclay and compatabilizer. *Arabian Journal of Chemistry.*
27. Manikantan, M. R., & Varadharaju, N., (2012). Preparation and properties of linear polyethylene based nanocomposite films for food packing. *Indian Journal of Engineering & Materials Sciences, 19,* 54–66.
28. Mohan, T. P., & Kanny, K., (2013). Melt blend studies of nanoclay-filled polypropylene (PP)–high density polyethylene (HDPE) composites. *J. Mater. Sci., 48,* 8292–8301, doi: 10.1007/s10853–013–7642–9.
29. Saba, N., et al., (2014). Review on potentiality of nano filler/natural fiber filled polymer hybrid composites. *Polymers, 6,* 2247–2273, doi: 10.3390/polym6082247, polymers–ISSN 2073–4360.
30. Alexandre, M., & Dubois, P., (2000). Polymer-layered silicate nanocomposites: preparation, properties and uses of a new class of materials. *Mater. Sci. Eng., 28,* 1–63.

31. Barboza, N. E. S., et al., (2013). Processing of a LLDPE/HDPE pressure vessel liner by rotomolding. *Materials Research,* São Carlos, Brazil, 17, pp. 236–241.
32. Karbhari, V. M., (2013). Non-Destructive Evaluation of Polymer Matrix Composites. *Techniques and Applications,* Woodhead Publishing, Cambridge.
33. Crawford, S. L., et al., (2010). Preliminary Assessment of NDE Methods on Inspection of HDPE Butt Fusion Piping Joints for Lack of Fusion With Validation From Mechanical Testing, ASME, PVP2010–25280, pp. 1039–1045. doi: 10.1115/PVP2010–25280.
34. Burt, V., (2015). *Corrosion in the Petrochemical Industry* (2nd edn.). ASME International, Ohio, USA.
35. Sood, S. C., (2014). Assimacopoulos, George, NDE welds inspection qualifications for pipework and pipelines. *Sensor Letters, 12*(8), 1243–1252(10), American Scientific Publishers, doi: 10.1166/sl.2014.3302.
36. Shi, J., et al., (2014). Ultrasonic Inspection of Large Diameter Polyethylene Pipe Used in Nuclear Power Plant, PVP2014–28609, V06BT06A042, p. 8. ASME Pressure Vessels and Piping Conference, doi: 10.1115/PVP2014–28609.
37. Andreucci, R., (2014). *Ensaio por Ultrassom.* ABENDI, São Paulo, Brazil, p. 6.
38. Tu, H., (2005). *Ultrasonic Attenuation Imaging and Analysis.* University of Wisconsin-Madison, USA.
39. Hellier, C., (2013). *Handbook of Nondestructive Evaluation* (2nd edn.). McGraw Hill, USA.
40. Laske, S., (2015). *Polymer Nanoclay Composite* (1st edn.). Elsevier, USA.
41. SIUI–Shantou Institute of Ultrasonic Instruments Co. Ltd. (2016). www.siui.com/enH/products_view_c5_9_21_52_i7664.html (Accessed on 25 July 2019).
42. Olympus, (2016). www.olympus-ims.com/pt/ndt-tutorials/thickness-gage/appendices-velocities/ (Accessed on 25 July 2019).
43. Velosa, J. C., et al., (2007). Development of a new generation of filament wound composite pressure cylinders. *Ciência e Tecnologia dos Materiais, 19*(1), 1–9.
44. Vela, M. C. V., et al., (2006). *Ciencia y Tecnología de Polímeros.* Universidad Politécnica de Valencia, Spain.
45. Silverstein, R. M., & Bassler, G. C., (1981). *Spectrometric Identification of Organic Compounds*, John Wiley, New York, USA.
46. Pavani, G. J., et al., (2015). Application of polymeric nanocomposites and carbon fiber composites in the production of natural gas reservoirs. *Journal of Nanomaterials* (Print), pp. 1–7.
47. Koo, J. H., (2006). *Polymer Nanocomposites: Processing, Characterization and Applications.* McGraw-Hill, USA.
48. McNally, T., et al., (2003). Polyamide-12 layered silicate nanocomposites by melt compounding. Recycling of Polymer Nanocomposites of Carbon Fiber. *Polymer, 44,* 2761–2772.
49. Chan, C. M., et al., (2002). *Polymer, 43,* 2981–2992.
50. Nath, D. C., et al., (2009). *J. Mater. Sci., 44,* 6078.
51. Modesti, M., et al., (2005). *Polymer, 46*(23), 10237.
52. Choudalakis, G., & Gotsis, A. D., (2009). Permeability of polymer/clay nanocomposites: A review. *European Polymer Journal, 45*(4), 967–984, Elsevier Science.

53. Manias, E., et al., (2001). Polypropylene/montmorillonite nanocomposites. Review of the synthetic routes and materials properties. *Chem. Mater, 13,* 3516–3523.
54. Kim, S. W., et al., (2001). Preparation of clay–dispersed poly(styrene-co-acrylonitrile) nanocomposites using poly(α-caprolactone) as a compatibilizer. *Polymer, 42,* 9837–9842.
55. Jiankun, L., et al., (2001). Study on intercalation and exfoliation behavior of organoclays in epoxy resin. *J. Polym. Sci. Polym. Phys., 39,* 115–120.
56. Liu, X., & Wu, Q., (2002). Polyamide 66/clay nanocomposites via melt intercalation. *Macromol. Mater Eng., 287,* 180–186.
57. Jin, Y. H., et al., (2002). Polyethylene/clay nanocomposite by in situ exfoliation of montmorillonite during Ziegler–Natta polymerization of ethylene. *Macromol Rapid Commun., 23,* 135–140.
58. Pandey, J. K., et al., (2005). An overview of the degradability of polymer nanocomposites. *Polym. Degrad, Stabil., 88,* 234–250.
59. MacDonald, R. W., & Huang, R. Y. M., (1981). Permeation of gases through modified polymer films. V. Permeation and diffusion of helium, nitrogen, methane, ethane and propane through gamma-ray crosslinked polyethylene. *Journal of Applied Polymer Science, 26*(7), United States.
60. Lange, J., & Wyser, Y., (2003). Recent innovations in barrier technologies for plastic packaging – a review. *Packag. Technol. Sci., 16,* 149–158.
61. Deepthi, M. V., et al., (2010). *Mater. Des., 31,* 2051–2060.
62. Adewole, J., K., et al., (2012). Transport properties of natural gas through polyethylene nanocomposites at high temperature and pressure. *J. Polym. Res., 19,* 9814.
63. Raharjo, R. D., et al., (2007). *J. Member. Sci., 306,* 75–92.
64. Klopffer, M., & Flaconnèche, B., (2001). *Oil Gas Sci. Technol. Rev., IFP, 56,* 223–244.
65. Utracki, L. A., (2004). *Clay-Containing Polymeric Nanocomposites.* Rapra Technology, England.
66. Tomova, D., & Reinemann, S., (2003). *Kunststoffe, 7,* 18–20.
67. ASTM D3985-05 – Standard Test Method for Oxygen Gas Transmission Rate Through Plastic Film and Sheeting Using a Coulometric Sensor. https://www.astm.org/Standards/D3985.htm (Accessed on 25 July 2019).
68. Walters, J. A., (2003). *Hoop-Wrapped Composite, Internally Pressurized Cylinders.* ASME Press, p. 69.
69. Ehring, R. J., (1992). *Plastics Recycling, Products and Processes.* New York: Hanser Publishers.
70. Kelen, T., (1983). *Polymer Degradation.* London: Van Nostrand Reinhold, p. 109.
71. Tammemagi, H., (1999). *The Waste Crisis: Landfills, Incinerators and the Search for a Sustainable Future.* Oxford University Press, New York – ISBN 0-19-512898-2.279.
72. Cruz, S. A., et al., (2008). Evaluation of rheological properties of virgin HDPE/recy-cled HDPE blends. *Polímeros, 18*(2), São Carlos, Brazil, abr./jun. http://dx.doi.org/10.1590/ S0104–14282008000200012 (Accessed on 25 July 2019).
73. ASTM D7209-06-Standard Guide for Waste Reduction, Resource Recovery, and Use of Recycled Polymeric Materials and Products. https://www.astm.org/Standards/D7209.htm (Accessed on 25 July 2019).

74. ABNT NBR 13230: 2008, (2016). Embalagens e acondicionamento plásticos recicláveis–Identificação e simbologia https://www.abntcatalogo.com.br (Accessed on 25 July 2019).
75. SPI – Society of the Plastics Industry, (2016). http://www.plasticsindustry.org/ (Accessed on 25 July 2019).
76. EPIC (2016). Environment and Plastics Industry Council. Plastics Recycling Made Easier with Resin Codes (pp. 3–6). Special news and views report: Different applications, different plastics. Mississauga, Ontario: www.plastics.ca/epic (Accessed on 25 July 2019).
77. Coltro, L., et al., (2008). Plastic materials recycling: The importance of the correct identification. *Polímeros, 18*(2), São Carlos, abr./jun., http://dx.doi.org/10.1590/S0104–14282008000200008 (Accessed on 25 July 2019).
78. Hollaway, L., (1993). *Polymer Composites for Civil and Structural Engineering.* Blackie Academic & Professional, Chapman & Hal, Glasgow, p. 164.

CHAPTER 12

Nanomaterial-Based Energy Storage and Supply System in Aircraft Systems

INDRADEEP KUMAR

Chairman of Bibhuti Education and Research, Bhagalpur, Bihar, India; Research Scholar, Department of Mechanical Engineering, Vels Institute of Science, Technology and Advanced Studies (VISTAS), Chennai – 600117, India

ABSTRACT

In the modern age, nanotechnology is found to be beneficial in different spheres as in few industries. The aviation industry is one of them. As the properties of materials vary a lot in nanomaterials (NMs) compared to the bulk material, the development of any field relies on the fact that these materials are much advantageous than bulk metals.

The aviation industry is one of the fast-growing industries in the world. Air transport is the only speedy way of shipping various products throughout the world. Numerous companies that are well developed or promising ones depend on the quality of air transportation. But at the same time, the environment is continuously threatened due to the emission of heat, carbon dioxide, water vapor, and other harmful gases, including black carbon. The expanding aviation industry is now becoming a crucial role in increasing carbon footprints on earth and the day by day competition of lowering the flight fare is at the cost of severe climatic change.

The aviation industry is one of the most foremost heavy industries in the world. Countless companies rely on the ability to ship products and people around the world with the speed that can only be achieved by air. Along with this massive economic value, however, comes enormous consumption and one of the largest carbon footprints on the earth relative to the size of the market. For this reason, the major drivers in recent aerospace R&D are towards lighter construction materials and more efficient engines - the overall aim being to reduce fuel consumption and carbon emissions associated with air freight and air travel. The remarkable interest in nanotechnology for the aerospace or aviation industry is justified by the potential of NMs and nanoengineering to help the industry achieve this goal. Nanotechnologies may impart vital improvement potentials in the development of both renewable energy sources and conventional energy sources. Nano-coated, storage device allows the optimization of lifespan and efficiency of systems in the development of a fuel storage tank and hence the saving of costs.

12.1 INTRODUCTION

For worldwide transportation of people in the shortest time and least hazard, the only means is the airplane. For business purpose too, a lot of companies are extremely dependent on the performance of the airplane. So the proper designing of the airplane is required to achieve maximum efficiency and performance. The efficiency can be increased by exploiting the excellent properties of the nanomaterials (NMs). NMs are cornerstones of Nanotechnology and Nanoscience. In recent years there is a tremendous research progress in the field of aviation. Nanoscience has the potential to revolutionaries the aerospace industry as it has offered lots of new materials with diversified functionalities. Many astonishing characteristics of the NMs have made it possible to commercialize many products which have changed the old concepts significantly.

Basically, nanoparticles are those which have at least one dimension with 100 nm where 1 nanometer (nm) = 10^{-9} m (Figure 12.1).

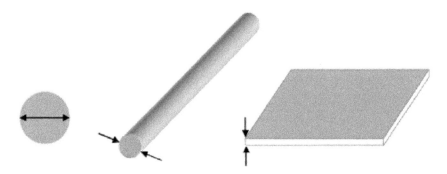

FIGURE 12.1 Scheme of various types of nanofillers or fillers with nanoscale dimensions.

Not all the systems of airplane can be revolutionalized by nanotechnology in a recent trend, but surely this new technology can be exploited to some extent to have a better system than before. This chapter deals with the possibilities of inclusion of NMs in certain parts, including the navigation system. For modern aviation, nanotechnology has a big prospective either in terms of enabling huge scale energy production process or designing efficient nanocoated energy storage material. But the extensive application is still hindered by the shortcomings like isolation of NPs, improved synthesis procedure, and critical application.

12.1.1 SIMPLE DEFINITION OF NANOTECHNOLOGY

Nanotechnology is the name of the application technology where nanoscience is applied in practical devices. Nanotechnology starts from atoms and molecules and their integration and manipulation to form the device. Tailoring of the properties of the materials in the nano range is possible now with nanotechnology especially for commercial and industrial applications. As material properties can be tailored in nanoscience and the devices made from materials, hence it can be said that the whole aviation industry depends on nanotechnology.

Two types of NMs can be defined:

- **Non-Intentionally-Made Nanomaterials:** This type of NMs is produced during volcanic eruptions, proteins, viruses, etc. or that produced due to unintentional human activity such as diesel combustion.
- **Intentionally-Made' Nanomaterials:** Deliberately produced NMs that are produced through a definite fabrication process or the NMs that are produced through some biological activity

Some good examples are:

Our nails increase in length of 1 nm in 1 second.
- The tip of a pin is about 1,000,000 nm in diameter.

The diameter of our hair is 80,000 nm.
- A DNA molecule is 1–2 nm wide.

12.2 WHY NANOMATERIALS (NMS)?

Anywhere within the airplane, NMs may be used. NMs in different forms like nanoceramics, nanocomposites, or polymeric NPs NMs can be used as required in the airplane system.

12.2.1 PROPERTIES OF NANO-MATERIALS

The structure of NMs is in between the bulk materials and those of atoms. The properties of microstructured materials mostly resemble with that of the bulk materials. But nanostructured materials have properties that are significantly different from bulk materials. Such difference of properties of the NMs is due to the facts:

- The nano-sized materials have a large number of surface atoms.
- Their surface energy is very high.
- Occurrence of spatial confinement.
- Reduced or no imperfection in the nanomaterial.

As the dimension of the material reduces to nano range, the surface to volume ratio of the material is increased, which leads to a large number of atoms available on the surface. That, in turn, increases the possibility of the material to interact with other atoms or group of atoms so that the surface-related properties are enhanced considerably. Especially if the sizes of NMs will be comparing to the length of the structure, the surface properties of nano-materials affects the entire material (Figure 12.2–12.4).

12.2.2 NOVEL PROPERTY

1. Very small number of molecules are contained in the material.
2. Confinement of the movement of the particles (Quantum confinement).
3. Nanoparticles, being of very small size, behave like wave-so wave-particle duality is observed.
4. Size being smaller, the small wavelength, high frequency, high energy properties are dominant.
5. Discrete energy levels are observed in NMs.
6. Quantum confinement.

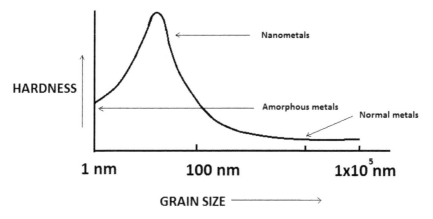

FIGURE 12.2 Grain size vs. Hardness graph.
(*Source:* Reproduced according to Ref. [23].)

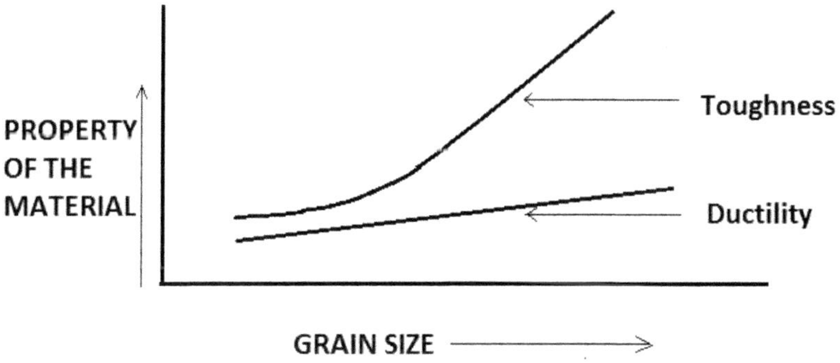

FIGURE 12.3 Grain Size vs. ductility and toughness curve.
(*Source:* Reproduced according to Ref. [24].)

12.2.3 HOW CAN NANOMATERIALS (NMS) BE FORMED?

NMs of different dimensions are produced in the manufacturing process; viz zero-dimensional NMs like atomic clusters, one-dimensional NMs like nanotubes, nanorods, two-dimensional NMs like buried layers and also three dimensional such as nanophase materials. NMs for aerospace are generally prepared by gas-phase synthesis process.

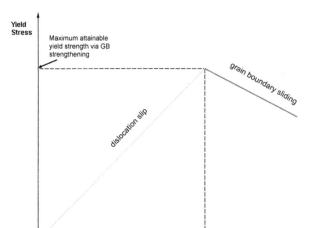

FIGURE 12.4 Hall-pitch strengthening limit.
(*Source:* From Wikipedia Ref. [24].)

12.2.4 CHEMICAL VAPOR DEPOSITION (CVD)

CVD is a popular process of synthesis of NPs where the substrate on exposure to the volatile gases, collect the precursors from the gases onto them, and the deposited layer is formed. This happens at high temperature in the CVD reactor. In a homogeneous deposition, particles are deposited on the cold surfaces by the thermophoretic forces. Nanopowders can be collected from the surface, or thin layer can be formed. This is what is called the homogeneous method. In this method, less reactive gases contain more reactive radicals at a higher temperature than the substrate.

In heterogeneous CVD, solid of any form deposits on the surface. Nucleation occurs on the surface itself, so that film structure is formed easily. To get film desirable film deposition, this the preferred synthesis method compared to homogeneous one.

CVD process is the preferred process of synthesis of NMs as this method has advantages over other methods. This allows the NMs to deposit into verities of forms like a powdered, thin film, thick film, coiled structure or desired shape. Their size, shape, crystallinity can be controlled excellently. Relatively pure particles are formed in large scale. The growth parameters such as temperature, carbon-rich gases, and catalysts can be controlled significantly in this method.

12.3 ENGINEERED MATERIALS AND STRUCTURES

Engineered materials are those which are not the same as produced but engineered or tailored as the need comes. The material properties of NMs can be changed according to the need. That is an important feature that cannot be achieved by the conventional materials. Now energy storage components can be accomplished with NMs and functionality is as a whole increased due to handling materials in the nano range. The obvious result is the improvement in performance and efficiency while we can use lightweight material with huge toughness, strength, and electronic properties.

Engineered materials and structures are divided into five areas:

i. **Lightweight Structures:** Structures having lightweight surrounds NMs with functional and structural properties which allow the decrease in the weight of aircraft by enabling lightweight NMs and higher efficiency aircraft components. These include long-lasting structure, lightweight and high-efficiency wiring, data cables, and devices.
ii. **Damage-Tolerant Systems:** Damage tolerance and control can be possible introducing NPs in the system that strengthens the interlaminar interfacial strength and repairs fast. Stiffness, toughness, strength, ductility, and smoothness should be achieved at a time. Damage should be controlled in unfriendly weather conditions, including high-impact events and vulnerability to high temperature.
iii. **Coatings:** on a substance is a thin engineered layer of some other substance spread over it to prevent environmental risk, icing, dust, and ionizing radiation. The coating can also protect the important parts from high-temperature damage.
iv. **Adhesives (Nanoadditive):** These are used in the aerospace industry in the interior and exterior or in the robotic inspection of spacecraft, satellite repairing orbital debris grappling and docking, low-precision rendezvous, astronaut extravehicular activity (EVA), and in-space assembly. These are used in the form of paste, liquid,

or film. These improve the durability, save weight, give challenging designs, and fuel efficiency for engines. Further high bonding quality can be achieved using nanotechnology in the future.

v. **Protection Against High Temperature:** Thermal management system gives the lightweight approach to protect systems to get damaged due to uncontrolled thermal cycling.

12.4 ENERGY STORAGE, POWER GENERATION, AND POWER DISTRIBUTION

The application of nanotechnology in the field of energy has three branches; generation, storage, and distribution. Atomic and molecular level interactions are mostly involved in energy generation and storage. Hence the materials developed from the atomic or molecular level must serve the purposes for the same.

These technologies are divided into three areas:

i. **Energy Storage:** The advancement in the energy storage systems is observed a lot by the application of nanotechnology. Lithium-ion batteries, ultracapacitors offer high power density and energy and are applicable in different missions. On the other side, NMs can be used for preparing electrodes with increased electrolytes and reactivity with better transport properties. These materials or composites should also be able to operate at high temperature.

ii. **Power Generation:** Photovoltaics (PVs) and thermo PV can provide huge power for aircraft and habitats. Piezoelectric and thermoelectric devices can convert vibration or thermopower into an electrical one. But the generated power is only a small amount. The efficiency of conversion is through by replacing the device components by nanocomposites and improving the mechanism of conversion of any power to electrical power.

iii. **Power Distribution:** Power distribution systems include buses, wiring, and harnesses for distribution and power management in spacecraft and aircraft. Nanotechnology can reduce the density of the storage systems and improved the durability by providing more durable conductive materials, lighter weight, and improve thermal management and insulation for wiring in energy distribution systems.

12.5 PROPULSION

In the propulsion system, NMs enable new or increased existing capabilities. This is possible in two different ways:

i. **Propellants:** Various propellant systems are hazardous, and it is necessary to handle the propellants with care. Cryo-propellants and hypergolic propellants are mostly used now which can be replaced by NMs derived propellants. These are less toxic and safe to use. By means of nanotechnology, the propellant transfer can also be made easy. This may greatly reduce launch costs, ground operations, and complexity.

ii. **Propulsion Components:** NMs can improve thermal conductivity, strength, and durability, which enable the development of more efficient energy storage, long-lasting, lighter components, and propulsion systems of aircraft.

12.6 NANOELECTRONICS FOR SENSORS AND ACTUATORS

NMs possess enhanced electronic properties along with their excellent mechanical properties. They can be exploited in making sensors and actuators. They require lower power and exhibit better performance. This has been possible due to their radiation shielding effect, and smaller volume helps them to get the benefit for greater packing efficiency.

Sensors and actuators are of wide use in the aerospace system. In the energy storage system, sensors are used for sensing and controlling of temperature, pressure, and strain. Checking of gas leakage and damage is also possible using such sensors. The sensors made by NPs are less invasive and very small so that they can be used in any device. The robotic systems contain sensors and actuators.

Other than sensor and actuator, communication and processing system depend mostly on logic devices. High-speed signal, memory devices, data storage system rely on the electronic properties of materials, and that can only be attained by NMs.

Nanoelectronics includes memory and logic devices for communication, processing systems, and data storage, that can enhance the performance of high-speed signal and control devices, like field emission based electronics. Nanotechnology has the proficiency to reduce power demand while improving the radiation hardness and performance of these devices.

12.7 NANOMATERIALS (NMS) AND NANOCOMPOSITES FOR STORAGE

12.7.1 COMPOSITES

i. **Carbon Nanotube (CNT) Composites:** CNTs have remarkable mechanical properties. They have high Young's modulus, tensile and compressive strength, extraordinary stiffness, very high toughness. Their thermal property is also astonishing. These materials have changeable conductivity. Though the excellent mechanical properties do not reflect when composites are prepared with them. The load transfer capacity is hindered for several reasons. The dispersion the CNTs and filling them in composites are difficult. But due to the fast progress in composite science and technology, CNT composites will certainly serve as storage material.

CNT reinforced polymer composites have the potential to be used in aviation due to their high strength and low weight. They can be also useful to resist fire and vibration.

Now the CNT-matrix interaction is of great interest, and it is proved that functionalization of the CNTs with different chemical groups mediated via defects inside them facilitate strong bonding between the nanotubes filler and matrix material.

ii. **Nanostructured Metals:** Nanostructured metals have high tensile and fatigue strength, high ductility, toughness, and hardness compared to the conventional materials. So these can be effectively used as a substitute of toxic materials in coating technology. They can replace chromium as these are toxic materials and also for the potential of the nano metals for using them in anticorrosive coating materials. Though these can be used in delicate use like landing gear or other parts of spacecraft, these involve complex synthesis method, and their ductility should be increased more before using them in a suitable place.

iii. **Nanomaterial/Polymer Nanocomposites:** Materials made up of two or more components which are in two or more phases are called composites. These two phases are one filler with one matrix material. Load transfer efficiency between the filler and the matrix determines the strength of the composite formed. Well-dispersion of the fillers in the matrix and prevention of the fillers after the dispersion from reaggregation are important for this purpose. For dispersion, or mechanical mixing is generally used, while different functionalization processes prevent reaggregation. For aircraft manufacturing nanofibers, graphene, CNTs or nanoclays are hugely used as fillers. Multi-walled CNTs or CNT ropes are now showing very satisfactory strength and stiffness while they are used with polymers. Multifunctional materials are now designed with nanotubes in composite form. The superb combination of high conductivity, low weight, and high strength has projected them to be ideal candidates in the aviation sector. Their improved resilience to vibration and fire makes them ideal for use in the aviation sector as energy storing devices.

It is now a well-established fact that a small percentage of nanomaterial fillers in the polymer matrix can increase the stiffness and strength in a large amount. So for low weight and high strength, now the aircraft manufacturing companies are deciding to replace some part of the airplane structure with polymer composites. This will save the cost as well as the fuel consumption in the near future.

12.8 NANO-COMPOSITE USED IN BUILDING AN AIRCRAFT ENERGY STORAGE DEVICE

12.8.1 *MATERIALS USED FOR THE MAIN STRUCTURE*

The materials to be used for the structure of the airframe must have the following properties:

- They should have high strength and stiffness;
- The materials should be lightweight;
- These should be anti-corrosive;
- Highly tough and durable material;
- They need little maintenance;
- Should be easily reparable and reusable.

12.8.2 NANO-MATERIALS BY WHICH THE REQUIREMENTS CAN FULFILL

Materials to be used should be NMs as they possess extraordinary mechanical, electrical, and thermal properties. Their properties can be tailored when needed by changing the manufacturing processes and parameters. Intelligent and smart materials can shape smart aircraft so that they are designed difficult to discoverable under radar and sonar. These eco-friendly, lightweight, and easily reparable materials are ideal for such applications.

12.8.2.1 CNT/POLYMER COMPOSITES

1. Carbon Nanotubes (CNT): Graphene sheets rolled along with the chiral vector form seamless cylindrical structures called CNTs. Single-walled, multi-walled CNTs and bundles or ropes of CNTs are formed. They have very high Young's modulus around 1 TPa and high tensile stress around 100 GPa. Crash resistance is very high too. They can bear a high temperature without failure. However, for the directional property of the CNT based polymer composites alignment of the CNTs in the matrix material is required. Still, Only 1% addition of CNTs in a matrix, its stiffness can increase between 36–42% and the strength by 25%

The probable reduction in the weight of aircraft components using composite materials reinforced with carbon nanotubes (CNT) can be as large as up to 70% compared to conventional carbon fiber reinforced polymers.

Some CNT-based composites which are used for Airframe structure are: CNT/Epoxy, CNT/Polyimide, CNT/PP, CNT/nylon, and CNT/PBO.

12.8.2.2 NANO-CLAYS REINFORCED POLYMER COMPOSITES

The major problem of using CNTs to prepare CNT/polymer composites is their alignment, as stated above. Good adhesion of the CNTs is necessary to achieve the goal. But this increases the production cost 10,000 fold. Still suitable industrial-scale production process is lacking. Compared to this, due to nontoxicity, easy abundance and easily processability, nanoclay composites, in the modern age have gained tremendous research interest. For example, a small amount of nanoclay can improve the properties of polyethylene composites considerably. When a small quantity of nanoclay is infused in between the filler and polymer, the overall strength of the composite increases. Thermal and Flame Retardant Barrier Properties are improved as well.

12.8.2.3 METAL NANOPARTICLES (NPS) INCORPORATED COMPOSITES

From hall-pitch relationship, it is obvious that with the decrease of grain size, the strength of the metals increases.

- Properties of metals are based on the Hall-Pitch relationship –that states as grain size decrease, strength increases. Nanocrystalline materials are characterized by sufficient increases in ultimate tensile strength, yield strength, and hardness. For example, the fatigue lifetime can be increased by up to 300% by using NMs with a sufficient reduction of grain size in comparison with conventional materials.
- Nanostructured metals, particularly titanium and aluminum alloys can improve the mechanical properties and enhance the corrosion resistance.
- Metals can be strengthened by ceramic fibers such as carbide, aluminum oxide, silicon, or aluminum nitride. Advantages of these so-called Metal matrix composites (MMC) are high thermal stability, high strength, high thermal conductivity, a low density, and a controllable thermal expansion. MMC has the potential to substitute aluminum and magnesium parts in the future

12.8.3 TRIBOLOGICAL COATINGS

Altitude, humidity, and changes in temperature, spacecraft suffer from corrosion and wear of many system parts. The tribological coating is effective where high pressure and sliding velocity is involved. The durability of the structure is required against these factors. Transition metal dichalcogenides powder composited with organic or inorganic matrix materials are used for aerospace applications. Magnesium alloys, in spite of their lightweight, cannot be used as a coating material, as they are more reactive and hence prone to chemical reactivity. Chromium can also be used as anti-corrosive material, but tribological coatings are used due to the toxicity of the material (Figure 12.5).

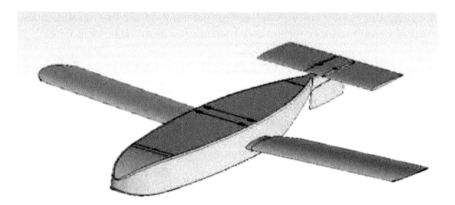

FIGURE 12.5 Coating of aircraft.

Silicon and boron oxides, cobalt-phosphorous nanocrystals are also used as anti-corrosion nanocoatings.

12.8.4 AERO-ENGINE PARTS

As the research on aerodynamics has been progressing fast, the consumption of fuel is diminished considerably compared to the past. But still, the air engines consume hydrocarbons as fuel. The visible progress is only in terms of more fuel-efficient engines and systems and also in term of reduction of weight. Further improvements are required for the engines, fuel consumption cost, and fuel quality.

Enhancement in aircraft engine efficiency can be reached by materials which allow lower engine weights, higher pressures, higher operating temperatures, and increased rotor operating stresses.

- The nanoscale materials that can bear high temperature, have the potential to improve the quality of the engine material so that thrust-to-weight ratio up to 50%. Using NMs the fuel can also be saved up to 25% compared to the conventional engines.

The engine blades are now coated with NMs. Today's research is focused on the tailoring the properties of the coating so that it can stick to the blades made of metal and help the engine run smoothly in a heated condition.

Thermal barrier coatings (TBC) can be applied on the metal blades to make the engines work at elevated temperature. TBC has common applications in gas turbines, aero engines, aerostructures, and chemical processing. TBC is a two-layered system. The inner layer is made of perovskite oxide ceramic, and it can resist strain. The outer layer, made of the same material should be chemically resistant. For TBC, nanophase ceramics can be used as when the NPs are in the range of 100 nm, they show extraordinary toughness.

- Airplanes move in such an environment that it has to pass through a harsh environment. The coatings can protect the surfaces and structure of the airplane from harsh climate, corrosion, erosion, scratching, or wearing of engine parts. For this, an effective coating made of nanostructured materials can be proposed. Friction can be avoided using nitrides, carbides, or metal nanoparticles.
- Yittria stabilized Nano-zirconia;
- SiC Nanoparticles in SiC-particle-reinforced alumina;
- TiN Nano-crystallites embedded in amorphous Si_3N_4 is used for Wear-resistant coatings.

The nanocomposite coatings are made of Diamond-like Carbide, crystalline Carbide, metal di-Chalcogenide, and TiN are used for low friction and wear-resistant applications of Airplane. Nanotube and NPs (Nano-graphite, Nano-Aluminum) containing polymer coating are used for EMI shielding low Friction and electrostatic discharge applications of Airplane surfaces.

12.8.5 NANO-MATERIALS FOR AIRPLANE ELECTRO-COMMUNICATION COMPONENTS

In an aircraft system, information and communication can be improved through the development of nanoelectronic systems. As the NPs have much improved electronic properties, they can be used for avionics and communication system.

- The main aim of the aviation sector is to offer a comfortable environment for the passenger. Improved flat screens, entertainment systems, and miniaturized energy-saving data storage systems can be designed for entertainment purpose.

Flat-screen monitors developed with CNTs can be less power consuming with a broader viewing angle. This can be an attractive display system for the passengers compared to LCD displays.
- "The home and office in the sky" can be envisaged with the help of integrated nano-electronic components.
- Weight savings of the aircraft frame is possible by replacing the body material by nanocomposites and also the wires for communication by nanotube improved plastic wires.
- Conductive plastics composited using nanotubes may be used in electronic devices for electrostatic dissipation.
- Polymer CNT composites or other NPs are to be used for making supercapacitors.
- An excellent fuel-management system can be developed using NEMS (Nano Electro Mechanical Systems) and MEMS (Micro Electro Mechanical Systems). Fuel consumption can be controlled using nanocoating.

12.9 ENVIRONMENT, HEALTH, AND SAFETY ASPECTS

1. **Health Risks:** We cannot imagine such a system where the human health is at stake. NMs brings with its superb properties a health issue and that to be solved first because passengers and Crew can spend a lot of time on aircraft. Coating of surfactants or biopolymer that are environment-friendly, can be used to nullify toxicity caused by NMs.
2. **Safety Risks:** Before future application of nanoscience and nanotechnology in an aircraft system, potential risks arising from long term applications should be studied well. Careful scrutiny is required whether he proposed model is prone to accidents or if it is falling or colliding with other objects in space.
3. **Environmental Benefits:** The components of an aircraft system should be eco-friendly and bring benefits of the environment. These benefits include onboard environmental control sensors, reduced fuel consumption, and more environmentally sound coatings.
4. **Safety Benefits:** The applications of NMs and nanocomposites in the airplane are really optimistic. Especially the safety of the people should be ascertained. Nanosensors should be smart

enough to imminent detection of any danger of the people. Safety of the people on the ground should be ensured too. The improved disaster management system utilizing improved satellite communication can be assured.

Safety of people on board can profit from the application of vibration and flame resistant NMs and nanosensor networks embedded in composite materials.

12.10 SOME OTHER APPLICATIONS

Ceramic nanocomposites, due to their low mass and high heat resistance, can be used in jet engines. Sealants and Gaskets for engine part can be prepared by nano graphite and nanosilica. Nano MMC are anti-corrosive and nonmagnetic with high strength and low weight. These are used in the aerospace industry.

Turbo engines can be protected using zirconia-based nanocomposites. Polymer nanocomposites have a great future for special design and other applications related to aerospace. Nano-chromium is used for the protection of the aluminum structure, and it also resists corrosion. Nano aluminum powder is used as a component of propellants and also modifies the burning rate of fuel. Nano copper serves as conductive plastic, lubricant, etc. Iron and iron oxide NPs may be used for conductive plastic

12.11 DISADVANTAGES OF NANO-MATERIALS

i. **Disposal of Nano-Waste:** Toxicity related issues are always alarming for NMs. Hence the disposal of nano waste may be harmful for the environment. Experimental evidences of exposure associated results are very few. Therefore the uncertainty associated with the effects of NMs is yet to be assessed in order to develop the disposal policies.

ii. **Internal Reaction Caused by Impurities:** Though NMs contain very lesser number of impurities in them, but NMs are so reactive that they interact with the impurities present in themselves.

iii. **Explosive Nature:** NPs due to their large surface area are highly reactive, and their combustion is exothermic. When it comes in direct contact with atmospheric oxygen, they act as strong explosives.
iv. **Necessity of Enclosure:** Kinetics of the reaction of NPs reveals that the reaction process is very fast. So working with NPs is most challenging. NPs are to be enclosed in different matrices to maintain the nano size of the particles.
v. **Production of Suitable Size of Nanoparticle is Difficult:** It is very tough to maintain the size of Nanoparticles when they are formed within the solution, and thus, the Nano-materials have to be enclosed in the matrix material.
vi. **Biologically Harmful:** NPs, are easily absorbed by skin, they become transparent to the cell dermis. Absorption of the materials, due to low mass is very high when inhaled. Once trapped inside lungs, these are impossible to breathe out. Large surface area of the NMs causes high reactivity as well as high toxicity. Nanomaterials have shown to cause of irritation, and have indicated to be carcinogenic.

12.12 CONCLUSION

The above discussion shows that the potential of the nanocomposite materials in the Aviation Sector. Using nanotechnology or nanocomposite in aviation gives the High Strength, Light Weight, Corrosion Resistant, materials with high toughness, and durability. Also, these materials need the least maintenance, and they are reusable. Cheaper and safer coating for the surface of the aircraft is easy with them. Unlike the conventional materials, these coating can protect the aircraft from a harsh environment more efficiently. Made with the NMs, the structure of the airplane can be repaired easily. In spite of having some disadvantages of nanotechnology, the overall gain can be counted indeed.

ACKNOWLEDGMENT

I would like to give thanks to my Parents Mr. Ramesh Chandra Lal Das and Mrs. Mira Devi and My brothers Mr. Dilip Das and Mr. Pradeep Das with my nephew D. Amay.

KEYWORDS

- composite material
- nano-composite material
- nanoelectronics
- nano-materials
- nanoscience
- nanostructure
- nanotechnology

REFERENCES

1. *EPA Nanotechnology White Paper,* EPA 100/B-07/001 February 2007.
2. http://www.nnin.org/nnin_edu.html (Accessed on 25 July 2019).
3. Alagarasi, A. *Introduction to Nano Materials* (pp. 1–3). http://www.azonano.com/article.aspx?ArticleID=3103 (Accessed on 18 October 2019).
4. http://www.workingin-nanotechnology.com (Accessed on 25 July 2019).
5. http://www.nanostudent.com (Accessed on 25 July 2019).
6. Will, S., (2003). *Nanotechnology In Aerospace Materials.* Published by VDI Technology Center, Future Technologies Division, Graf-Recke-Str. 8440239, Düsseldorf, Germany, Future Technologies No. 47, Düsseldorf, ISSN 1436–5928.
7. *Space Development and Systems* (Vol. 1, pp. 1262–1266 .). VDI Technology Center.
8. www.nanoforum.org (Accessed on 25 July 2019).
9. www.nasa.gov (Accessed on 25 July 2019).
10. Zweck, A., & Luther, W., (2003). *Application of Nanotechnology in Aerospace Corporation.* http://www.aero.org// (Accessed on 25 July 2019). Nanotechnology For Aerospace by Boeing.
11. Mark, D., (2006). *Taczak: A Brief Review of Nanomaterials for Aerospace Applications: Carbon Nanotube-Reinforced Polymer Composites* (pp. 39–46). MITRE Nanosystems Group, Virginia.
12. Lincoln, D. M., Vaia, R. A., Brown, J. M., & Tolle, T. H. B., (2000). *IEEE Aerospace Conference Proceedings, 4,* 183.
13. Kim, B. K., Seo, J. W., & Jeong, H. M., (2003). *Eur. Polym. J., 39,* 85.
14. Patton, R. D., C. U. Pittman Jr., Wang, L., & Hill, J. R., (1999). *Compos. Part A., 30,* 1081.
15. Smith, J. G. Jr., Watson, K. A., Thompson, C. M., & Connell, J. W., (2002). *34th Inter. SAMPE Technical Conference, Maryland* (pp. 97–106).
16. Sandler, J., Werner, P., Shaffer, M. S. P., Demchuk, V., Altstädt, V., & Windle, A. H., (2002). *Compos. Part A., 33,* 1033.

17. Andrews, R., Jacques, D., Qian, D., & Rantell, T. A., (2002). *Acc. Chem. Res., 35*, 1008.
18. McCullough, R. L., (1990). In: Whitney, J. M., & McCullough, R. L., (eds.), *Delaware Composites Design Encyclopedia* (Vol. 2, pp. 49–90). Lancaster, PA: Technomic Pub. Co.
19. Odegard, G. M., Gates, T. S., Nicholson, L. M., & Wise, K. E., (2001). *NASA/TM-2001–210863*.
20. Star, A., Stoddart, J. F., Steuerman, D., Diehl, M., Boukai, A., Wong, E. W., Yang, X., Chung, S. W., Choi, H., & Heath, J. R., (2001). *Angew. Chem. Int. Ed., 40*, 721.
21. Odegard, G. M., Gates, T. S., & Wise, K., (2002). *Structural Dynamics and Materials Conference* (pp. 127–139). Denver.
22. Zhang, S., Sun, D., Fu, Y., & Du, H., (2003). Recent advances of superhard nanocomposite coatings: A review. *Surf. Coat. Technol., 167*, 113–119.
23. Kawasaki, M., Ahn, B., Kumar, P., J-iJang, L., & Langdon, T. G. Nano- and micromechanical properties of ultrafine-grained materials processed by severe plastic deformation techniques. *Advanced Engineering Materials, 19*(1).
24. https://en.wikipedia.org/wiki/Grain_boundary_strengthening#/media/File:HallPetchLimit.png (Accessed on 25 July 2019).

Index

A

Absorption coefficient method, 200, 202, 203
Acceptor molecules, 6
Acoustic waves, 262
Acrylonitrile-butadiene-styrene (ABS), 13
Actuators, 55, 66, 99, 291
Adenosinetriphosphatase (ATPase), 166
Aero-engine parts, 296
Aerosol deposition method, 54
Aerospace applications, 125, 134, 295
Aerostructures, 297
Air sanitization, 145
Aircraft energy
　storage device, 293
　　main structure, 293
　　materials used, 293
　system, 121, 122, 124, 125, 128, 129, 131, 136
　　modeling, 129
　　simulation, 129, 136
　　simulators, 136
Airplane electro-communication components, 297
Algorithms, 122, 124
Alkali, 28, 55
Aluminum
　nitride, 12, 33, 295
　oxide, 12, 33, 295
Ambient temperature, 79, 81
Ameliorate, 6
Amino acid, 167–169
Aminopropyltriethoxysilane (APTES), 9
Ammonia, 28, 98
Amorphous
　carbon, 7
　silicon, 22, 102, 189
　zone, 269
Anaerobic digestion, 109

Analytical models, 129
Anatase, 6, 173
Anode, 25, 27, 70, 103, 104, 146, 148
Antenna complexes, 162
Antimicrobials, 17
Antireflection, 43, 54, 55
Archaealrhodopsins, 163
Archaeans, 163, 167
Arginine, 167
Aromatic amino acids, 167
Artificial intelligence (AI), 122, 124, 216
Aspartic acids, 167, 169
Atom-by-atom process, 44
Atomic
　clusters, 287
　force microscopy (AFM), 45, 46, 57, 175, 220
Attenuated total reflectance (ATR), 219
Au/Pt/Ag colloidal nanocomposites (Au/Pt/Ag TNCs), 216–218, 220–226
Automatic
　cooling device (ACD), 65, 75
　energy transport device (AETD), 75, 78
Azimuthal angle, 201

B

Bacteria cells, 17
Bacterial pigments, 177
Bacteriorhodopsin (BR), 161, 163, 165–180
Bacterioruberin, 177
　sensitizers, 177
Bandgap, 23, 24, 46, 47, 55, 96, 103, 106, 165, 185, 189, 191, 192, 204–208
　assumption method, 204, 206
Bernoulli's equation, 73, 74
Bimetallic nanoparticles (BNPs), 217
Bioenergy, 108
Biofuel, 4, 21, 94, 108
Biogas, 108, 109

Biomass, 21, 28, 31, 108, 109, 146
Biomolecular electronics, 171
Biomolecules, 161, 162
Bio-nano hybrid complexes, 178–180
Bio-nanometal-cluster composites, 215, 230, 231
Bio-photonic devices, 161
Bio-physical kinetics, 215, 230
Biopolymer, 298
Biosensors, 67, 178
Black
  box models, 138
  carbon, 283
Born-Oppenheimer approximation, 239
Boron nitride, 12, 33
Bulk materials, 25, 94, 122, 125, 187, 286

C

Cadmium-telluride (CdTe), 22, 102, 189, 191
Calorimetry, 265
Carbon
  black, 15
  cloth, 234, 235
  cryogels, 7, 33, 96
  dioxide, 28, 101, 147, 148, 268, 283
  emission, 4, 32, 108, 151
  fiber, 255–258, 260–262, 264–266, 268, 274, 276, 277, 294
  monoxide, 109, 256
  nanofibers (CNFs), 18, 29
  nanostructures, 6
  nanotube (CNT), 7, 12, 15, 29, 30, 33, 97, 122, 133, 139, 140, 148, 150, 152, 153, 155, 177, 189–191, 234, 235, 251, 292–294, 298
  tetrachloride, 220
Catalytic activity, 216, 217, 230
Cathode, 25, 50, 146, 148, 150, 153, 234
Cell membrane, 163, 166, 169
Cellular automata, 122, 124
Central processing units (CPUs), 76
Ceramics matrix nanocomposites (CMNC), 19
Chemical
  bonds, 108
  reactants, 146
  vapor deposition (CVD), 19, 49, 54, 190, 288
Chemisorption, 235
Chlorophyll, 162, 169, 170
Chromophore, 162, 166–169, 172, 173, 179
Chronoamperometry (CA), 227, 230, 231
Circumferential stress (σc), 265
Colloidal
  dispersion gels (CDG), 26
  nanoparticles, 161
  suspension, 19, 66, 68
Composite material, 30, 140, 205, 208, 257, 266, 301
  application, 96
  development, 101
Compressed natural gas (CNG), 255–262, 265, 267, 268, 272–277
Concentrating solar power (CSP), 72
Conductive transparent oxide (CTO), 23
Conductivity, 12, 13, 22, 30, 33, 70, 75, 98, 148, 150, 154, 190, 250, 251, 292, 293
Coolant, 67, 78, 82
Copper indium gallium
  dieseline, 22
  selenide (CIGS), 102, 189
Corrosion, 152, 153, 217, 256, 257, 295–297, 299
Coulometric method, 271
Coupling agent, 18
Covalent bonds, 15
Critical volume fraction, 71
Crystalline, 13, 22, 23, 45, 48, 50, 54, 95, 102, 103, 163, 164, 186, 220, 234, 269, 297
  substance, 45
Crystallinity, 51, 54, 57, 96, 289
Crystallographic
  orientation, 50
  structure, 44
Cyanobacteria, 163
Cyclic voltammetry (CV), 57, 227–229, 231
Cytoplasm, 179

Index

D

Damage-tolerant systems, 289
Dark-adapted (DA), 168
Density
  DOS (density of state), 225, 228
  functional theory (DFT), 192, 193, 208, 225, 227, 228, 233, 236, 237, 239–241
Department of Science and Technology (DST), 82
Dielectric, 6
  constant, 6, 43, 44, 48, 49, 55, 56, 200, 234
  loss, 43, 51, 55, 56
Diffusion coefficient, 268–270
Direct
  absorption solar collectors (DASC), 72
  current (DC), 22, 50
Dispersion, 6, 9, 10, 18, 26, 33, 122, 124, 216, 262, 266, 267, 292, 293
Dopant, 55
  density, 204, 205
Doping, 51, 55, 106, 152, 189, 190, 235
Drug permeation, 15
Dye-sensitization, 103
Dye-sensitized, 21, 95, 110, 164, 191
  solar cell (DSSC), 21, 23, 95, 103, 164, 180
Dynamic
  light scattering (DLS), 219–221
  model, 131
  random access memory (DRAM), 55

E

EDX analysis, 224
Elastomers, 29
Electric
  double-layer supercapacitors, 7
  field, 48, 237, 238, 241, 244
  vehicle (EV), 25, 27, 70, 148
Electrical
  conductivity, 8, 17, 18, 98, 233, 235, 250, 251
  transformer, 82
Electrocatalysts, 5, 33, 255
Electrocatalytic activity, 190, 226, 227, 231
Electrochemical
  capacitors (ECs), 29
  deposition, 49
  properties, 7
Electrodeposition, 19
Electrodes, 5, 7, 25, 27, 30, 33, 49, 50, 96, 98, 99, 103, 150, 152, 153, 164, 173–175, 190, 191, 235, 237, 290
Electrolyte, 5, 25, 27, 30, 32, 98, 99, 103, 108, 146, 153, 164, 174, 191, 234, 290
Electromagnetic
  energy, 68
  field, 46
  force, 68
  transduction, 68
Electromotive force, 70
Electron
  acceptor, 225
  beam evaporation process, 57
  density, 193, 217, 238, 240, 241, 244
  difference potential (EDP), 239
  diffusion, 164
  donor, 177, 225
  excitation, 22
  hole pairs, 24, 236
  paramagnetic resonance (ESR), 7
Electronic
  effect, 217
  kinesis, 164
  transition, 47
Electrons, 7, 17, 23, 27, 45, 50, 95, 102, 104, 106, 164, 191–193, 204, 205, 216, 217, 224, 236, 238–241
Electro-optic devices, 55
Electrostatic
  energy, 237, 238, 244, 251
  forces, 14
  potential, 173, 238, 244, 247
Energy
  application, 34, 99
  change, 147
  dispersion, 148
  dispersive x-ray (EDX), 11, 219, 220, 224
  flux, 197, 203, 205, 208

harvester, 56, 65, 68, 69, 81, 82
stockpiling, 148, 149
storage, 6, 7, 22, 29, 44, 56, 57, 65, 67, 70, 72, 81, 97, 101, 124–130, 139, 146, 147, 151, 162, 233–235, 237, 241, 285, 289–291
utilizations, 149
Epoxy resin, 13, 255, 258, 261, 265, 266, 276
Equilibrium, 241, 262, 268
Equivalent series resistance (ESR), 7, 30
Ethylene
propylene-diene monomer, 9
vinyl acetate (EVA), 13, 289
Evaporation method, 54

F

Fermi level, 204, 205, 247, 249
Ferroelectric
oxide thin films, 43
properties, 56
Ferrohydrodynamic (FHD), 66, 73, 74
Fick's first law, 268
Field
effect transistor (FET), 236, 251
emission scanning electron microscopy (FESEM), 46, 57, 220, 224
Filament winding (FW), 255, 257, 258, 261
Finite element simulation (FEM), 264
First generation (1G), 102
Flexural strength, 8, 18
Fluid
magnetization, 72, 73, 75, 78, 79, 81
velocity, 73
Flywheel energy storage (FES), 98
Förster resonance energy transfer (FRET), 179
Fossil fuels, 20, 27, 28, 31, 91, 92, 102, 146, 147
Fourier transform infrared (FT-IR), 219, 265
Frequency-specific absorption, 205
Fullerene, 7, 150
Functionalized carbon nanotubes (FCNTs), 122

G

Gas permeability, 5, 255, 260, 261, 266, 268, 270, 272, 277
Gasification, 108, 109
Genetic algorithm, 122 124
Geometrical effect, 217
Ghost atoms, 239
Global
electricity demand, 29
warming, 20, 94
Glutamic acids, 167
Gold (Au), 216
nanohybrids, 235
G-protein coupled receptor (GPCR), 167
Gradient magnetic field, 66
Granulometry, 267, 275
Graphene, 7, 12, 27, 29, 30, 33, 97, 99, 150–153, 176, 186, 189–191, 234, 235, 293
catalysis, 151
lithium-ion batteries (LIBS), 150
nanotechnology, 150
oxide (G-O), 56
solar cells (SCS), 150
supercapacitors, 151
Graphite, 5, 7, 12, 33, 71, 98, 99, 235, 297, 299
Gravitational force effects, 130
Green
algae, 163
catalyst, 231
Greenhouse
effects, 31
gases, 27, 28, 31, 32, 94, 101, 108

H

Hall-Pitch relationship, 295
Halobacterium, 163, 166
Halogen, 28
Halophiles, 163
Halophilic bacteria, 165
Halorhodopsins, 163
Hard carbon technology, 27
Heat transfer
coefficient, 70
devices, 82

Heterogeneous, 7, 105, 176, 262, 288
  photo-catalysis, 7
Hetero-nanohybrids, 234
Heterostructure, 152, 162, 171, 192, 236
High
  combustion enthalpy, 107
  density polyethylene (HDPE), 13, 261, 264, 270–273
  resolution
    scanning electron microscopy (HR-SEM), 219, 220
    transmission electron microscopy (HR-TEM), 219
Highest molecular-lowest unoccupied molecular orbital (HOMO-LUMO), 225, 227
Homogeneity, 51
Homogeneous
  materials, 276
  transfer films, 16
Hot spot, 79, 80
  temperature (HST), 79, 80
Humidity sensor, 56
Hybrid, 5, 145, 150
  electric vehicles (HEVs), 27
  electrodes, 174
  nano-bio solar cells, 171
  structures, 233, 251
Hybridization, 8, 106
Hydraulic subsystems, 132
Hydrocarbon, 24, 105, 260, 296
Hydrodynamic diameter, 219, 220
Hydroelectric energy, 101
Hydrogen
  accumulating point, 149
  evolution, 105
  gas, 28, 95, 107, 108
Hydrolysis, 7, 20
Hydrophilic and lipophilic polysilicon (HLP), 26
Hydrophobic surface, 162
Hydrostatic test, 257, 259, 260, 264–266, 272, 274–276
Hydroxyl (OH), 20
Hypersaline water, 163
Hyperthermia, 67

I

Indium tin oxide (ITO), 52, 54, 171, 190
Industrialization, 91, 92, 109
Inert gas, 50
Inhomogeneous electron density, 222
Inorganic
  composites, 18
  hybrid, 6
Input pedigree results, 138
Instrumentation, 122, 124
Interfacial
  adhesion, 10
  tension (IFT), 25, 26
Intermolecular spaces, 269
International
  Council on Systems Engineering (INCOSE), 133
  Standard Organization (ISO), 185, 255, 258, 259, 265, 274, 276, 277
Iodide ions, 108
Isomerization, 173, 179
Isotropic properties, 257, 258

K

Kelvin body force, 75
Ketoprofen, 15
Kinetic energy, 74, 77, 193, 194, 240, 241, 262
Kohn Shyam equation, 240

L

Lambert-Beer's law, 202
Large-scale simulation, 126, 137
Lattice parameters, 45, 51
L-glycerol, 167
Life cycle phases, 139
Light
  absorption spectra, 219
  emitting diodes (LEDs), 99, 146, 154, 155, 236, 251
Linear
  combination of atomic orbitals (LACO), 239
  low-density polyethylene (LLDPE), 264
Lithiation, 70

Lithium
  hexafluorophosphate (LiPF), 5
  ion, 5, 70, 150, 233
    batteries (LIBs), 5, 7, 25, 70, 99, 150
    metal oxide, 5
Local density approximation (LDA), 193, 195, 208, 233, 239
Longitudinal stress (σl), 265
Low density polyethylene (LDPE), 13, 271
Lysine, 167, 169

M

Magnetic
  field, 50, 66, 69, 71, 73, 74, 76–79, 81
  fluid, 65–74, 76–82
    applications, 66, 68, 70–72, 76, 78
    column, 70
    droplets, 69
    lubrication, 69
  flux, 68, 70
  forces, 66
  memories, 54
  properties, 66, 82, 122, 125, 153
Magnetization, 73–75, 78–80
Magnetocaloric materials, 66
Magneto-electronic devices, 66
Magnetohydrodynamic (MHD), 77
Magnetron, 50, 56
Maleic acid, 260, 272
MATLAB, 121–123, 125, 132, 139
Matrices, 8, 29, 149, 241, 274, 300
Matrix, 10, 12, 14, 16, 18, 19, 33, 200, 202–204, 235, 241, 274, 292–295, 300
Maxwell's equations, 125
Mechatronics, 132
Memristive cells, 56
Memristors, 56
Metal
  matrix composites (MMC), 295, 299
  nanohybrids, 234
  nanoparticles (NPS), 14
    incorporated composites, 295
  oxides, 107, 108
  sulfides, 107, 108
Metallic nanocomposites (MMNC), 19

Metalloporphyrins, 162
Methane, 7, 257, 260, 268, 270, 275
Methanol oxidation reaction (MOR), 225, 226, 228, 230
Micro
  nano fly ash (MNFA), 109
  composites, 19
  electro mechanical systems (MEMS), 56, 69, 298
Microbial
  biofilm, 17
  life, 163
Microcracks, 267
Microcube, 224
Microelectronics, 76
Microfluidic devices, 66
Micromagnetics, 125
Micromorph silicon, 102
Microorganisms, 31, 163
Micro-pitting, 29
Microwave, 43, 44, 48, 56
Mid-scale simulation, 136
Miller indices, 238
Miniaturization, 25, 30, 43, 44, 54, 76, 233
Mitochondria, 166
Mitutoyo caliper, 264
Model-based system engineering (MBSE), 133–135
Modular energy, 22
Modulus, 5, 8, 9, 16, 98, 187, 261, 266, 292, 294
Molecular
  dynamics, 122, 133, 134, 139
    simulation, 122, 139
  scale, 123
  semiconductors, 102
  sensitizers, 177
Monometallic
  nanoparticles (MNPs), 217, 220, 221
  surfaces, 217
Montmorillonite clay, 16, 263, 266, 267
Mulliken
  charge, 238, 242, 251
  population (MP), 239, 243, 251
Multiple walled nanotubes (MWNTs), 98, 99

Index

N

Nano-applications, 147
Nano-assemblers, 122, 125
Nano-bio hybrid photocatalyst, 161
Nanoadditive, 289
Nanoaluminum, 299
Nanocapacitors, 237
Nanocatalysts, 109
Nanoceramics, 286
Nanoclay, 10, 255, 257, 260, 261, 263, 267, 270–272, 275, 276, 293, 294
　selection, 267
　reinforced polymer composites, 294
Nanocluster, 145
Nanocoatings, 121, 148, 296
Nanocomposite, 3–14, 17–25, 27–30, 32–34, 91, 92, 96, 121, 145, 149, 151–154, 185, 187, 192, 215, 216, 218, 220, 222, 225, 227, 231, 246, 255, 260, 266, 267, 270, 272, 273, 276, 277, 286, 290, 292, 297–300
　different types, 19
　　preparations methods, 19
　energy industries application, 20
　　energy storage in supercapacitors, 29
　　enhancing batteries efficiency, 25
　　hydrogen energy production, 27
　　other energy related systems and applications, 30
　　petroleum industry, 24
　　solar energy harvesting, 20
　dye-sensitized solar cells (DSSC), 21, 23, 103, 105, 164, 177
　photo-catalysis, 24
　photovoltaic (PV) cells, 22
　solar collectors, 21
　　wind power production, 28
　materials, 3, 12, 32, 33, 91, 300, 301
　preparation, 18
　properties, 5
　　mechanical property, 8
　　thermal property, 11
Nanocrystal, 23, 177
Nanocrystalline, 43, 51, 54, 164, 173, 295
　film, 54
　morphology, 23, 103
　thin films, 54
Nanocrystals, 145, 177, 186, 237, 296
Nanodevice, 237, 239
Nanodimensions, 19
Nanodiscs, 121
Nanoelectromechanical systems (NEMS), 69, 298
Nanoelectronics, 291, 301
　actuators, 291
　devices, 123, 165
　sensors, 291
　systems, 297
Nanofibers, 145, 178, 293
Nanofilms, 23
Nanofluid, 22, 70–72, 75, 109
　geothermal energy, 109
Nanoforest, 164
Nanogenerator, 69
Nanohybrid, 175, 179, 233–237, 241–251
Nanointerface, 123
Nanolayers, 267
Nano-lubricants, 29
Nanomaterials (NMs), 5, 6, 19, 20, 23, 25, 27, 29, 31, 92, 94, 95, 97, 98, 101, 102, 109, 110, 121, 140, 145–151, 154, 155, 161, 165, 180, 185–189, 209, 233, 236, 283–296, 298–301
　properties, 286
Nanometer, 122, 123, 125, 179, 186, 187, 198, 216, 266, 284
Nanooscillators, 122
Nanoparticle (NPs), 3–6, 8, 9, 11, 14–19, 22–26, 32–34, 66, 69–73, 94, 95, 100, 102–104, 106–109, 121, 122, 124, 145, 150, 152, 153, 155, 165, 171–180, 186–189, 198, 209, 216–217, 220, 222, 224, 234, 235, 251, 266, 267, 269, 284–286, 288, 289, 291, 295, 297–300
　beneficial effects, 15
　　electrical beneficial effects, 17
　　health beneficial effects, 17
　　mechanical beneficial effect, 16
　effects, 14
　　composites, 18
　energy applications, 234
　matrix interaction, 14
　photo-catalysis, 104

Nanopatterning, 49
Nanophase materials, 287
Nanophotocatalysis, 32
Nanoplates, 100, 276
Nanopolymers, 256, 260
Nanopowders, 186, 288
Nanorods, 95, 100, 145, 186, 287
Nanoscale, 14, 18, 20, 23, 30, 44, 46, 98, 121, 122, 124, 125, 131, 132, 165, 171, 185–187, 192, 284, 296
    devices, 171
    science, 20
Nanoscience, 20, 21, 31, 32, 140, 152, 284, 285, 298, 301
Nanosensor, 152, 298, 299
Nanosheets, 24, 150, 164, 186, 234
Nanosilica, 299
Nanosphere, 100
Nanostructure, 6, 27, 91, 92, 95, 97, 100–102, 106, 110, 136, 149–151, 161, 186, 189, 191, 192, 301
    application, 96
        materials development, 101, 107
    modeling, 140
Nanosystems, 122–124, 139
Nanotechnology, 21, 24, 25, 27–33, 43, 66, 101, 102, 121–125, 140, 147–152, 162, 192, 215, 216, 235, 283–285, 290, 291, 298, 300, 301
    simple definition, 285
    solar energy, 101
    systems, 122, 124
Nano-thin films, 102
Nanotips, 164
Nanotubes, 5, 9, 10, 30, 95, 99, 121, 164, 174, 186, 287, 292–294, 298
Nanotubular films, 175
Nanowires, 24, 95, 104, 121, 186, 236
National power grid, 4
Near-infrared (NIR), 57, 177
Negative capacitance (NC), 236, 245, 246, 251
Neutralization, 50
Neutrally wet polysilicon (NWP), 26
Nitrogen-boron species, 7
Non-destructive test technology (NDT), 262

Nonlinear, 56, 69, 130, 138, 171
    cavitations, 130
    photonics, 56
Non-magnetic fluid, 73
Nontoxicity, 165, 294
Non-volatile
    memory applications, 56
    random access memory technology, 56
Novel property, 286
Nucleation, 288
Nucleophilic count, 20

O

Optimization, 23, 51, 123–125, 140, 284
Orbital hybridization, 8
Organic
    composites, 18
    degradation, 107
    hybrid, 6
    hydrogels, 7
    materials, 108
    matter, 108
    nanofilms, 23
    phase, 19
    photovoltaic (PV), 189
        (OPV), 189
    pollutants, 24
Oxide thin film, 49–51, 54, 56, 57
    applications, 54
    characterization, 44
        optical properties, 46
        structural properties, 44
    deposition techniques, 49

P

Palladium (Pd), 216
Parallel plate capacitor, 49, 239, 243–246
Particle-particle interaction, 14
Passive solar, 21
Pechini method, 54
Perovskite solar cells (PSCs), 102, 164, 180
Phase
    change materials (PCM), 71, 154
    type process, 135
Photoanode, 23, 95, 164, 174, 175
    redox, 164

Index

Photocatalysis, 6, 21, 24, 28, 95, 104–107, 110, 162, 230, 231, 234
Photocatalytic
 action, 145
 activity, 100
 hydrogen, 107, 145
 materials, 145
 processes, 7
 purifications, 145
Photochemical reaction, 100
Photocurrent
 density, 171, 173–176
 generation, 174, 177
Photocycle, 161, 168–170, 172, 173, 176–178, 180
Photodetectors, 236
Photoelectrochemical cells, 161
Photoelectrode, 164, 177, 178
Photoenergy, 167, 180
 conversion schemes, 161
Photoexcitation, 23, 179
Photo-generated catalysis activity (PCA), 24
Photo-isomerization, 169
Photoluminescence, 179, 188, 191
Photolysis, 7, 107
Photon, 47, 98, 102, 166, 168, 169, 176, 179, 180, 188, 196, 197, 199, 200, 202, 203, 205, 206, 208
 energy, 47, 188, 196, 197, 199, 200, 202, 203, 205, 206, 208
 flux, 196, 202, 207
Photophosphorylation, 169
Photophysical kinetics, 231
Photo-reaction mechanism, 24
Photoreceptor, 162, 163
Photoresponse, 162
Photosensitizer, 173
Photosensory
 pigments, 163
 receptors, 163
Photosynthesis, 5, 21, 28, 95, 162, 164, 166, 169
Photovoltage, 171, 173
Photovoltaic (PV), 21–23, 95, 97, 98, 102, 103, 108, 110, 150, 152, 155, 161, 163–165, 171, 176, 177, 180, 185, 187–189, 191, 192, 195, 205, 208, 209, 290
 applications, 191
 background, 163
 cells, 22, 23, 110, 164
Phylogenetic range, 163
Physical vapor deposition (PVD), 19, 49
Picosecond, 169
Piezoelectric
 devices, 54
 energy, 68
 oxide films, 43, 44
 properties, 55
Planar waveguides, 24
Plasmonic field, 171–173, 176, 188, 189, 216
Platinum (Pt), 216
Poisson's equation, 204
Pollutants, 24, 256, 257
Poly(ethyleneterephthalate) (PET), 13
Polyamide, 13
Polyamidoamines (PAMAM), 15
Polycarbonate (PC), 13, 219
Polycondensation reactions, 7
Polydispersity, 100
Polyetheretherketone (PEEK), 13
Polyethylene, 13, 256, 257, 260–263, 267, 271, 272, 276, 294
 blend, 267, 270, 271
Polyforms, 6
Polyisobutylene (PIB), 13
Polylactic acid (PLA), 9, 10
Polymer, 5, 6, 8, 10, 12–15, 19, 29, 33, 92, 99, 102, 121, 123, 152, 153, 188, 190, 236, 246, 255–262, 264, 266–270, 272–274, 276, 277, 292–295, 297
 composites, 29, 153, 260, 262, 276, 292–294
 gas permeability, 268
 layered silicates (PLS), 5
 liner, 256–258, 260, 261
 matrices, 8
 matrix, 10, 19, 33, 267, 270, 293
  nanocomposites (PMNC), 19
 nanocomposites, 6, 19, 29, 266, 273, 292
  recycling, 273
 reservoir, 256, 257, 260, 261

characterization, 261, 264
description, 261
hydrostatic testing, 264
ultrasound assay, 262
statics, 5
Polymeric
materials, 33
nanocomposites, 4, 18
Polymethylmethacrylate (PMMA), 13
Polyoxymethylene (Homo) (POM), 13
Polypeptide chain, 167
Polyphenylene sulfide (PPS), 13, 16
Polypropylene (PP), 13, 294
Polystyrene (PS), 13, 16, 17, 219
Polysulfone (PSU), 13
Polytetrafluoroethylene (PTFE), 13
Polyvinyl chloride (PVC), 13
Polyvinylidenedifluoride (PVDF), 13
Porphyrin pigments, 162
Power
conversion efficiency (PCE), 103, 161, 188, 191, 208, 209, 237, 251
distribution, 201, 290
generation, 56, 66, 69, 77, 92, 164, 290
Protein sensitized solar cell (PSSC), 165, 180
Proteobacteria, 163
Proteoliposomes, 171
Proton, 163, 166, 169, 179
pumping activity, 174
pumps, 163, 165, 168, 171
Protonable groups, 169
Prototype, 76, 81, 128, 135, 243
Pseudocapacitance, 234
Pseudocapacitors, 234
Pulse laser deposition (PLD), 54
Pure
montmorillonite (P-MMT), 10, 11
unsaturated polyester (UPE), 10, 11
Pyrolysis, 7, 19, 96, 108, 109, 260, 274

Q

Quantum
devices, 76
dot (QD), 23, 102, 121, 152, 179, 180, 186, 188, 191, 209
sensitized SCs (QDSSCs), 191
efficiency, 165, 169
imprisonment, 145
mechanical effects, 125
mechanics, 239
Quaternary ammonium salts (QAS), 10

R

Radio frequency (RF), 49–52, 55–57
Radiographic tests, 262
Reactive framework, 124
Reaggregation, 293
Real-time execution, 137
Recycling, 31, 256, 260, 265, 273, 274, 276, 277
Redox
electrolyte, 174
reaction, 7, 24, 96, 228, 234
Refractive index, 46, 47, 55, 200
Renewable
energy, 5, 20, 27, 33, 34, 71, 91–99, 101, 110, 146, 151, 154, 180, 192, 215, 284
solar energy, 215, 231
Repulsion forces, 14
Reservoir layers, 265
Resonators, 49
Retinal chromophore, 168, 179
Rheological properties, 273
Rhodium (Rh), 216
Rhodopsin, 162, 163, 167
Room temperature (RT), 52, 53, 78, 99, 218, 262, 264
Rotomolding, 257, 261, 264, 268, 275

S

Scanning electron microscopy (SEM), 10, 11, 45, 46, 57, 215, 219, 220, 224, 230
Scattering, 45, 99, 104, 149, 188, 217, 219
Second-generation (2G), 102
Selected area electron diffraction (SAED), 223, 224
Self-monitoring analysis and reporting technology (SMART), 25
Semiconductor, 6, 7, 22, 23, 28, 32, 95, 100, 102, 103, 105, 107, 121, 145, 148,

Index 313

164, 165, 173, 186, 189, 191, 195, 236, 246
Sensor
  applications, 56
  elements, 54
Short fiber-reinforced polyphenylene sulfide (SFPS), 16
Silicon
  nitride, 12, 33
  solar cells, 163
Silver (Ag), 216
Simulation, 77, 122–140, 185, 193, 195, 237, 239, 261, 264, 275
  S&M (simulation and modeling), 140
Simulator, 133, 136
Single
  phase prototype transformer, 81
  walled nanotubes (SWNTs), 98, 99, 178
Soft carbon technology, 27
Solar
  cells (SCs), 5, 6, 21–23, 72, 93–95, 98, 99, 102, 103, 110, 145, 150, 155, 161–165, 171, 173, 175, 177, 178, 180, 188–192, 215, 230, 234–236, 246
  collectors, 21, 22, 71, 72, 99
  desalination, 22
  drying devices, 22
  energy, 5, 20–24, 32, 54, 70–72, 81, 92, 95, 96, 101, 103, 104, 108, 145, 151, 161–165, 231
  harvesting, 20, 163, 165
  light-harvesting systems, 162, 166
  membrane, 180
  power plants, 22
  proton, 23
  radiation, 21, 23, 92, 96, 101, 176
  spectrum, 103, 205
  thermal
    applications, 21
    system, 72
Solenoid, 73–75
Sol-gel, 19, 20, 49, 51, 56, 235
  method, 19
  technique, 51, 54
  process, 19, 20, 51
Solid-state lighting, 146

Spectrophotometer, 219
Spectroscopy, 46, 173, 215, 230, 265
Split post dielectric resonator (SPDR), 48, 49, 57
Sterilization, 24, 105
Stoichiometric ratio, 260
Styrene-acrylonitrilecopolymer (SAN), 13
Substantial magnetic forces, 66
Substrates, 24, 54, 171, 217
Supercapacitor, 30, 57, 70, 96, 98, 99, 101, 151, 215, 234, 235, 298
  electrode, 235
  fabrications, 56
Super-hydrophilicity, 24, 105
Surface
  enhanced
    infrared spectrum (SEIRS), 219
    Raman scattering (SERS), 178, 216, 217, 219
  photovoltage spectroscopy (SPS), 171
  plasmon resonance (SPR), 152, 188, 189, 209, 216
Sustainable energy, 4, 21, 32, 91, 95, 109, 110, 215, 231
Swarm intelligence, 122, 124
Systems engineering (SE), 133, 134

T

Tandem cell, 164
Temperature
  gradient, 66, 78, 79, 81
  sensitive magnetic fluid (TSMF), 65, 73, 76–81
Tensile
  test, 266–268, 272, 273, 275
  strength, 8, 9, 16, 98, 265, 266, 268, 272, 275, 295
  modulus, 9, 16
Terawatts (TW), 94–96
  scale solar energy, 32
Tetrapyrrole complexes, 162
The-lens detector (TLD), 220
Thermal
  barrier coatings (TBC), 297
  conductivity, 8, 12, 13, 22, 32, 33, 70, 71, 78, 98, 99, 152, 291, 295

engineering, 76
expansion, 12, 295
loops, 76
properties, 4, 11, 12, 34, 154, 294
stability, 12, 17, 98, 173, 266, 295
transition temperature, 12
vents, 163
Thermalization, 205
Thermodynamics, 75
Thermoelectric change, 148
Thermo-magnetic effect, 78
Thermo-mechanical properties, 17
Thermophoretic forces, 288
Thermoplastics, 12, 29, 256
Thermopower, 290
Thermosets, 12, 13, 29
Thin-film electrodes, 30
Third-generation (3G), 102
Thylakoids, 176
Titanium
   aluminides, 148
   oxides, 27
Transmembrane, 167, 179
   electron movement, 17
   lipids, 167
Transmission
   electron microscopy (TEM), 100, 215, 219, 222, 223, 230, 231
   spectra, 46, 233, 247–251
Transmissivity, 72
Tribological
   advantages, 16
   coatings, 295
   properties, 16
Trimetallic nanoparticles (TNPs), 216, 217
Turbidity, 24

U

Ultra-capacitors, 29, 290
Ultrahigh molecular weight polyethylene (UHMWPE), 13
Ultrasonic
   inspection, 262, 263

   propagation, 263
   wave, 262, 263
Ultrasound, 262–264, 277
Ultraviolet (UV), 24, 106, 107, 180, 215, 219–221, 230
   visible spectroscopy, 219, 220
Upconversion nanoparticles (UCNPs), 177
Urbanization, 91, 109

V

Vacuum devices, 66
Vanadium pentoxide, 7
Velocity, 78, 263, 264, 295
Verbal models, 128
Verification and validation (V&V), 137, 138
Versatile gadgets, 148
Vertebrate retinas, 166
Viscosity, 25, 26, 75, 78
Visual rhodopsins, 163
Vitamin A aldehyde, 162
Voltaic cells, 25

W

Wavefunction, 193, 239–241
Wavelength, 45, 47, 165, 219, 220, 263, 286
Wettability, 25, 26
Wide bandgap semiconductor, 23

X

X-ray, 44–46, 219
   diffraction (XRD), 44, 45, 57
   photoelectron spectroscopy (XPS), 225, 226
   analysis, 225

Z

Zeolites, 149
Zeta
   deviation analysis, 222
   potential (ZP), 222
Zetasizer software, 219